Atlas of Astronomical Discoveries

Govert Schilling

Atlas of Astronomical Discoveries

Springer

Author
Govert Schilling
Bloemendalsestraat 32
3811 ES Amersfoort
Netherlands
mail@govertschilling.nl

Originally published in 2008 as 'Atlas van Astronomische Ontdekkingen' by Fontaine Publishers,
The Netherlands Translation: Andy Brown

ISBN 978-1-4419-7810-3 e-ISBN 978-1-4419-7811-0
DOI 10.1007/978-1-4419-7811-0
Springer New York Dordrecht Heidelberg London

Library of Congress Control Number: 2011922049

Printed on acid-free paper

Springer is part of Springer Science+Business Media (www.springer.com)

Preface

Four hundred years is little more than a blink of an eye in terms of cosmic time. In 400 years the Sun rises nearly 150,000 times, it has been Wednesday 20,000 times, and a Full Moon more than 5,000 times. These are big numbers, but most celestial phenomena evolve much more slowly. The Earth may go around the Sun 400 times in as many years, but Saturn completes less than 14 orbits and Halley's Comet a little more than 5, while the dwarf planet Eris will have completed only 70% of a single orbit.

The night sky hardly changes in 400 years. The fastest moving star in the sky – which is only visible with a telescope – will have moved a little more than 1° further north, and to see any real changes in the shapes of the constellations you need to wait a hundred times longer. In 400 years, the Sun completes less than a millionth of its orbit around the center of the Milky Way. That is like walking around the Place de l'Etoile in Paris and only moving 1.5 mm.

Large numbers are the trademark of astronomy, but if we look at it in the right perspective, we see that the universe hardly changes in 400 years. The Sun may convert 50 quadrillion tons of hydrogen into helium in that time, but that is only a trillionth part of its total mass. And although the Andromeda galaxy has moved more than a trillion kilometers closer to the Milky Way, that only means that the light it emits takes a month less to reach us than the two and a half million years that we are used to.

In the age of the universe, 400 years is about the same as one minute in the life of an old person. In our Milky Way, a few hundred new stars may have seen the light, and in the vast cosmos with its countless galaxies, a few billion supernovas will have exploded. But generally speaking, the universe looks exactly the same today as it did at the start of the seventeenth century. In that period of time the universe has just blinked.

That makes it all the more remarkable when we look at what has changed in our understanding of the universe. From their rather inferior place in space, on a small planet at the edge of a spiral galaxy, astronomers have succeeded in penetrating to the depths of the universe and into the vaults of cosmic history. Our knowledge of the universe has undergone a revolutionary development, largely thanks to the invention of the telescope in 1608.

This *Atlas of Astronomical Discoveries* offers a spectacular review of the past 400 years of telescopic astronomy. In one hundred breathtaking snapshots, it looks at the most important astronomical discoveries since the invention of the telescope. Short texts tell familiar and less well-known stories behind these discoveries – stories of amazement, curiosity, perseverance and luck, but above all, stories of the unstoppable process of unraveling the secrets of the universe in which we all live.

In the next 400 years, the cosmos will again change very little. But we are likely to have to wait much less for new revolutionary developments in astronomy. A few orbits of the Sun, at the most.

Govert Schilling

page 1-4 **The Dawn of Astronomy**

page 5-32 **1608 - 1708**
New Vistas and Cosmic Laws
Mountaineering on the Moon 1609
Children of Jupiter 1610
Tarnished Blazon 1611
Cosmic Order 1619
Blowing in the Solar Wind 1619
Changeable Behavior 1638
Cosmic Hula Hoops 1656
Planet Mapping 1659
Stormy Weather 1664
Planetary Surveyors 1672
Universal Power of Attraction 1687
Recurring Visitors 1705

page 33-52 **1708 - 1808**
Swarms of Stars on a Three-Dimensional Stage
Star Trek 1718
Subtle Swings 1728
Veiled Sister 1761
Fallen Planet 1781
Route du soleil 1783
Deadly Beauty 1790
Celestial Vermin 1801
Rocks from Space 1803

page 53-88 **1808 - 1908**
Paving the Way for Major Theoretical Breakthroughs
Deceptive Lines 1814
Distant Suns 1838
Moving Waves 1842
Inconstant Sun 1843
Swirling Veils 1845
Newton's Triumph 1846
Enervating Explosions 1859
Stellar Runt 1862
Elementary Puzzles 1868
Fear and Dread 1877
Extraterrestrial Waterways 1877
Celestial Fireworks 1892
Minimum Temperature 1893
Luminous Thunder 1901
Almost Empty 1904
Color Coding 1905

page 89-134 **1908 - 1958**
A Speck of Cosmic Dust in an Evolving Cosmos
Heavenly Messengers 1912
Cosmic Yardstick 1912
Nearest Neighbor 1915
Sprinter in the Night Sky 1916
Galactic Dimensions 1918
Einstein Proved Right 1919
Island Universes 1924
Rotating Disk 1927
Inflated Space 1929
Lilliputian Planet 1930

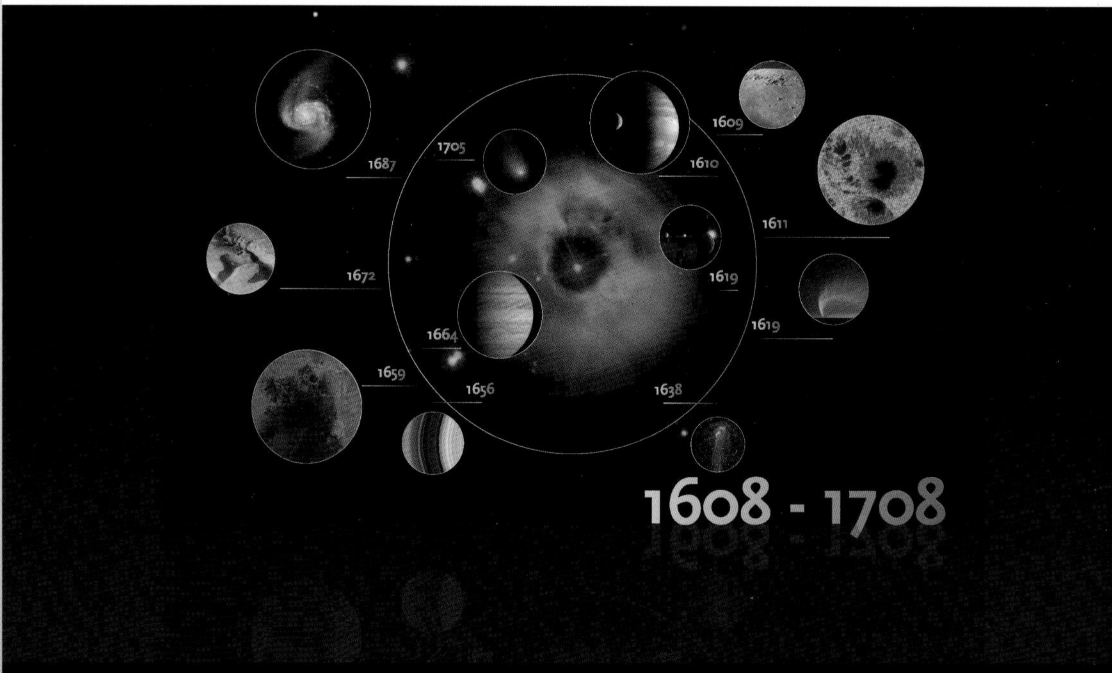

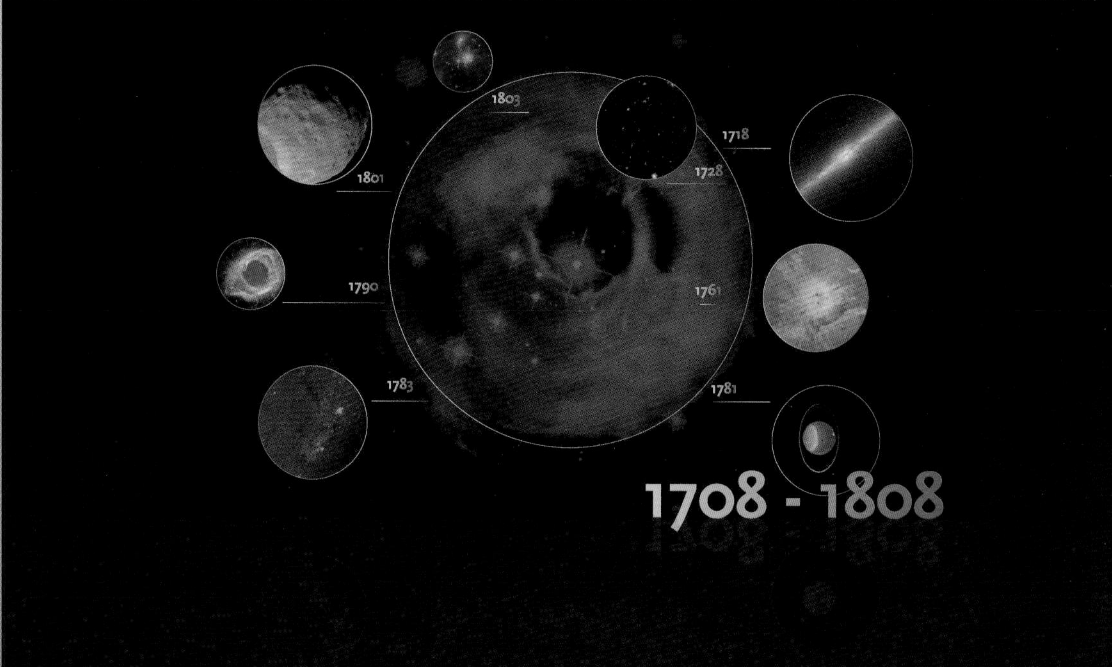

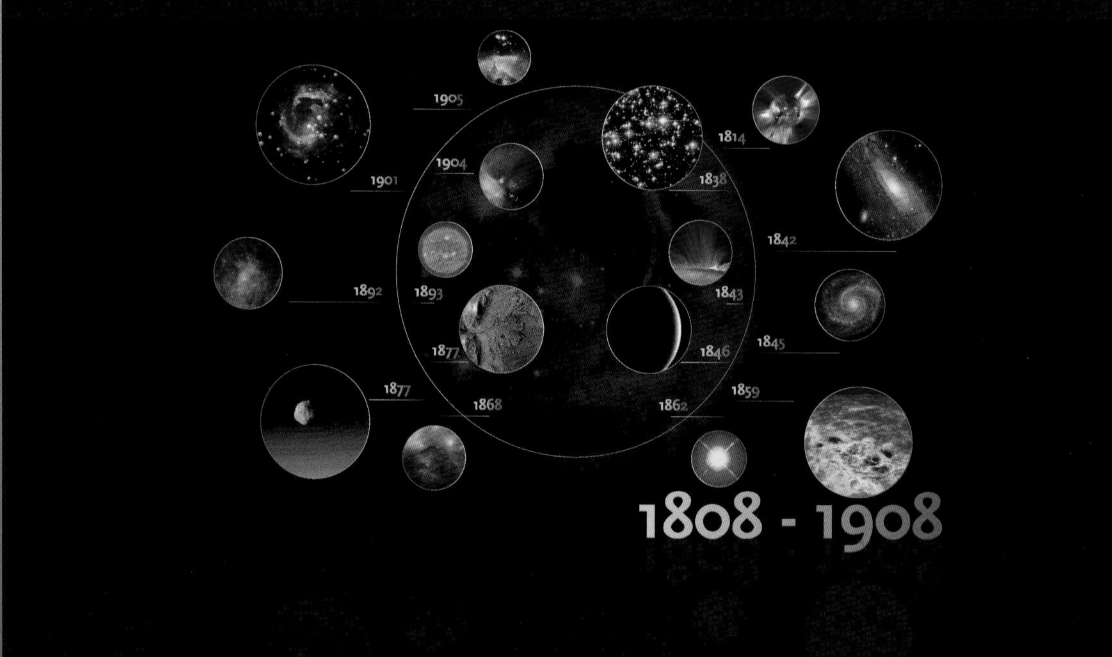

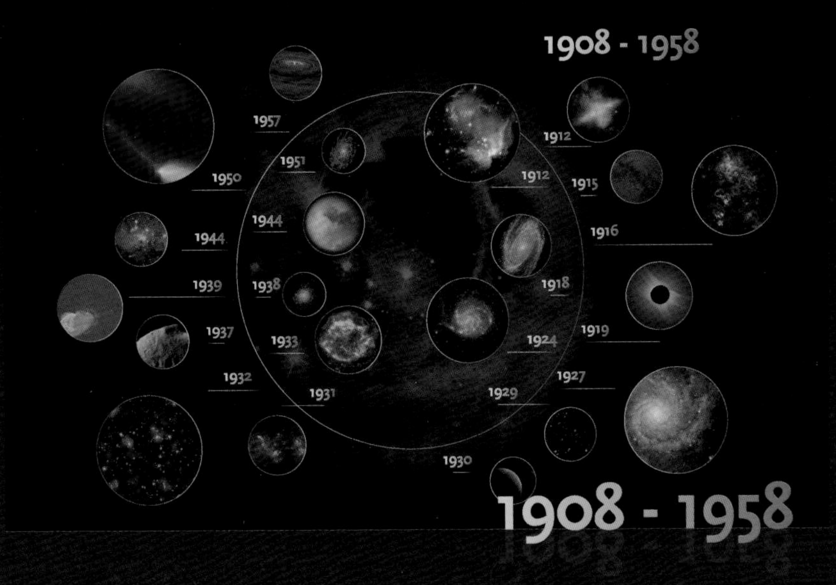

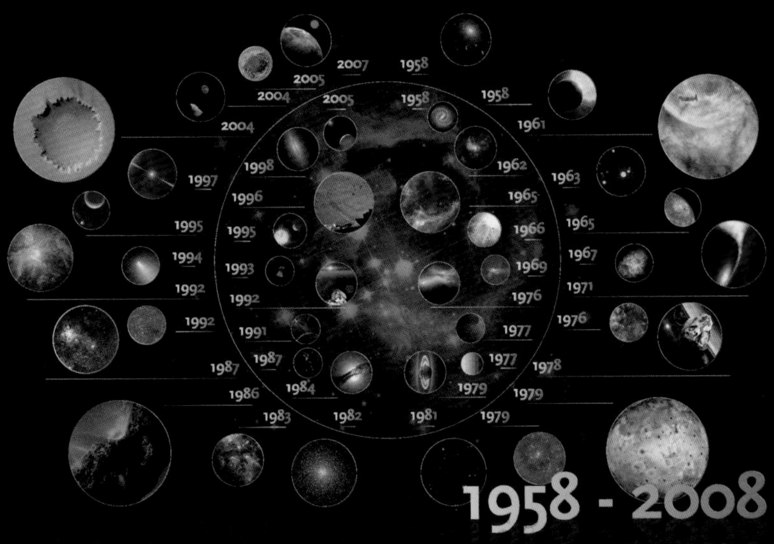

1931 Galactic Podcast
1932 Invisible Stuff
1933 Terminal Explosions
1937 Near Miss
1938 Sunny Train Ride
1939 Celestial Radio Beacon
1944 Discoveries in Wartime
1944 Veiled Moon
1950 Icy Cloud
1951 Hydrogen Hiss
1957 Elementary Stardust

1958 - 2008 page 135-224

Revolutionary Discoveries in a Mysterious Universe

1958 Cosmic Gathering
1958 Majestic Spiral
1958 Interplanetary Weather
1961 Cosmic Evolution
1962 Penetrating Look
1963 Far-Off Shores
1965 Ancient Light
1965 Radar Revelations
1966 Optical Illusion
1967 Extraterrestrial Beeps
1969 Military Spin-Off
1971 Bright Shell, Black Core
1976 Asymmetric Expansion
1976 Striking Similarity
1977 Hybrid Intruder
1977 Skinny Hoops
1978 Double Planet
1979 Dusty Belt
1979 Sulfurous Surprise
1979 Curved Light
1981 Empty Space
1982 Nervous Beacon
1983 Heat Surplus
1984 Fatal Attraction
1986 Gushing Iceberg
1987 Elongated Galaxies
1987 Mysterious Explosion
1991 Hostile Environment
1992 Baby Photo
1992 Beyond Pluto
1992 Cosmic Nursery
1993 Unsightly Little Moons
1994 Collision Course
1995 Other Worlds
1995 Failed Star
1996 Fossil Remains
1997 Cosmic Fireworks
1998 Accelerating Universe
2004 Wet History
2004 Target Earth
2005 Warrior Princess
2005 Icy Landing
2007 New Earth

In the outer regions of a spiral galaxy, somewhere on the edge of the Virgo cluster, a star explodes. Shock waves from the supernova explosion cause a thin cloud of interstellar gas and dust particles to condense. The balance is disturbed, and gravity grabs its chance: the cloud contracts under its own weight. A hundred thousand years later, a new sun shines here.

The birth of a star is a violent, messy and incomplete process becuase not all of the matter in the collapsing cloud ends up in the star. One percent ends up in a flat, rotating disk which, over a period of ten million years, concentrates into a handful of planets – cold spheres of ice, stone and metal.

The Dawn of

One of these planets is a small, wet oasis with warm inland seas and a protective atmosphere. Cosmic hydrocarbons find a fertile medium for organic chemistry- Amino acids and proteins. Nucleic acids. DNA. The first living cell. Evolution.

More than four billion years later, *Homo erectus* looks skywards from the African savanna. Intrigued by the alternation of day and night, the phases of the Moon, and the twinkling stars in the night sky. A child of the universe, he stands eye-to-eye with his own origins. Astronomy is born.

The universe hardly changes in a few hundred thousand years. What does change though, is the way we humans look at the infinite and impalpable world beyond the Earth. The world of celestial bodies, with their own cycles and orbits. Marking the seasons, announcing the arrival of the monsoon, and seeming to control life on Earth.

Ursa Major and Orion are beacons for navigating the night sky. The equinoxes and solstices define the seasons, and the heliacal rising of Sirius presages the annual flooding of the Nile. Hunters, farmers, seamen – no one can live without the stars, and anyone who keeps a close eye on the firmament naturally becomes curious about the mysterious drama being played out among the stars.

Wandering stars pass through the Zodiac. Meteors shoot through the sky; a spectacular comet arcs above the horizon. Sometimes a new star appears, only to be extinguished again some months later. Every now and again the Sun is eclipsed. All of nature is in the magic grip of the cosmos.

Astronomy

Babylonians discover the regularity of celestial phenomena. The Chinese note the appearance of 'guest stars' and comets. The Greeks philosophize on the nature of things, on order and regularity, on *kosmos* versus *kaos*. In his geocentric model, Aristotle of Athens sees the planets as supported by crystal spheres. Eratosthenes of Cyrene measures the circumference of the Earth. Hipparchus of Rhodes catalogs the stars. Ptolemy of Alexandria draws and calculates, traces circles and epicycles, and explains the seemingly unpredictable movements in the firmament in a 13-volume mathematical treatise published in the middle of the second century.

The geocentric worldview, with the Earth at the center and the celestial bodies around it in complex, multiple circular orbits, spreads throughout the Old World. In the ninth and tenth centuries, via India, it reaches Persia where mathematicians and astrologers at the palaces of the sultans and caliphs develop measuring instruments and determine the positions of celestial bodies in the sky. It then passes via North Africa to Spain, where King Alfonso X commissions the most accurate planetary tables ever produced, and from where the old Aristotelian ideas conquer the rest of Europe – a continent that is just starting to wrest itself free from the obscurity of the Middle Ages.

It is not until the mid-sixteenth century that the Polish canon and astronomer Nicolaus Copernicus drops a pebble into the still waters of Greek cosmology. The resulting circular ripples no longer place the Earth at the center of the universe, but the Sun at its center. The crystal spheres are shattered, the division of the cosmos into sublunar and superlunar dissipates like soap bubbles. In his life's work *On the Revolutions of the Heavenly Spheres*, Copernicus puts the Earth on the same level as the planets; it is not the geometric center of the universe but a world orbiting the Sun.

The Renaissance marks the cautious beginnings of modern science, where the worldview is based not on myth and tradition but on observation and experiment. Medieval natural philosophy shows its first hairline cracks and will break into two within a couple of hundred years. Astronomy and astrology part company, and science and faith no longer walk hand in hand.

While naturalists and philosophers embrace Copernicus' heliocentric worldview, observers like Tycho Brahe determine the positions of the stars and planets using gigantic quadrants, and theologians debate the implications of these new insights, a God-fearing spectacle-maker in Middelburg, the Dutch Republic, grinds two convex lenses and places them at both ends of a cardboard tube. Here, Hans Lipperhey builds the first working telescope and gives a public demonstration of his 'tube to see great distances' to Prince Maurits at the end of September 1608, heralding the advent of modern astronomy.

Supernova explosions still occur everywhere in the universe. New stars and planets are continually being born. There may be countless hidden oases in the universe where life has evolved because the cosmos is infinitely large and will last forever. But, in at least one place in that unimaginably large and immeasurable vastness, in the outer regions of a spiral galaxy, somewhere on the edge of the Virgo cluster, an irreversible step has been taken. Here, on our small Earth, we have started exploring and unraveling the secrets of the universe, and astounding discoveries are coming together to form a breathtaking picture of the world we live in.

1705
1687
1672
1664
1659
1656

1608 - 1708

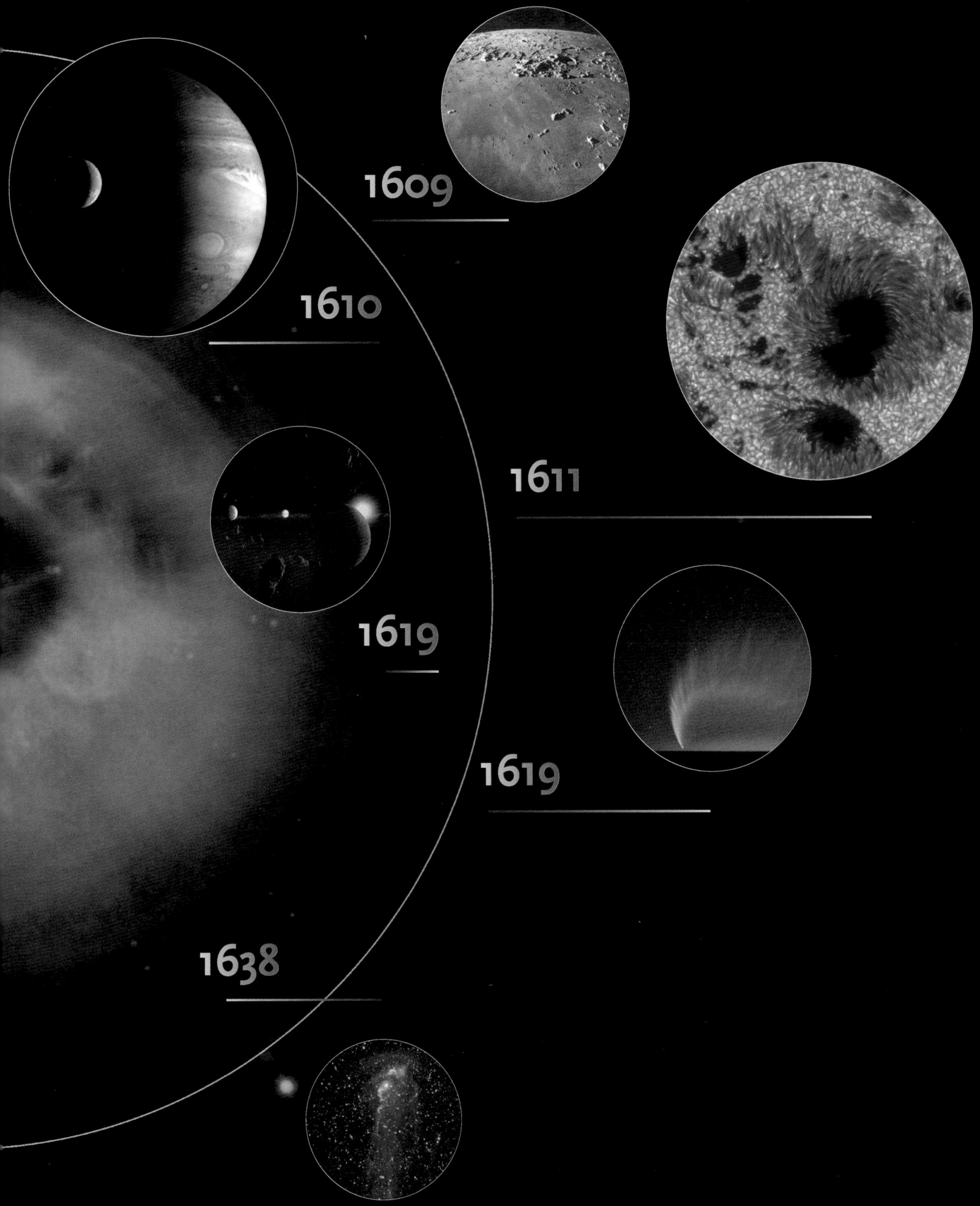

The invention of the telescope unleashes a revolution in astronomy. (until 1608 the only way we can look at the universe is through our own eyes, and they are not the most sensitive instruments imaginable). But now the human eye has a helping hand, and a complete new world opens up.

New Vistas and

The first telescopes that astronomers aim at the night sky have poor quality lenses and a minuscule field of view, and suffer from chromatic dispersion and imaging errors, but even with these small, simple instruments they are able to make revolutionary discoveries. Wherever they look, astronomers observe new objects and details.

The seventeenth century is the age of the great voyages of discovery, of new horizons on the other side of the world's oceans. In science, too, new vistas are opening up. It is an age of astonishment and as yet not understood connections, especially in astronomy. Are there seas on the Moon? Does Saturn have handles? Are there distant suns in the Milky Way? How far away are all these objects? And how do they influence each other?

Cosmic Laws

But it is also the age of new laws. Mathematical laws that describe the paths of the planets; physical laws that determine the motions of celestial bodies. Newton's laws of gravity make it possible to explain, calculate and predict the orbits of planets and comets, although we still know little about their physical properties.

Astronomers finally get the cosmos in their sights. The telescope is their weapon; the hunt for the secrets of the universe has begun.

1609

Mountaineering on the Moon

Galileo Galilei Discovers Mountains on the Moon

The Greek sage Democritus was already writing about mountains and valleys on the Moon in the fifth century BC, but prejudices and misconceptions are tenacious. For many centuries the Moon, like the Sun, was considered a heavenly body in the most literal sense – a perfect, almost divine object in no way at all comparable to the pock-marked Earth.

The telescope is hardly a year old, however, when that idea has to be jettisoned. On July 26th, 1609, the English scientist Thomas Harriot has already made the first drawings of the Moon, as seen through a simple telescope. Harriot's sketchbook shows blemishes and craters. But he does not publish his observations or a physical description of what he has drawn.

A few months later, on November 30th in Tuscany, Galileo Galilei directs his self-built telescope at the Moon. Galileo's latest telescope has a magnification of 20, showing details never seen by anyone before. He sees an irregular shadow line. Large and smaller craters. Mountains. A landscape full of light and shadow, peaks and troughs. The Moon looks like the Earth!

Galileo's Moon drawings are published in early 1610. *Sidereus Nuncius (Starry Messenger)* unleashes a revolution in astronomy. The telescope offers a new perspective on the cosmos. Planets are small disks that reveal details, like phases and moons. Not everything in the universe rotates around the Earth, and the Moon is not a divinely perfect sphere, but a world like ours. Can there be water there? Or life?

The topography of the Moon is first charted accurately at the end of the eighteenth century. Johann Schröter makes wonderful drawings of deep craters, long mountain ridges, winding valleys and high, isolated mountain peaks that cast long shadows over the Moon landscape. Just how high are the mountains? How steep and impenetrable are their slopes?

Astronomers try to capture the extreme lunar landscape in maps, drawings and plaster models. Writers and artists go even further; in science fiction and space art the Moon becomes a world of high, sharp peaks, deep ravines and vertical rock faces.

It is the absence of an atmosphere that causes this new misconception. On the Moon, shadows remain sharp and pitch black, even if the Sun is low and they are tens of kilometers long. In reality the mountains on the Moon are not much steeper than those here on Earth.

With stereo photography and altitude meters the topography of almost the entire surface of the Moon is charted. Space probes, like NASA's Lunar Reconnaissance Orbiter, have done that in even greater detail. Four hundred years after Galileo's discovery we know the mountains on the Moon as well as we do the Rocky Mountains and the Himalayas. How long before we see the first lunar mountaineers?

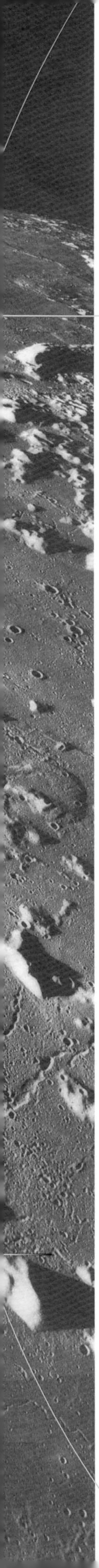

➋ Geologist Harrison Schmitt is the only scientist to have set foot on the Moon. He was the lunar module pilot of the Apollo 17 mission, which made the last manned flight to the Moon in December 1972. (NASA)

➌ Craters, mountains and 'rilles' dominate the lunar landscape in the area of the crater Euler. The origin of the small, elongated string of craters has still not been determined conclusively; they may be related to volcanic activity at some time in the Moon's past. (NASA)

Children of Jupiter

Galileo Galilei Discovers Moons Around Another Planet

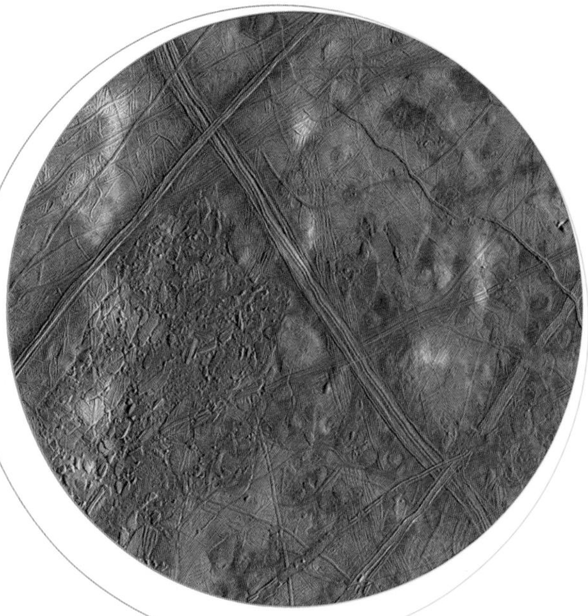

It is January 1610 and, on every clear night, Galileo Galilei can be found outside with his self-built telescope. The winter sky above the hills of Tuscany is breathtakingly beautiful; one just can't get enough of it. On Thursday the 7th, Galileo points his telescope at Jupiter, shining brightly in the constellation of Taurus, about halfway between the Pleiades and the almost Full Moon. That night, in a single, short glance, he discovers three new worlds.

Close to the bright planet, Galileo sees three stars, two to the east and one to the west. Do they belong to Taurus? Galileo is not sure. The three stars are surprisingly bright, and they are all neatly lined up. He makes a sketch of their position.

The next night, Galileo observes Jupiter again. The planet has moved a little in the night sky, but the three stars seem to have moved along with it, although all three are now to the west. Night after night Galileo keeps a close watch on the mysterious objects. And, then, on Wednesday January 13th, he discovers a fourth star.

Galilei describes his discovery in the *Sidereus Nuncius* (*Starry Messenger*), published in March 1610. Jupiter has four companions, four moons. From the Earth we see their orbits edge-on, which makes it look as though they move back and forth, each at its own pace. The solar system has four new celestial bodies.

The discovery is further evidence of the validity of Copernicus' heliocentric worldview. In his book *De Revolutionibus Orbium Coelestium* (*On the Revolutions of the Heavenly Spheres*) Nicolaus Copernicus writes in 1543 that the Earth is not the center of the universe, but orbits the Sun, just like the other planets. The book encounters fierce resistance, especially by the Vatican. How can there be other centers around which celestial bodies rotate other than the Earth? And anyway, if the Earth moves, wouldn't the Moon get left behind?

Galileo realizes that the four moons he has discovered prove that Copernicus was right. They orbit Jupiter, which means that the Earth is not the only center of motion. In addition, they move through space along with their parent planet, which he surmises must also be the case for our Moon and planet Earth.

In 1614, at the suggestion of Johannes Kepler, the German Simon Marius (who also claims to have seen the moons in January 1610) gives them their current names: Io, Europa, Ganymede, and Callisto – four lovers of Zeus, the king of the gods. Galileo's proposal to name them after Tuscan nobles is not adopted.

In 1979 the American space probe Voyager 1 takes the first close-up pictures of Jupiter's four large moons, four complete worlds with volcanoes, craters and subterranean oceans. At the end of the 1990s, another planetary probe charts them in great detail. The probe is called Galileo, after their Italian discoverer.

↑ The icy surface of Jupiter's moon Europa is crisscrossed with cracks and fissures. It is likely that there is an ocean of liquid water under the ice. Europa is smaller than the Moon, but probably contains more water than the Earth. (NASA/JPL/University of Arizona)

→ The moon Io contrasts with the night side of the giant planet Jupiter. This photograph was made by the space probe New Horizons, on its way to Pluto. On the rim of Io, the eruption of a sulfur volcano can be seen. (NASA/JHU-APL)

1610

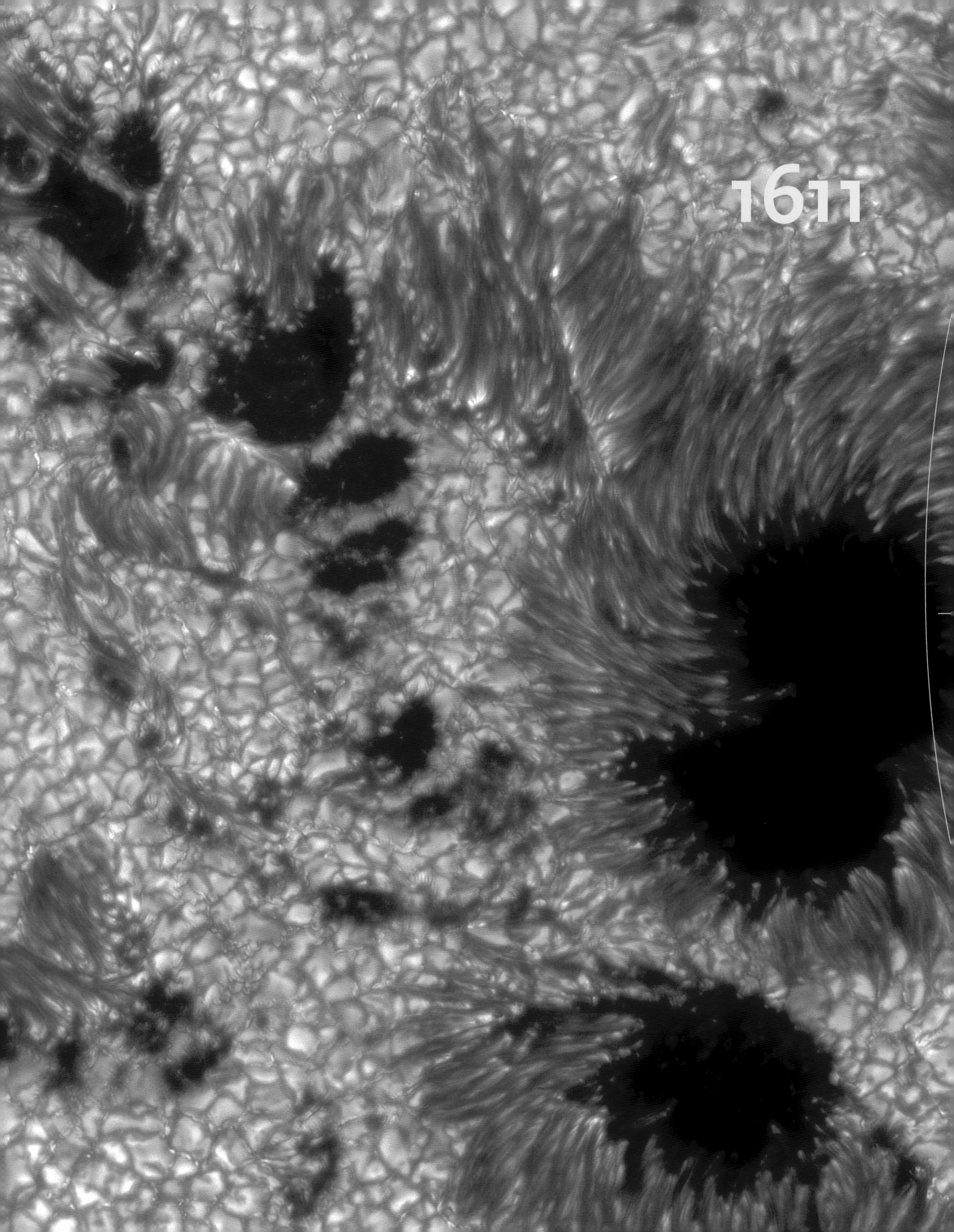
1611

Tarnished Blazon

Johannes Fabricius Discovers Spots on the Sun

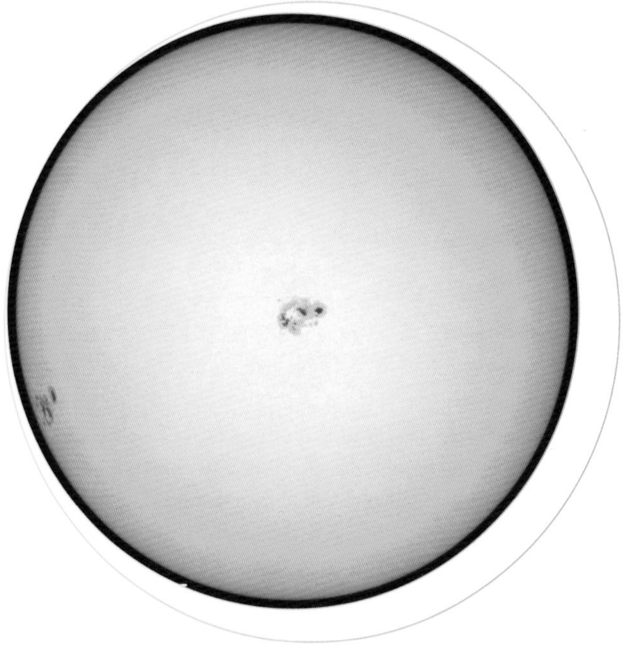

Johannes Fabricius inherits his interest in the night sky from his father David, who is a preacher in Eastern Friesland and a fanatical amateur astronomer. In August 1596, David discovers a star in the constellation of Cetus that varies in brightness. He corresponds with leading astronomers like Tycho Brahe, Michael Maestlin and Johannes Kepler.

His son, Johannes, studies in the Netherlands, where the telescope was invented a few years previously. Early in 1611, around his 24th birthday, he returns to his home in Osteel and takes a telescope with him. Together with his father, he observes the stars in the night sky.

On Wednesday March 9th, on his father's 47th birthday, Johannes is up early. It is a beautiful day, and shortly after sunrise he points the telescope cautiously at the Sun. The bright light hurts his eyes but he is not imagining things: there are dark spots clearly visible in the blinding surface.

Father and son keep a close eye on the Sun in the days that follow. The spots move from east to west- the Sun is rotating on its axis! Johannes has also learned that you have to publish scientific discoveries, so that you can lay claim to them later. On June 13th, his essay *De Maculis in Sole Observatis, et Apparente earum cum Sole Conversione Narratio* (*Narration on Spots Observed on the Sun and Their Apparent Rotation with the Sun*) appears in print.

It is not until the following year that Christoph Scheiner and Galileo Galilei publish their own observations of sunspots. Galileo spends a lot of time looking at the sun, but only near sunrise and sunset. Later he uses a safer method: he projects the image of the Sun on a white card. A similar technique is used by Scheiner. Learned European astronomers begin to conduct a heated debate on the true nature of the black spots, and the contribution of the two amateurs from Eastern Friesland is forgotten. Johannes Fabricius dies in 1616 at the age of 29. His father is murdered a year later by a farmer from his own parish who he had accused of stealing a goose.

Many sunspots are larger than the Earth, and a sizeable group of spots can be a few hundred thousand kilometers across. Special telescopes, including the Dutch and Swedish solar telescopes on La Palma and Tenerife, now make detailed photos and movies of sunspots almost daily.

Sunspots are not black but look dark because they are a few thousand degrees cooler than the rest of the Sun. They are caused by distorted magnetic fields that prevent gas from bubbling up from the interior of the Sun. This creates a cooler area on the surface that is also a little lower than its surroundings.

➊ On October 24th, 2003, a large group of sunspots was visible in the middle of the Sun. A second group appeared on the left rim. The increased sunspot activity of the Sun in the fall of 2003 was accompanied by powerful solar flares. (NASA/ESA)

➋ This close-up of active region 10030 was made with the Swedish solar telescope on La Palma. Sunspots are many hundreds of degrees cooler and lie deeper than the area around them. (Royal Swedish Academy of Sciences)

Cosmic Order

Johannes Kepler Discovers the Laws of Planetary Motion

It is July 19th, 1595. Johannes Kepler is 23 years old when he suddenly has a revelation while giving a mathematics lecture in Graz. He describes it a year and a half later in his book *Mysterium Cosmographicum (The Cosmographic Mystery)*. The brilliant German mathematician and astronomer is convinced that the dimensions of a planet's orbit display a regularity that is related to the five Platonic solids. Kepler is on the tracks of a divine plan.

Johannes Kepler has one foot in the past and the other in the future. He marks the turning point between intellectual darkness and scientific enlightenment. Kepler is a mystic and astrologer, but also discovers the laws of planetary motion and improves the Dutch telescope.

In the classical Greek worldview the planets move at constant speed in perfectly circular orbits, and the Earth is at the center of the cosmos. Nicolaus Copernicus dethroned the Earth in 1543, but held on to the concept of uniform circular motion; Kepler finishes the job Copernicus had started.

In 1600 Kepler moves to Prague, where he assists the great Danish astronomer Tycho Brahe. Tycho gets him to tackle a problem he himself cannot solve: the motion of the planet Mars, which defies explanation in terms of circles. After Tycho's death in 1601 (Kepler succeeds him as court astronomer to Emperor Rudolf II), Kepler has his second revelation: Mars does not move in a circle but in an ellipse, and its orbital speed varies gradually.

These are exciting years and Kepler writes a standard work on optics. In October 1604, he discovers a new star in the constellation Serpens, and in 1609 he publishes his *Astronomia Nova (The New Astronomy)*, describing and explaining the properties of elliptical orbits.

Kepler's first law says nothing more than planets follow elliptical orbits, with the Sun at one focus. His second law, also discussed in *Astronomia Nova*, describes how a planet's orbital speed varies according to its distance from the Sun. But Kepler's book of cosmic laws was not completed until 1619. In *Harmonices Mundi (Harmony of the Worlds)* he describes how the dimensions and orbital periods of the planets are related – his third law.

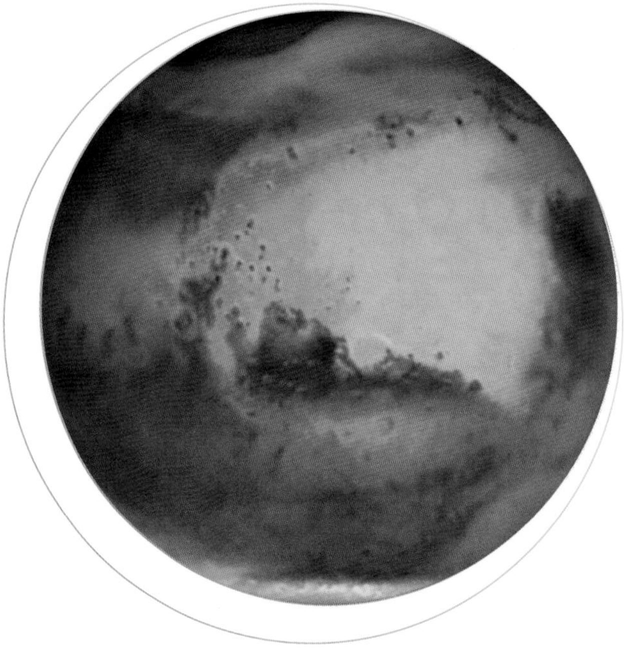

Kepler never takes complete leave of his mystical ideas. In *Harmonices Mundi*, he refers again to the Platonic solids, and to divine harmonics created by the planets. For someone living on the cusp of two ages, there is perhaps not such a great difference between the mathematical order of planetary motions and the geometry of the cosmographic mystery.

Kepler's laws are universal, and they apply not only to our solar system, but also to other planetary systems, binary stars and distant galaxies. What was revealed to Kepler may not have been a divine plan, but it was certainly a cosmic order.

Johannes Kepler discovered his laws of planetary motion when he was investigating the planet Mars. This photo of Mars was made by the Hubble Space Telescope and shows a dust storm in the northern polar region of the planet. (NASA/ESA/Hubble Heritage Team)

Planets and asteroids follow elliptical orbits and all obey Kepler's laws of motion. This illustration shows a planetary system around another star, where three Neptune-like planets and numerous smaller lumps of rock have been discovered. (ESO)

1619

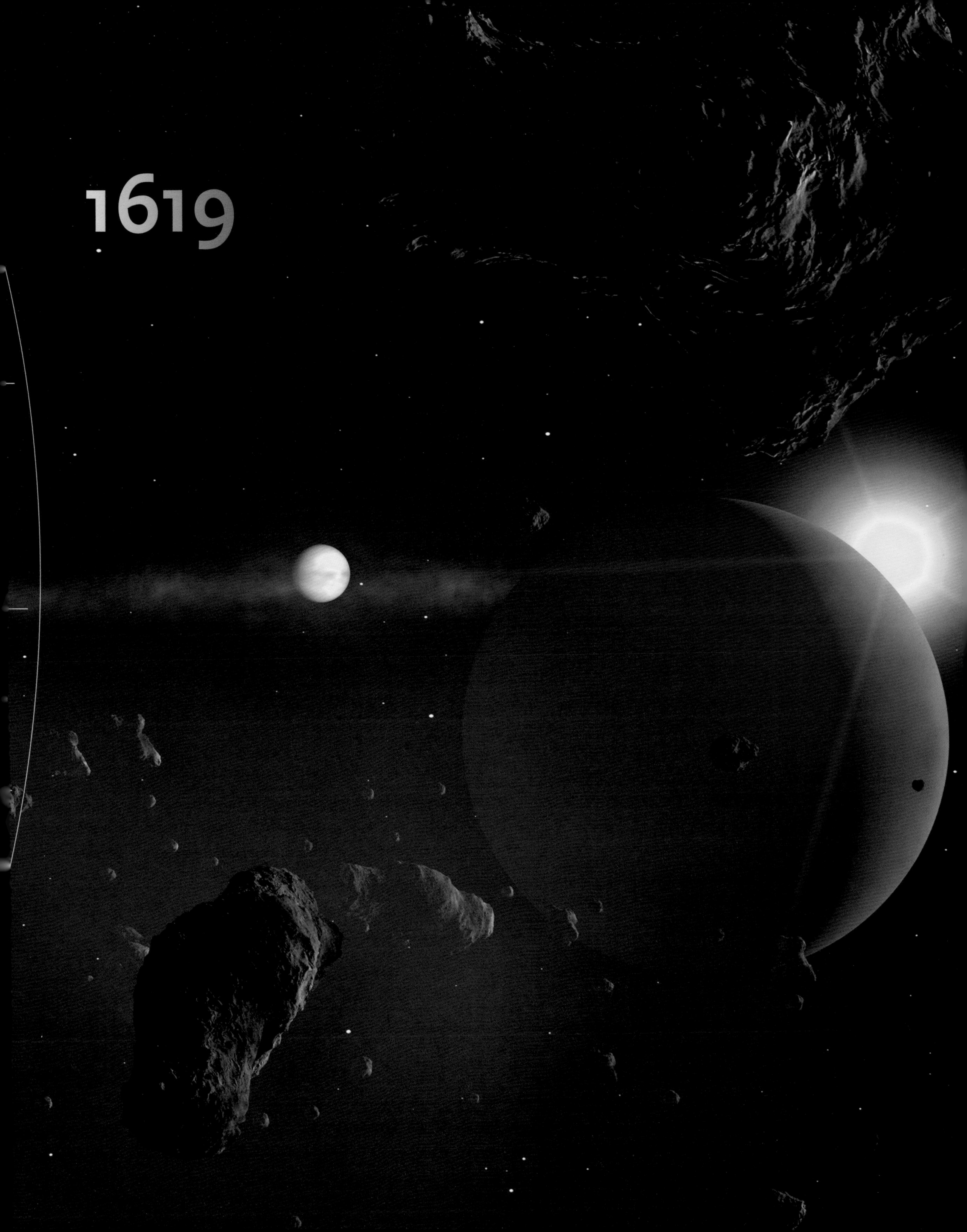

1619

Blowing in the Solar Wind

Johannes Kepler Discovers Why Comet Tails Point Away from the Sun

Katharina Guldenmann sees the Great Comet of 1577 as a bad omen. Her husband, Heinrich Kepler, has left her. Her prematurely born son Johannes has had smallpox and is regularly ill. And yet she takes the 6-year-old outside to see the comet when it appears early in 1578. Two years later they watch a total lunar eclipse together, and Johannes' interest in the universe has been awakened for good.

Comets are impressive celestial phenomena that appear suddenly without warning and then disappear again a few weeks or months later. Since time immemorial they have been linked to disaster and catastrophe and are generally considered to be part of the 'sublunar' regions; after all, they can hardly have anything to do with the perfect permanence of the cosmos.

The spectacular comet of 1577, with its long curved tail, puts an end to that belief. Measurements by Tycho Brahe show that the comet must be further away than the Moon. The question is, though, do comets move in circular paths through space, like the Moon and the planets, or are they one-off visitors that plough a straight course through the cosmos?

Johannes Kepler firmly believes that comets follow straight courses (how else can it be explained that each comet only appears in the firmament once)? Around 1607, when a bright comet once again appears in the sky, Kepler sets down his ideas on paper for the first time. They originate in a condensation of the etheric aura, and are accompanied by a spirit. They reflect the light of the Sun, and exercise great influence on Earthly events.

In the seventh century, Chinese astronomers had already discovered that the tail of a comet is always turned away from the Sun. Girolamo Fracastoro and Petrus Apianus reached the same conclusion on the basis of their observations of the comet of 1531. But Kepler is the first to come up with an explanation: the sunlight passes through the etheric substance of the comet, carrying comet particles with it.

The appearance of three bright comets in 1618 rekindles Kepler's interest in these unpredictable celestial bodies. A year later, he publishes a book entitled *De Cometis Libelli Tres* (*Comet Trilogy*), containing his explanation.

Kepler's mystical and astrological ideas on the comets moving in straight lines are wide off the mark, but his explanation of the 'antisolar' nature of comet tails was not far from the truth. Gas and dust particles from a comet are blown away by the electrically charged particles in the solar wind and by the slight pressure of the light from the Sun.

For Kepler's mother Katharina Guldenmann, the comets of 1618 bring little good fortune. The herbal healer is tried for witchcraft and, in August 1620, is sent to prison for over a year.

In 2014, the Philae lander of the European comet probe Rosetta will make a soft landing on the icy core of comet Churyumov-Gerasimenko. Rosetta will study the comet's composition and outgassing activity. (ESA/AOES Medialab)

Comet McNaught was one of the most spectacular comets of recent years. Unfortunately it was only clearly visible from the southern hemisphere. This photograph was taken just after sunset in early 2007; the comet's tail points away from the Sun. (ESO/Sebastian Deiries)

Changeable Behavior

Jan Fokkes Discovers the Variability of Mira

Jan Fokkes is not the first to see a new star in the constellation Cetus, but he is the first to discover the regularity in its brightness. Partly because of that discovery, he is given the name *Lumen Frisiae (Light of Friesland)*. He cannot enjoy his fame very long, though: he dies of tuberculosis in 1651, at the age of 32.

In 1572, Tycho Brahe sees a new star flare up in the constellation Cassiopeia. This *Stella Nova* is the final nail in the coffin of the classical Greek notion that the cosmos is eternal and unchangeable. No wonder then that David Fabricius, an East Frisian preacher and amateur astronomer, contacts Tycho in August 1596 when he discovers a *Stella Nova* in the constellation Cetus. Thanks to Fabricius' good connections, the star is included a couple of years later in Johann Bayer's astronomical atlas *Uranometria*, under the name Omicron Ceti.

In August 1600, Amsterdam mapmaker Willem Janszoon Blaeu discovers a new star in Cygnus, and in the autumn of 1604 Johannes Kepler sees a new object in Serpens. All of these stars are visible for a short time and then disappear again from sight.

Jan Fokkes from Holwerd studies astronomy, philosophy and medicine at the University of Franeker and is very familiar with the night sky. On December 21st 1638, when Fokkes is 20 years old, he sees a new star in Cetus, in exactly the same position that Fabricius had seen his *Stella Nova* more than 40 years previously. A few weeks later the star has disappeared, but Fokkes keeps a close eye on Cetus. At the end of 1639 he sees the star again, and again 11 months later.

This is not a new star but one which periodically changes in brightness. Most of time it is too faint to be seen without a telescope, but once in 11 months it is possible for a few weeks. Fokkes – who Latinized his name to Johannes Phocylides Holwarda – has discovered the first periodic variable star.

The Polish astronomer Johannes Hevelius calls the star Mira (Wonderful). Later astronomers discover that many more stars regularly vary in brightness. For example, Algol, in the constellation Perseus, becomes fainter every few days when a cooler companion passes in front of it. The brightness of Delta Cephei oscillates because the star swells in size and then contracts again.

Mira is the prototype of a long-period variable star. These are old giant stars that pulsate very slowly and emit enormous quantities of stellar gas. Mira is the only one that is visible – now and again – with the naked eye.

In 1923 Robert Grant Aitken discovers that Mira is a binary star. NASA's ultraviolet satellite Galex discovers in 2007 that the star trails a 13 lightyear-long gas tail behind it as it moves through the Milky Way at high speed, and even today Mira continues to surprise astronomers.

As can be seem from this X-ray photo, taken by the Chandra X-ray Observatory, Mira is a binary star. Gas blown away by the slowly pulsating main star accumulates in the disk around the companion. (NASA/CXC/SAO/M. Karovska)

NASA's ultraviolet satellite Galex made this photograph of the variable star Mira in the constellation of Cetus. The star moves through the Milky Way at high speed, dragging a 'tail' of thin, hot gas behind it. (NASA/JPL)

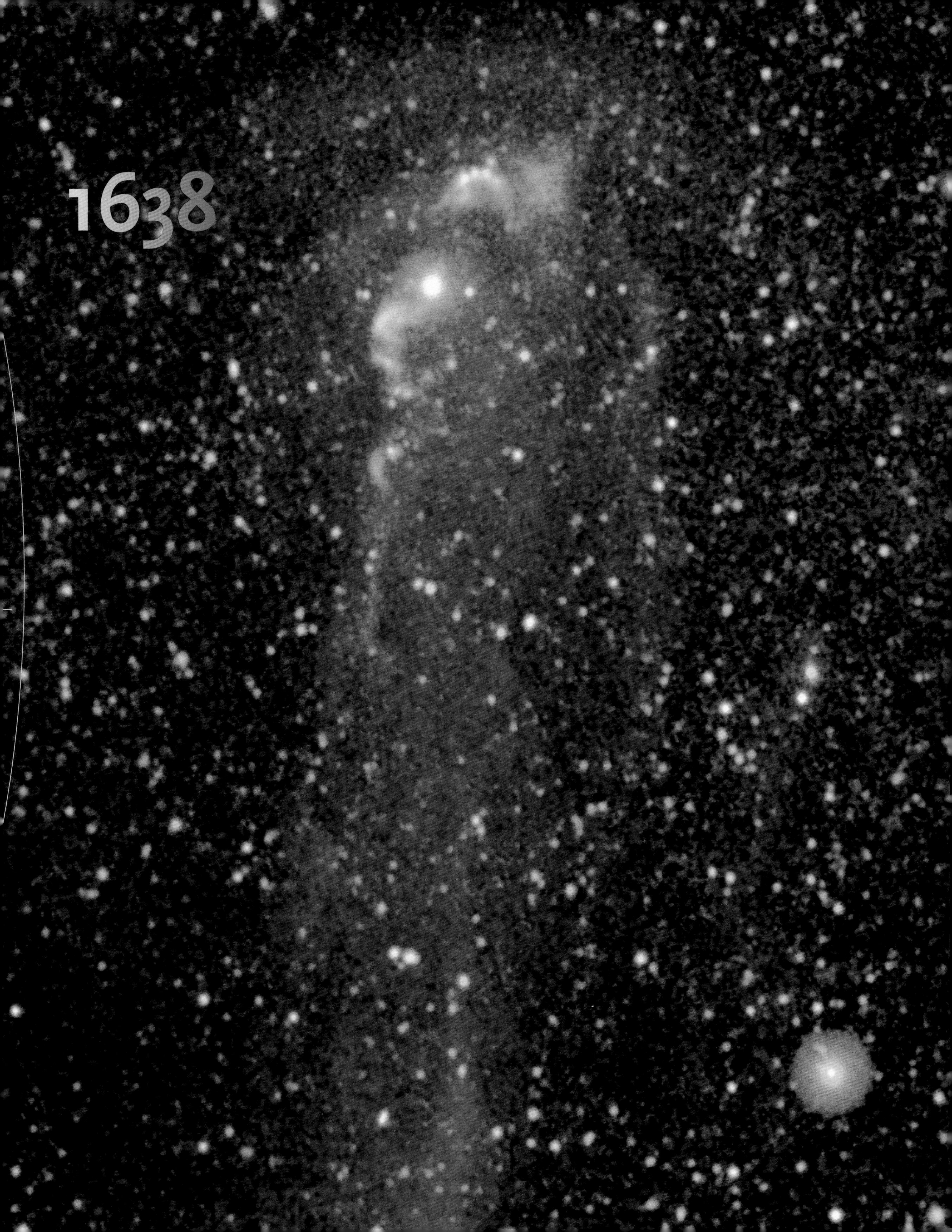
1638

1656

Cosmic Hula Hoops

Christiaan Huygens Discovers the True Nature of Saturn's Rings

For nearly half a century astronomers have been puzzled by the mystery of the planet Saturn. It sometimes seems to have strange protuberances, like the handles of a Roman vase, and some observers think they are two large moons that almost touch the planet. What makes it even more of a riddle is that once every 15 years or so, Saturn's 'ears' disappear, only to reappear a few weeks later.

The Dutch physicist and astronomer Christiaan Huygens solves this puzzle for good. Huygens is one of the greatest scholars of his time. One of the many things he studies is the wave-like behavior of light, and he invents the pendulum clock. He also grinds lenses and builds telescopes of unprecedented quality.

In 1655, Huygens discovers the large Saturnian moon Titan, and in early 1656, he keeps a close eye on the planet's 'handles.' They get smaller and smaller and, by March, they have disappeared completely. Suddenly 26-year-old Christiaan realizes that the changes in Saturn's shape are caused by a thin, flat ring which we continually see from a different angle.

Huygens needs more time to confirm his idea. He wants to record his discovery without giving away too many details too soon. In 1656 he publishes the anagram *a a a a a a a c c c c c d e e e e e h i i i i i i i l l l l m m n n n n n n n n n o o o o p p q r r s t t t t t u u u u u*, which stands for *Annulo cingitur, tenui, plano, nusquam cohaerente, ad eclipticam inclinato* (Surrounded by a ring, thin, flat, touching nothing, at an angle to the ecliptic). Three years later, Huygens publishes a small book entitled *Systema Saturnium*, containing everything that is known at the time about the planet, its moon and its ring.

Two centuries after Huygens' revolutionary discovery, scientists realize that Saturn's ring cannot be a solid object but must instead consist of countless small particles that orbit the planet – material which for some reason or another has never solidified into a full-fledged moon.

Space probes later make detailed pictures of the ring system with its hundreds of separate rings and intermittent spaces, its temporary variations in density and spiral wave formations, caused by the gravity of small moons. Research into the dynamics of the ring system also provides insights into other flat, rotating structures in the universe, such as spiral galaxies and gas and dust disks around stars which may in the future evolve into planets.

The giant planets Jupiter, Uranus and Neptune also have ring systems, but they are by no means as majestic as the one around Saturn. This ringed planet has become an icon of astronomy, as familiar as a five-pointed star or a crescent moon; Christiaan Huygens would have been delighted.

On this photograph, taken by the planetary explorer Cassini, the southern hemisphere of Saturn is visible through the planet's semi-transparent rings. Above right, the rings cast their shadow on Saturn's northern hemisphere. (NASA/JPL/SSI)

The dark section in the middle of this photo is the Cassini Division, named for Jean-Dominique Cassini, who discovered it in 1675. Many narrow rings have since been discovered in this 'empty' zone between the bright B Ring (*left*) and the somewhat darker A Ring (*right*). (NASA/JPL/SSI)

Planet Mapping

Christiaan Huygens Discovers Surface Details on Mars

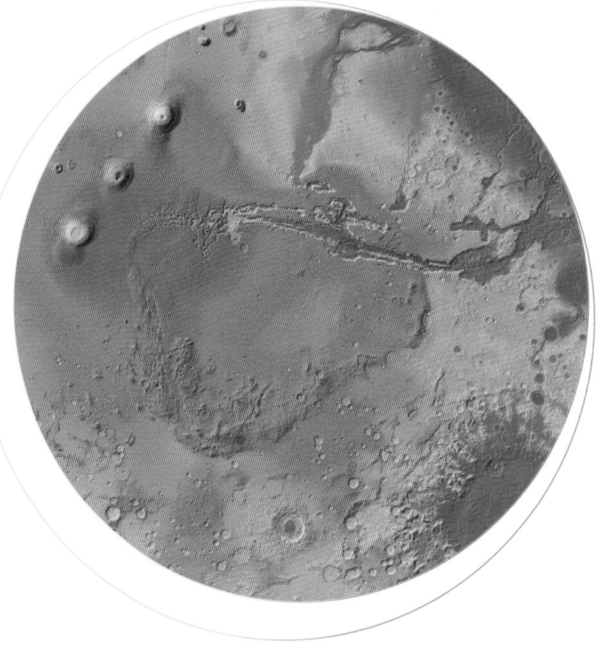

Christiaan Huygens discovers Saturn's moon Titan and the true nature of the planet's rings, but Saturn itself remains without structure, even through the high-quality telescopes Huygens builds together with his brother Constantijn. This is remarkable since, like the Earth, the planets orbit the Sun, and you would expect them to have mountains and seas too.

In the autumn of 1659, Huygens aims his telescope at the planet Mars, which rises in the evening as a bright red-orange star between the horns of Taurus. Huygens looks intensely at the small planetary disk. Every now and again, if the Earth's atmosphere is calm for a while and the telescope image is less shaky, he can make out a spot on the surface, a dark triangle on Mars. The sketch Huygens made that night is the first 'map' of another planet.

During the night the spot moves- Mars is rotating on its axis! The next night the dark triangle is back in the middle of the planetary disk. Mars appears to have the same rotational period as the Earth: 24 hours.

Yet Huygens is not finished with Mars. He has invented a micrometer, with which he can measure the apparent diameter of a planet in arcseconds. Kepler's laws tell him roughly at what distance Mars is from the Earth which means he can calculate Mars' true diameter: 60% of the Earth's.

In 1666, Huygens' Italian-French colleague, Jean-Dominique Cassini, discovers that Mars has polar caps. (Huygens sees them himself a few years later.) Cassini also makes a more accurate calculation of the planet's rotation period: 24 hours and 40 minutes. Huygens measures the angle of tilt of Mars' axis, which proves to be almost the same as that of the Earth. The planet is so similar to ours that there almost has to be life there. In his popular book *Kosmotheoros* (1698; published in English as *The Celestial Worlds Discover'd: or, Conjectures Concerning the Inhabitants, Plants and Productions of the Worlds in the Planets*) Huygens lets his imagination run wild.

The spot that Huygens recorded in 1659 is later called Syrtis Major, after the Gulf of Sidra off the Lybian coast. It is one of the most striking features of the planet. A large plain more than 1,000 kilometers across, its dark color caused by the presence of volcanic basalt.

Mars is the only planet on which surface details can be seen from the Earth. With Jupiter and Saturn we can only see bands of clouds in the atmosphere. Venus is also hidden behind a dense covering of clouds, and Mercury is too small and too close to the Sun. Huygens is the first in a long line of Mars observers who use increasingly sensitive telescopes to chart the Earth's neighboring planet.

The surface of Mars has now been recorded in almost as much detail as that of the Earth. Satellites orbiting the planet can photograph details only one or two meters wide. With altimeters and stereoscopy the images are made threedimensional. Syrtis Major holds no more secrets.

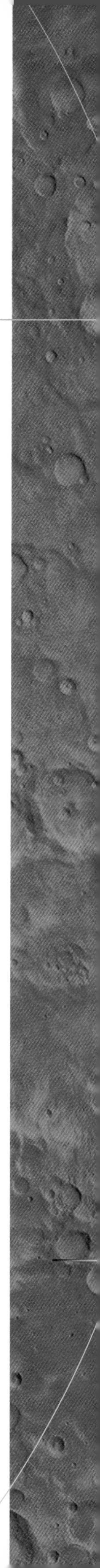

On this topographical map of Mars, the colors indicate differences in altitude. The white spots top left are large, high shield volcanoes. The blue-green horizontal formation to the right is Valles Marineris, a deep canyon system. (Google Mars)

Syrtis Major Planitia – the dark spot first recorded by Christiaan Huygens in 1656 – has now been mapped in detail by planetary explorers in orbit around Mars. (NASA/JPL/USGS)

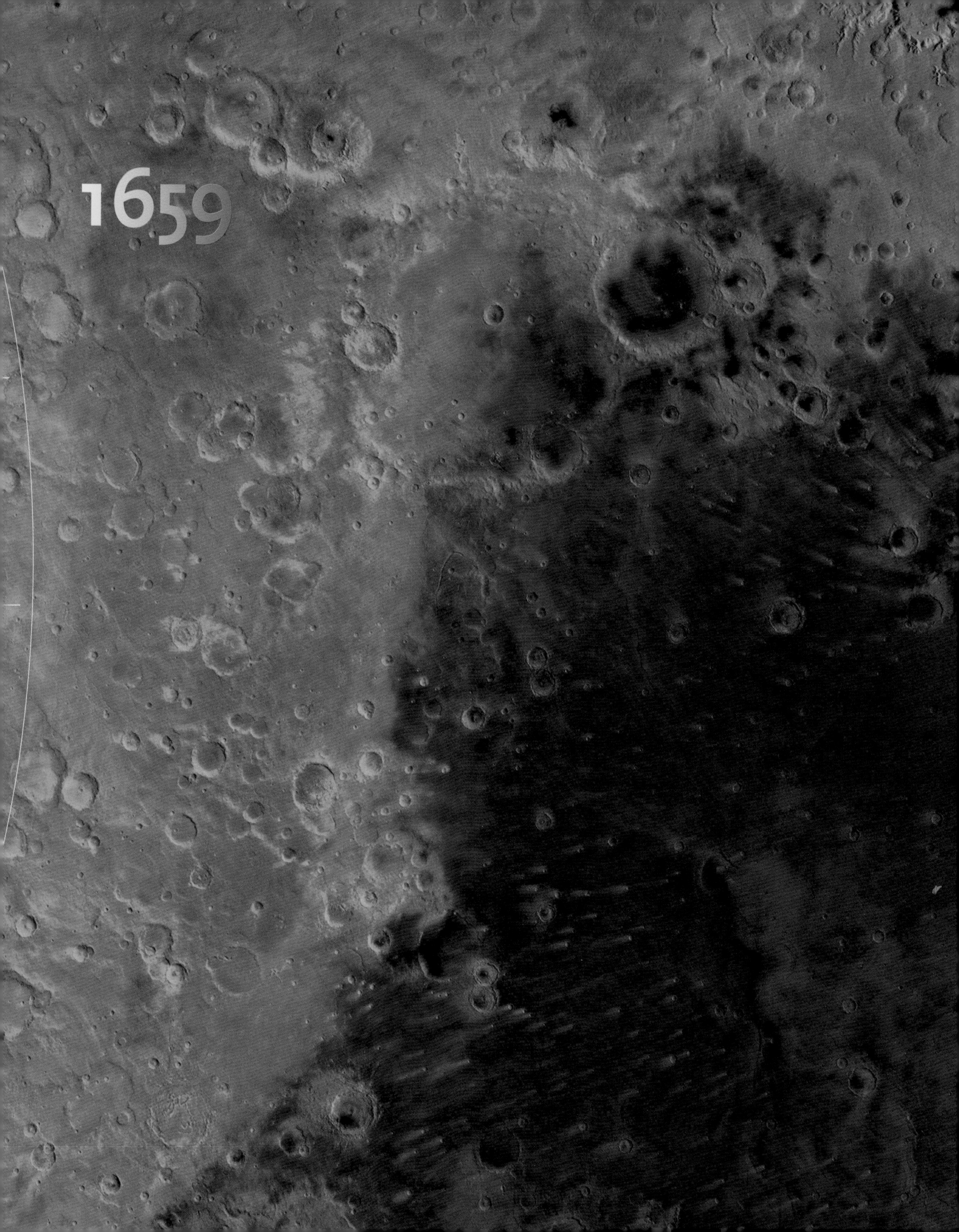
1659

1664

Stormy Weather

Robert Hooke Discovers the Great Red Spot on Jupiter

Robert Hooke may be in weak health, but there is nothing wrong with his eyes, and practically every day he can be found glued to the eyepiece of a microscope or a telescope. Both instruments are relatively new, and there is a whole new world to be discovered in both the micro-cosmos and the macro-cosmos.

Hooke is a real all-rounder. He assists Robert Boyle in building his air pumps, develops a pocket watch, observes cells under the microscope, is an expert mathematician and architect, makes drawings of the Moon and the planets, and is working on a theory of gravity. At the age of 26 he is appointed curator at the newly established Royal Society.

Almost immediately Hooke finds himself at loggerheads with practically everybody. He is jealous and suspicious, lays claim to other people's inventions, and becomes embroiled in a heated argument with Isaac Newton about who first developed the theory of gravity.

His sharp eyesight stands him in good stead in May 1664; the planet Jupiter is in the constellation Sagittarius and does not rise high above the horizon. The image is unclear, but Hooke sees a spot on the planet and determines Jupiter's rotational period at ten hours. A year later the same spot is described by Jean-Dominique Cassini.

There are no more recorded observations of the Great Red Spot, as it is now called, until after 1830, although it is visible in a painting in the Vatican by Donato Creti, dated 1711. But it is not altogether certain that the current spot is the same as the one Hooke saw.

In the twentieth century, the Great Red Spot is extremely striking. Is it a disturbance in the planet's atmosphere, caused by a high volcano on the surface? Or is there perhaps a large lake under the spot? More accurate observations and better theories put an end to all the speculation. The Great Red Spot is a gigantic anticyclonic storm, and its color comes from organic compounds bubbling up from deeper layers in the atmosphere.

Space probes produce detailed images of the spot for the first time at the end of the 1970s. At the start of the twenty-first century, two new red spots appear in the giant planet's restless atmosphere, and no one knows how they will develop further or what will eventually happen to the original spot.

Hooke and Newton never quite bury the hatchet. In 1676, in a letter to Hooke, Newton writes the famous sentence: 'If I have seen further [than others] it is only by standing on the shoulders of giants.' This is undoubtedly a new sneer: Hooke is small in stature and has a hump. Shortly after Hooke's death in 1703, his portrait at the Royal Society is irretrievably misplaced when the venerable club, presided over by Sir Isaac Newton, moves to new premises. Consequently, no one knows what the discoverer of the Great Red Spot looks like.

On this infrared photograph of the Great Red Spot, the different colors indicate the altitude of the cloud tops. Blue clouds are reasonably deep, while pink and white ones are high. The image was made in 1996 by the Jupiter probe Galileo. (NASA/JPL/Cornell University)

Jupiter's Great Red Spot is the largest storm system in the solar system. The giant planet's dynamic atmosphere was photographed in detail at the end of 2000 by the Cassini space probe, which flew past Jupiter on its way to Saturn. (NASA/JPL/SSI)

Planetary Surveyors

Jean-Dominique Cassini Discovers the Scale of the Solar System

Astronomy is the science of large numbers. But how do you measure the distance to a celestial body? Despite all kinds of inventive tricks, little is known for certain about the scale of the universe until the middle of the seventeenth century. Around 1600, Tycho Brahe still assumes that the Sun is eight million kilometers from the Earth, yet the real distance is nearly 20 times greater.

Kepler's third law, on the connection between the distances and orbital periods of the planets, offers some solace. The orbital periods are known, and Kepler's law tells us that Mars, for example, is one and a half times more distant from the Sun than the Earth. That does not tell us the actual distance, though, and Kepler draws a map of the solar system, but the scale is unknown.

Jean-Dominique Cassini solves the riddle in 1672. The Italian astronomer (his name is originally Giovanni Domenico) becomes the first director of the Paris Observatory in 1671 and together with his assistants Jean Picard and Jean Richer he measures the distance from the Earth to Mars. This enables him to determine the scale of the solar system.

In the autumn of 1672, Mars is in opposition with the Sun. The planet is opposite the Sun in the sky, and at its closest to the Earth. Cassini sends Richer to Cayenne in French Guyana, remaining in Paris with Picard. At previously agreed upon times, they measure the position of Mars as accurately as possible in relation to the stars.

Cassini applies a classical method used by land surveyors for measurement. If you measure the direction of a distant church tower from two points at a known distance from each other, you can calculate the distance to the tower. The same applies to Mars. The distance between Paris and Cayenne is known; the distance to the planet can then be calculated from the measured difference in position.

The astronomers publish their results in 1673. Mars is 70 million kilometers from the Earth. That means that the distance from the Earth to the Sun (the astronomical unit) is 140 million kilometers. At last we know the size of the solar system: according to Cassini the furthest planet, Saturn, is an average of 1,330 million kilometers from the Sun.

Similar parallax measurements, for example of transits of Venus and of the asteroid Eros, later result in a more accurate measurement of the distance between the Earth and the Sun of 150 million kilometers. Cassini's figure, based on the observations of Mars, is inaccurate by only 7%. The most precise value for the astronomic unit is now 149,597,870.691 km.

Cassini also discovers four of Saturn's moons: Iapetus (1671), Rhea (1672) and Tethys and Dione (in 1684). In 1675 he discovers the dark division in Saturn's rings that bear his name. Yet at a great age, he loses his eyesight; the first man to see the scale of the solar system dies in 1712, completely blind.

The Earth is a small, pale blue dot on this enlargement of a photograph of the solar system taken by the American planetary explorer Voyager 1 from a distance of four billion kilometers. Voyager 1 was the first spacecraft to leave the solar system. (NASA/JPL)

Measurements of Mars once made it possible to determine the scale of the solar system. Today the planet itself has been measured in detail in three dimensions, including by the European Mars Express, which supplied this 3D image of Echus Chasma. (ESA/DLR/G. Neukum)

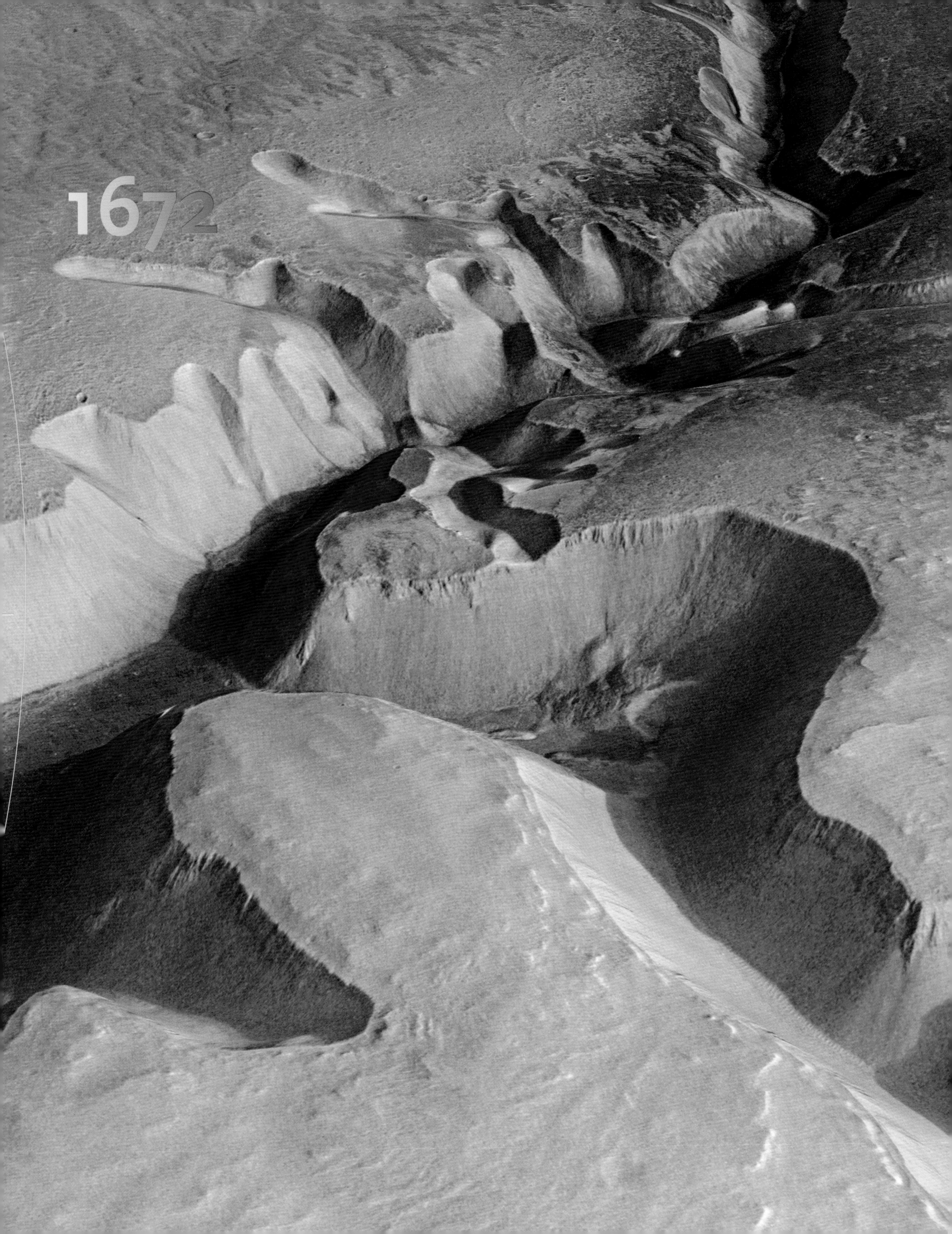
1672

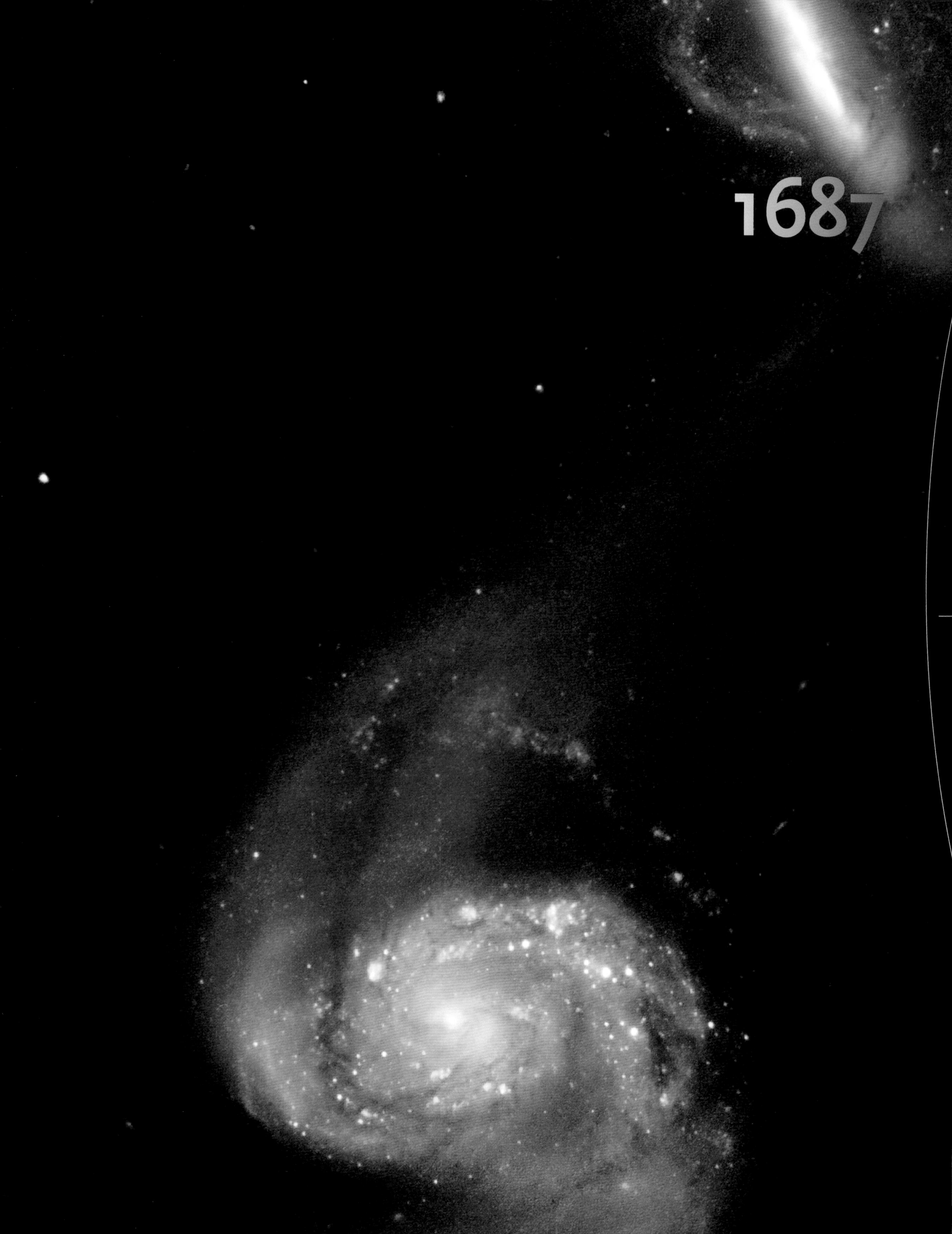

1687

Universal Power of Attraction

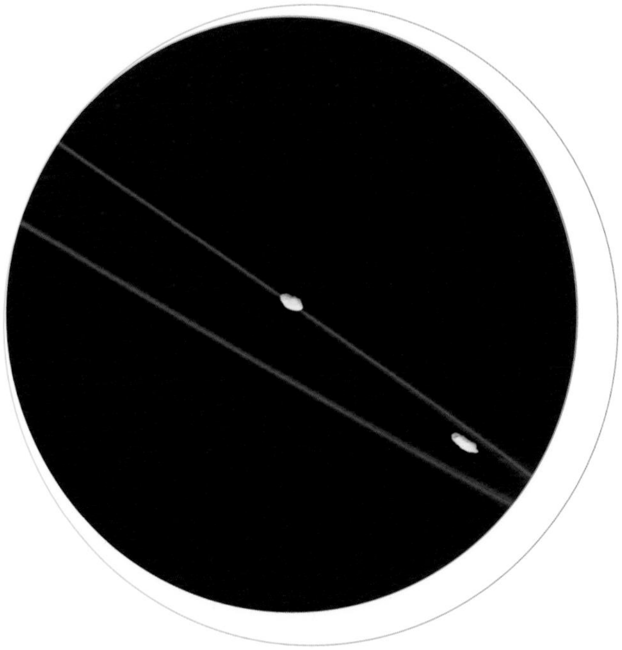

Isaac Newton Discovers the Law of Gravity

Isaac Newton is 22 years old when the bubonic plague breaks out in England in 1665. The University of Cambridge, where he has just completed his studies, is closed as a precautionary measure. Newton returns to his parents' house in Woolesthorpe where a falling apple in his mother's orchard sows the seed of his theory of gravity. If the Earth's mysterious power of attraction can reach the branch of a tree, then why not the Moon?

It is another 20 years before the classic book *Philosophiae Naturalis Principia Mathematica (Mathematical Principles of Natural Philosophy)* sees the light. Newton is hyperintelligent, possibly autistic, chaotic, and does a whole range of things at the same time. He studies Galileo's laws of motion and Kepler's laws, circular motion and centrifugal forces, develops differential calculus at the same time as Gottfried Leibniz, writes a standard work on optics, invents the reflector telescope, and also immerses himself in alchemy and theology.

In 1684, Robert Hooke – who had corresponded with Newton ten years earlier on gravity, inertia and elliptical motion – claims that he is the discoverer of the 'inverse-square law.' The law states that attraction between two celestial bodies is inversely proportional to the square of the distance between them. In the summer of the same year, the influential astronomer Edmund Halley pays a visit to Newton to ask him what the truth of the matter is. Newton digs out his old notes and Halley urges him to publish his ideas on gravity in book form.

Two and a half years later, on July 5th 1687, Newton's *Principia* is published. Many see it as the most influential book in the history of natural science. The publication is funded by Halley because the Royal Society's book budget for that year had been allocated to a book on fish.

Newton's theory of gravity provides a mathematical basis for Kepler's empirically deduced laws, and finally lays to rest the division between sublunar and superlunar: the same laws apply in the cosmos as here on Earth.

Only two centuries later Newton's body of theory acquires its first small cracks. Albert Einstein shows that the classic laws of motion and gravity no longer apply under extreme circumstances. Yet if you were on a space flight to Mars, you would not need Einstein's Theory of Relativity: Newton will do just fine.

Incidentally, the title of Newton's *magnum opus* is very appropriate. Physicists know the mathematical principles of gravity, but they are still in the dark regarding its true nature. Almost 350 years after the apple fell in Woolesthorpe, gravity remains one of the greatest mysteries in physics.

➋ The two small Saturnian moons Prometheus (right) and Pandora (left) perform a complex gravitational dance with each other and with the dust particles in the planet's thin F Ring. The two 'shepherd' moons ensure that the ring does not get any wider. (NASA/JPL/SSI)

➌ Two galaxies in close proximity feel each other's gravity and tidal forces, causing wisps of gas and stellar flows to be dragged outwards. Gravity is by far the most important force of nature in determining the large-scale structure of the universe. (NASA/ESA/Hubble Heritage Team)

Recurring Visitors

Edmund Halley Discovers the Periodicity of Comets

Edmund Halley is 25 when he sees the comet that will later bear his name, in the autumn of 1682, the year in which the multifaceted scientist marries Mary Toone. This is one comet that does not bring bad luck.

At the end of the seventeenth century astronomers are preoccupied with the mystery of comets. It is clear that they follow paths through the solar system, and Newton's theory of gravity leaves no doubt that their motion is affected by the Sun. But accurate positional measurements are rare, making it difficult to determine the mathematical properties of the paths followed by historical comets.

Edmund Halley is an all-rounder like Robert Hooke. At the age of 20 he travels to St. Helena to study the constellations of the southern hemisphere where he also studies the weather and climate, invents the diving bell, and lays the basis for actuarial science. He encourages Newton to write and publish his *Principia*, and later conducts an extensive correspondence with the genius in Cambridge on how to determine the orbital properties of comets.

In 1695 Halley discovers that the comet of 1682 moves through the solar system in virtually the same way as the comets of 1531 and 1607, which were observed by Petrus Apianus and Johannes Kepler. Could this be the same object? Do comets move in long, elongated elliptical orbits around the Sun?

Halley does not immediately elaborate on his ideas. From 1698 to 1700, he leads two long scientific expeditions to the southern Atlantic Ocean and is the first to accurately map out the Earth's magnetic field. After being appointed a professor at the University of Oxford in 1703, he again applies himself to the mystery of the comets.

In 1705 Halley publishes his *Synopsis Astronomiae Cometicae (Synopsis on Cometary Astronomy)*, in which he claims that the comets of 1531, 1607 and 1682 are indeed the same object and predicts that it will return in 1758.

Gravitational distortions caused by the giant planets Jupiter and Saturn delay the comet's return. It is not until December 25, 1758 that the German farmer and amateur astronomer Johann Palitzsch first catches a glimpse of the comet. A few years later, it is named after Edmund Halley.

Halley's Comet is the most famous in history, and is the only short-period comet visible with the naked eye. When it returns in 1986, Halley's Comet is visited by a small armada of unmanned space probes, including the European Giotto, which makes the first close-ups of a comet's nucleus: an irregular clump if ice and rubble that, under the influence of the heat of the Sun, spews geysers of gas and dust into space.

The comet is scheduled to return to the inner solar system in 2061. For the time being, it keeps the memory of Edmund Halley alive.

The famous Halley's Comet was visible from Italy in 1301 and served as a model for the Star of Bethlehem in Giotto di Bondone's fresco 'Adoration of the Magi' in de Scrovegni Chapel in Padua. (ESA)

Comet NEAT, discovered by the Near-Earth Asteroid Tracking program, is one of the many short-period comets in the inner solar system. The actual nucleus of the comet consists of a small lump of ice and debris. The photograph clearly shows the comet's rarefied coma and tail, both of which consist of gas and dust particles. (WIYN/NOAO/AURA/NSF)

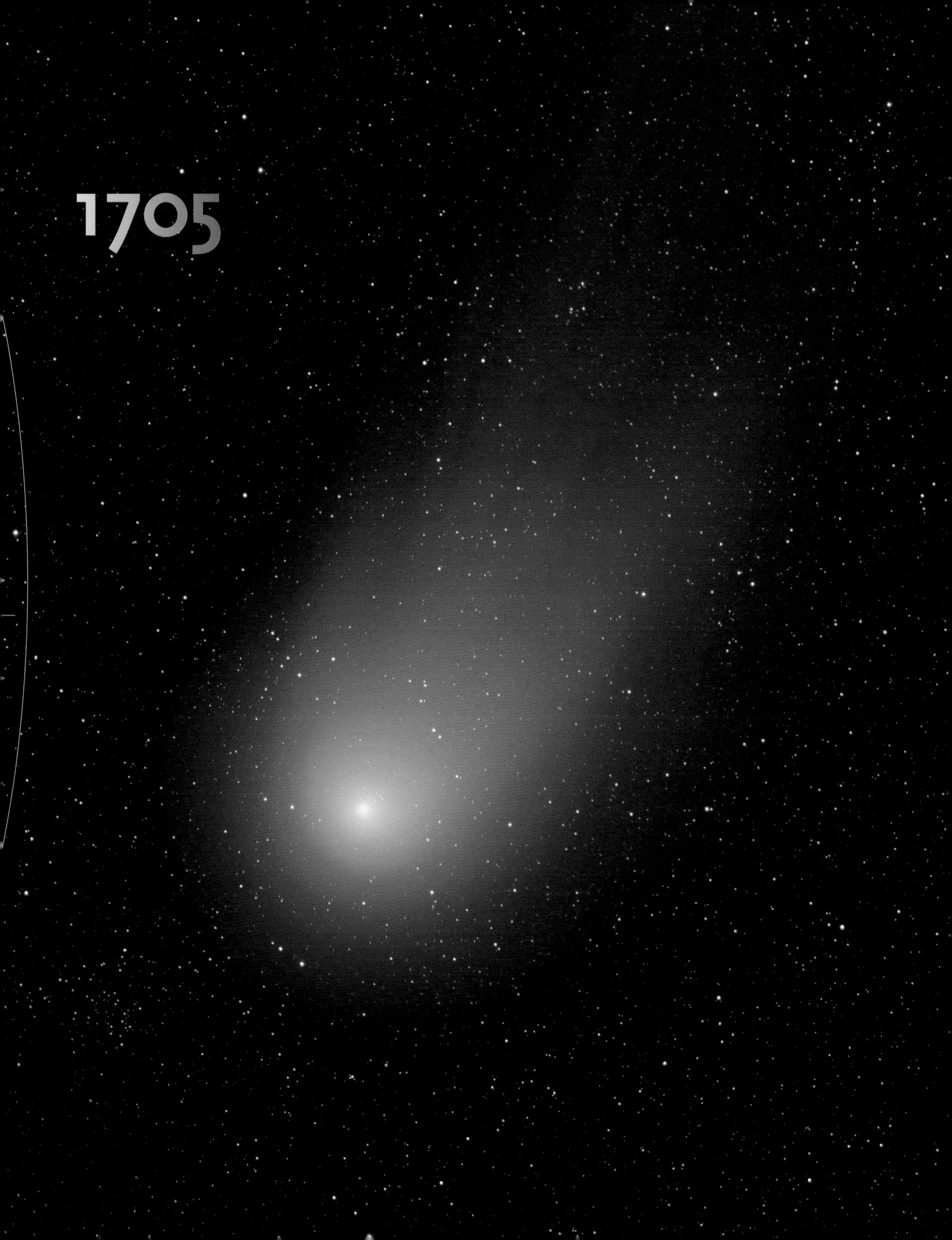
1705

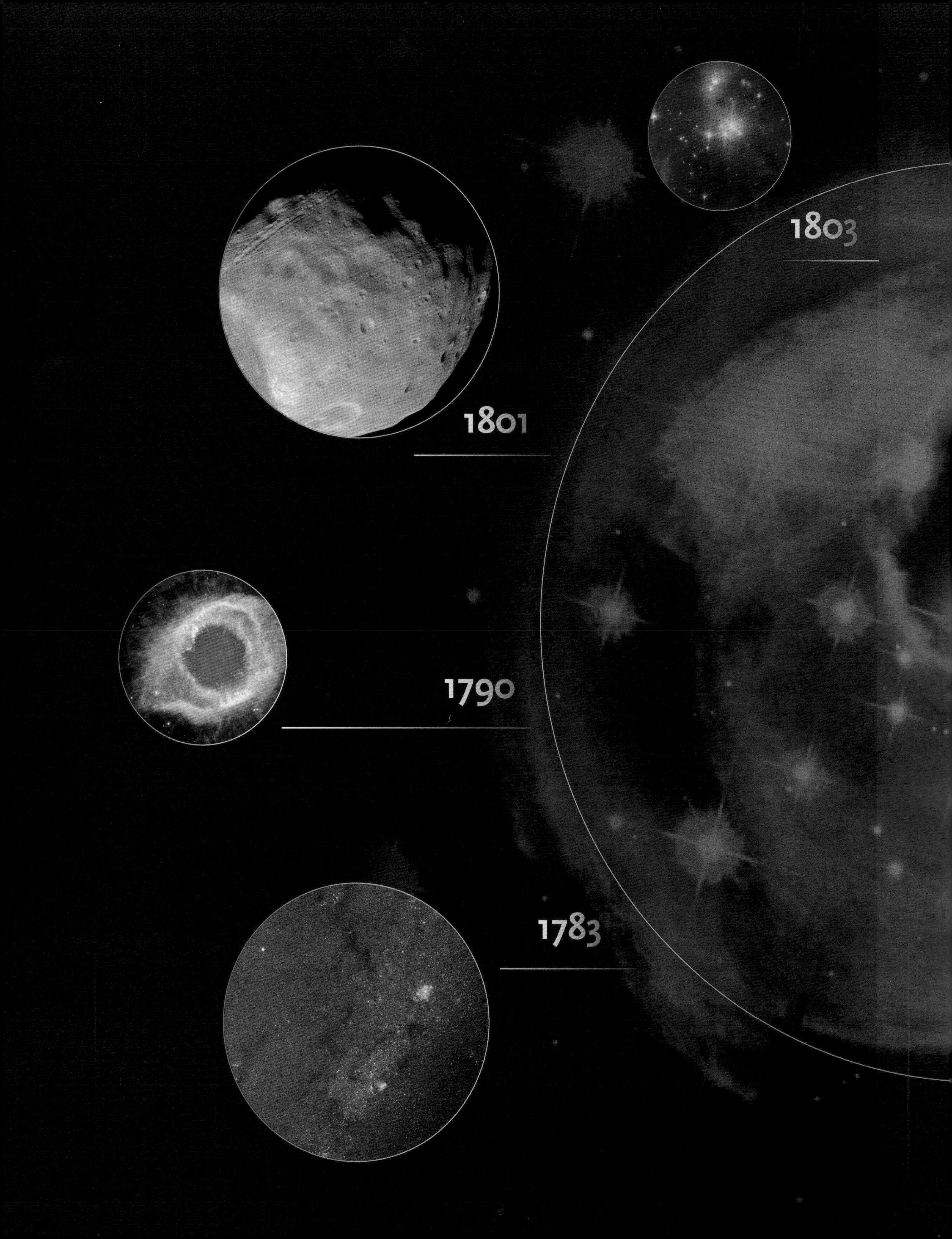
1803
1801
1790
1783

1708 - 1808

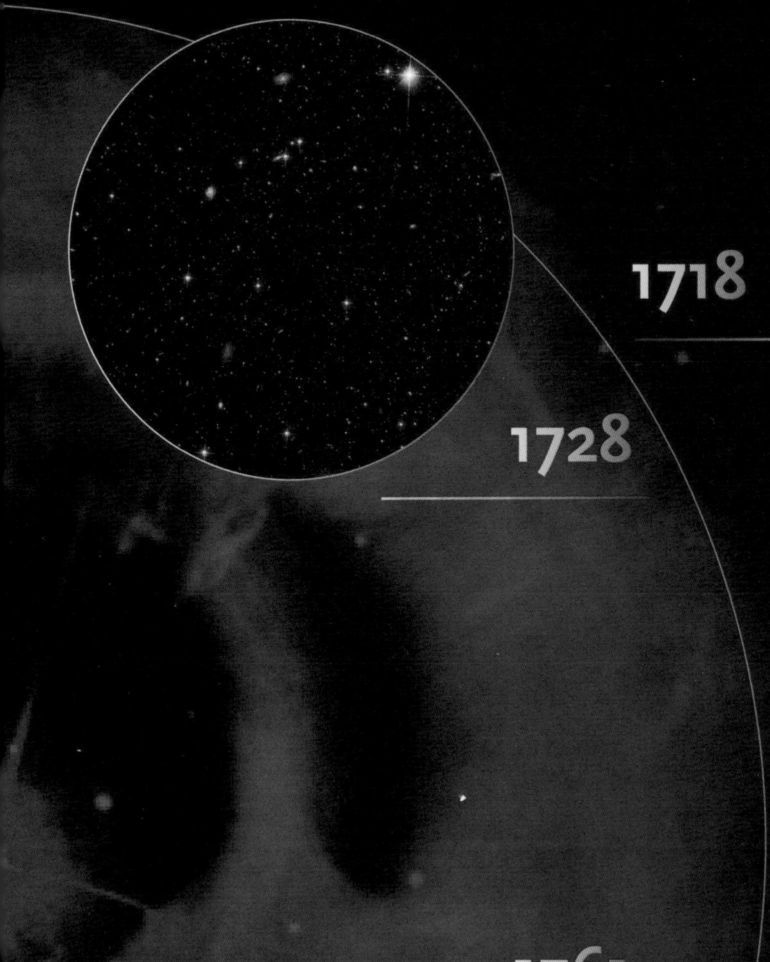

1718

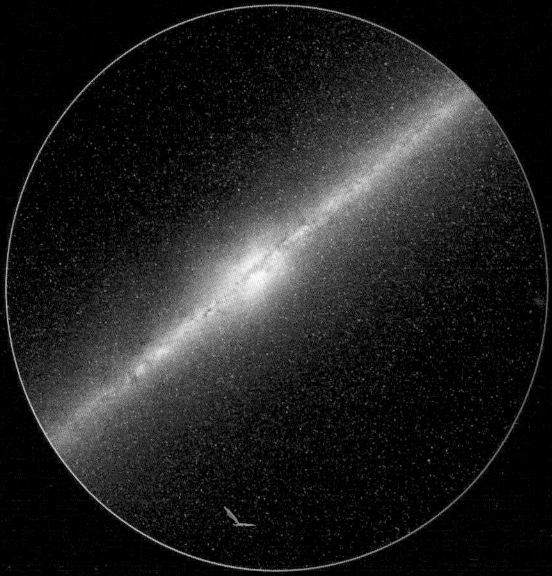

1728

1761

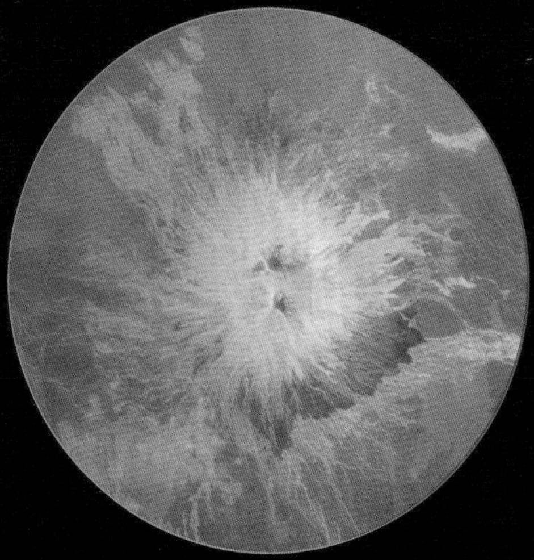

1781

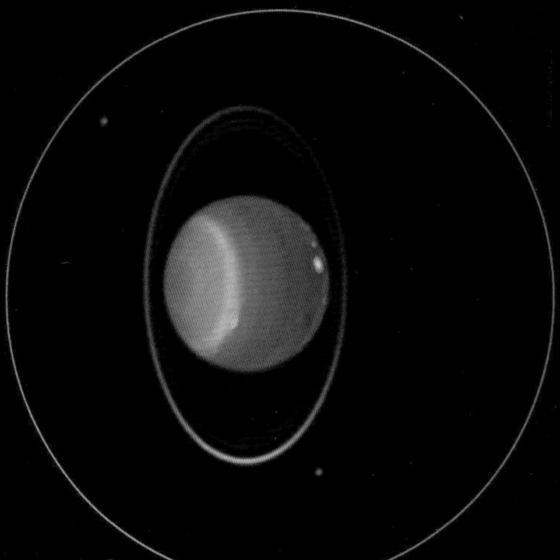

Observing everything that appeared in the eyepiece of the telescope at random – that pretty much typified astronomy in the first half of the seventeenth century. No wonder: everything was new, and new discoveries were there for the picking; astronomers were like children in a newly opened candy shop. But in the eighteenth century, this haphazard method slowly made way for a more

Swarms of Stars on a

systematic approach that ultimately led to important new insights into the nature of the universe.

One of the greatest surprises was perhaps the discovery that the night sky is not static and unchanging, as had been taken for granted for thousands of years. Instead it proved to be a dynamic, three-dimensional stage, swarming with myriads of stars. The fact that we see the same Big Dipper as the cave men says more about our transience than the assumed permanence of the firmament.

In the eighteenth century the systematic study of the night sky is taken in hand by William Herschel, who catalogs countless nebulae and binary stars and caps it all in 1781 when he discovers a new planet far beyond the orbit of Saturn. Besides all this, Herschel also explains the motion of the Sun and explores the three dimensional structure of the Milky Way.

Three-Dimensional Stage

None of these discoveries would have been possible with the simple telescopes built by Hans Lipperhey and Galileo Galilei. Yet telescope-building also takes great steps forward in this period; the telescopes built by Herschel – who is in fact an amateur astronomer – are the largest ever made.

1718

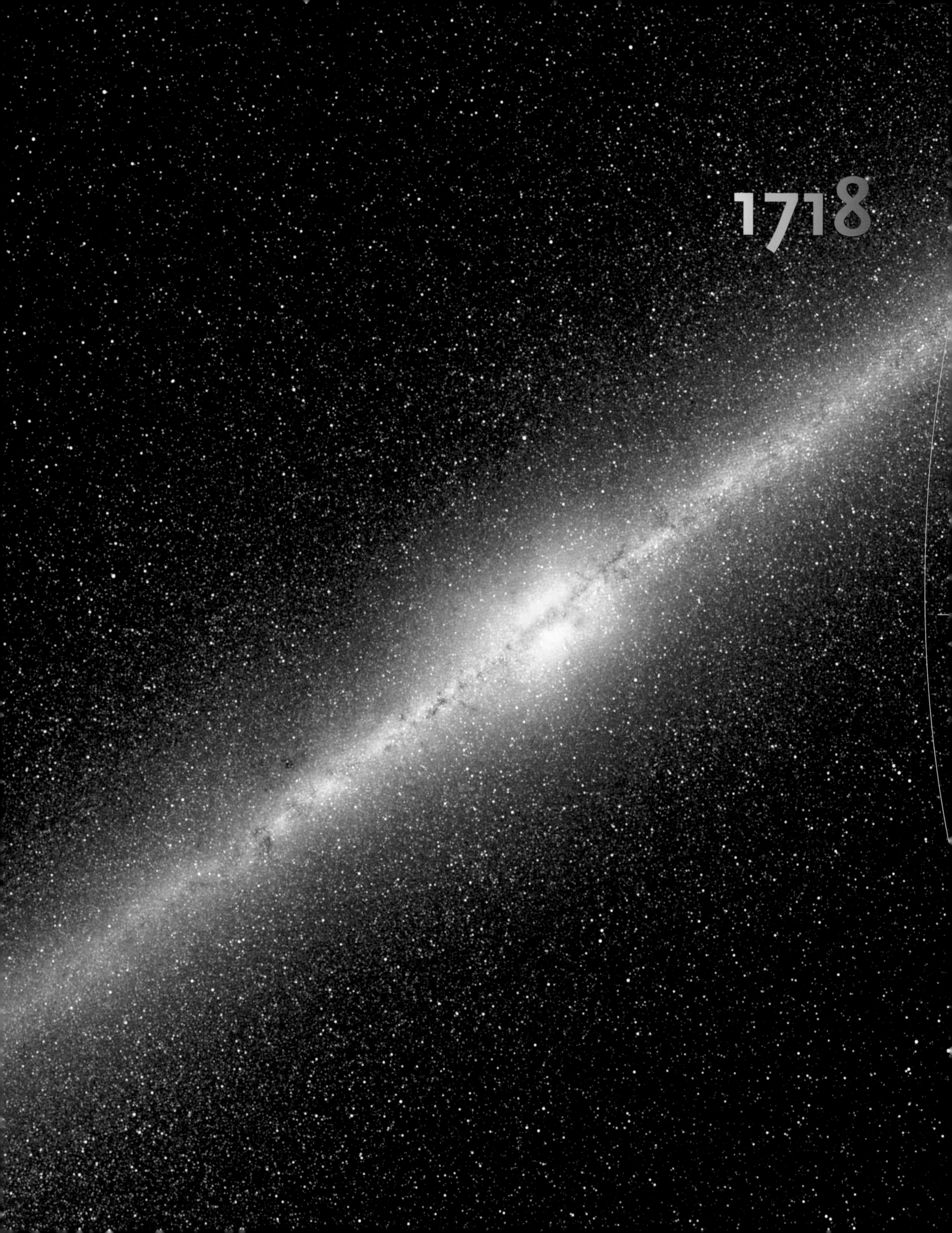

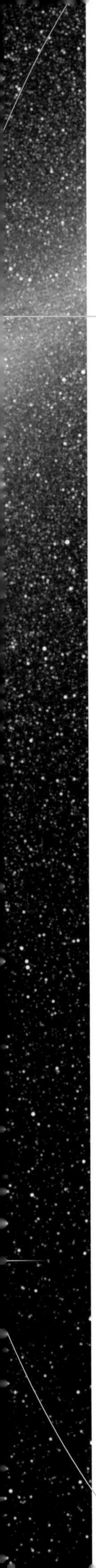

Star Trek

Edmund Halley Discovers the Proper Motion of Stars

The appearance of the night sky is constantly changing. In the course of the night all celestial bodies rotate around the Pole Star. New stars rise in the east and others disappear below the horizon in the west and because the Earth is orbiting the Sun, we also see a different part of the night sky at different times of the year.

But the relative positions of the stars do not appear to change. The Big Dipper and Orion always look the same, and astronomers have referred to 'fixed stars' for many centuries, with good reason.

In the second century BC, the Greek astronomer Hipparchus measures the positions of the bodies in the night sky. More than 200 years later, Ptolemy incorporates most of Hipparchus' star catalogue in his standard work, the *Almagest*, which contains the positions of 1,022 stars.

Edmund Halley conducts much more accurate positional measurements, using telescopes, micrometers and accurate clocks. Halley compares his star positions with those of Ptolemy, which were based on measurements made more than eighteen centuries previously. The big question is, though, has anything changed in the meantime?

In 1718 Halley observes a significant change of position for three bright stars. Sirius, the brightest star in the sky, is more than half a degree further south than it was in the time of Hipparchus. Arcturus, the main star in the constellation Boötes, has even moved by more than a full degree. And Aldebaran in the constellation Taurus is no longer in exactly the same place in the firmament.

The invariability of the night sky is clearly an illusion. Anyone who waits long enough will eventually see the Big Dipper and Orion change. The proper motion of stars that are relatively close to the Earth, like Arcturus, Sirius and Aldebaran, is quite easy to observe; more distant stars seem to move less quickly.

In 1916 Edward Emerson Barnard discovers that a faint star in the constellation Ophiuchus moves no less than 10.3 arcseconds a year – a record. Barnard's Star is a red dwarf at less than six lightyears distance, and can only be seen with a telescope. Since Hipparchus' time, it has moved a distance of more than 5°.

Stars not only display proper motion in the sky, they also move closer to or further away from the Earth. If this radial velocity and the distance of a star are known, its actual spatial velocity can be calculated.

Between 1989 and 1993, the European satellite Hipparcos (named after the Greek astronomer but also an abbreviation for HIgh-Precision PARallax COllecting Satellite) conducts precision measurements of the positions, distances and motions of more than a hundred thousand stars in the Milky Way. The Hipparcos catalog is the most accurate star catalogue ever.

Arcturus is the brightest star in the constellation of Boötes. The illustration here comes from Johann Bayer's star atlas *Uranometria*. Arcturus was one of the first stars whose proper motion in the sky was measured. Arcturus possibly originated in another galaxy.

Five hundred million stars are visible in this image of the Milky Way, based on infrared photographs made by the Two Micron All Sky Survey (2MASS). The future European satellite Gaia will measure the proper motion of nearly all of them. (2MASS)

Subtle Swings

James Bradley Discovers the Aberration of Starlight

The stars move, and they are at different distances from the Earth. But how do you determine those distances? Parallax measurements offer the solution. For example, the Earth orbits the Sun so in January you see a star in a slightly different direction than in July. That difference in position is known as a star's parallax. Using the parallax, it is easy to measure a star's distance from the Earth.

In the early eighteenth century, astronomers begin to hunt for the parallax, but without success. The positional measurements have to be conducted with extreme precision, and refraction of light by the Earth's atmosphere makes that even more difficult.

British astronomers James Bradley and Samuel Molyneux concentrate on the star Gamma Draconis. The star is reasonably bright and is probably relatively close to Earth. Furthermore, in London, it moves exactly through the zenith – the point directly above your head, meaning that there is no problem with atmospheric refraction.

At the end of 1725, the two astronomers set up a large telescope alongside the chimney of Molyneux' house in Kew. Over several days in December they measure the position of Gamma Draconis, which proves to vary somewhat. In March 1726 the star reaches its most southerly position in the sky; six months later it is forty arcseconds more to the north.

But this cannot be caused by the parallax; the star is shifting in the wrong direction. Bradley and Molyneux do not understand it at all. Is there perhaps a slight, periodical shift in the Earth's axis? If so, a star on the other side of the celestial pole would show the same change in the opposite direction. However, that proved not to be the case.

The positional measurements conducted by Bradley and Molyneux are the most accurate ever made. Other telescopes show similar shifts in the positions of 200 other stars and all of those stars have the same maximum deviation from their average position in the sky: twenty arcseconds. There is an unexplained aberration – a deviation in both the literal and figurative sense.

Molyneux dies in April 1728 without ever knowing the solution to the aberration mystery. A couple of months later Bradley suddenly sees the light; abberation is caused by the motion of the Earth around the Sun and the finite speed of light. Because the direction of the Earth's motion is continually changing, the direction from which starlight reaches us is also always slightly different. It is like when you walk in the rain and have to hold your umbrella tilted forward a little, even though the rain is falling straight downwards.

The accidental discovery of the aberration of starlight provides definite proof that the Earth is in motion. At a later age, James Bradley finally discovers nutation – the small, periodic change in the position of the Earth's axis, which has a cycle of 18.6 years. It will be more than a century, however, before the parallax of a star is determined for the first time.

Gamma Draconis was the first star for which the aberration of starlight was discovered. This is the periodic shift in position of a star in the sky as the result of the Earth's movement around the Sun and the finite speed of light.
(Digitized Sky Survey)

The 'deepest' photograph of the cosmos ever made shows more than 300,000 stars in the halo of the Andromeda Galaxy, the nearest big neighbor to our own Milky Way. Every star in the sky shows an annual aberration in its position.
(NASA/ESA/T.M. Brown)

1728

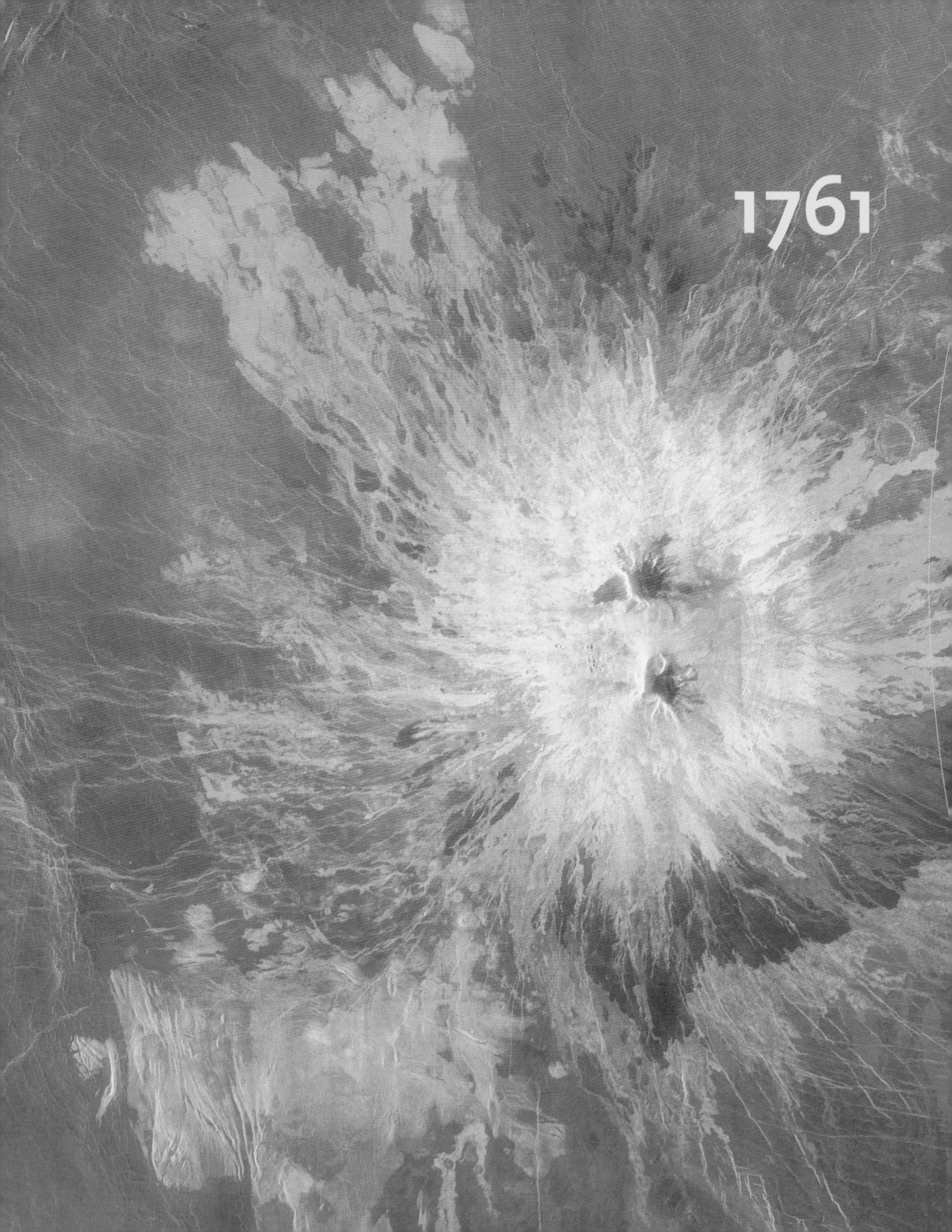
1761

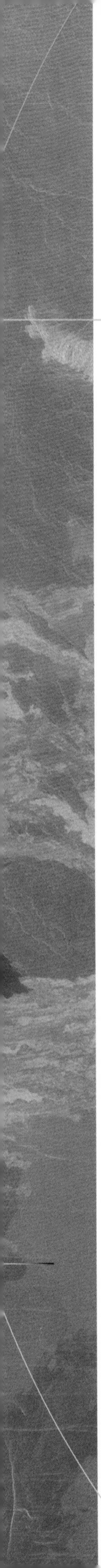

Veiled Sister

Mikhail Lomonosov Discovers the Atmosphere Around Venus

The planet Venus moves around the Sun in a smaller orbit than the Earth. Every now and again it passes between the Earth and the Sun and you can see a round black spot move across the face of the Sun. Transits of Venus occur in pairs, with about 8 years between them, and the next pair of transits then take place a little over a century later.

In 1627, Johannes Kepler predicts that there will be a transit of Venus at the end of 1631, but unfortunately no one observes it. The English astronomers Jeremiah Horrocks and William Crabtree do, however, see the transit of December 4, 1639. Astronomers realize that accurate timing measurements of a Venus transit will make it possible to determine the size of the solar system. This method will be much more accurate than the one applied by Jean-Dominique Cassini in 1672, based on parallax observations of Mars. In 1716, with this in mind, Edmund Halley calls on astronomers to closely observe the transit of Venus on June 6, 1761.

One of these observers is the Russian genius Mikhail Vasilyevich Lomonosov. Lomonosov is born in 1711, the son of a poor fisherman in a small village near Arkhangelsk, in the far north of Russia. Amazingly, he walks to Moscow, completes a 12-year course of study in 5 years, and eventually becomes a professor in St. Petersburg.

Lomonosov is the Einstein of Russia. He is a physicist, astronomer, philologist, chemist, geophysicist, philosopher, poet and artist. He improves on Robert Boyle's gas theory, predicts the existence of Antarctica, and discovers the law of the conservation of matter. In 1655 he founds Moscow State University.

On June 6, 1761, Lomonosov observes the transit of Venus together with his colleagues Andrey Kasilinkov and Nikolay Kurganov of the St. Petersburg Observatory. At the start of the transit he sees a ring of light around the silhouette of the planet which lasts less than a second. There is only one explanation: like the Earth, Venus must have an atmosphere which refracts and scatters sunlight. That same year Lomonosov publishes his discovery in Russian and German.

The discovery that Venus has an atmosphere is world news. Who knows, perhaps Venus is not only the Earth's neighbor but also its sister! In the course of the twentieth century, however, it becomes clear that Venus bears very little resemblance to the Earth. In 1932, Walter Adams and Theodore Dunham discover carbon dioxide in the planet's atmosphere, and on December 14, 1962, the American space probe Mariner 2 passes close to Venus and measures a surface temperature of almost 500 degrees Celsius and air pressure of more than 90 atm.

However in 1761 Lomonosov still toys with the idea that there might be life on Venus, and in the mid-twentieth century the planet is seen as a tropical oasis. But we now know that Venus is the Earth's scorched, infernal step-sister and not her true sister as originally thought.

➋ The surface of Venus is permanently shrouded in a closed blanket of clouds. The cloud blanket, which reflects the light of the Sun, gives Venus its extreme brightness. The air pressure on the surface of the planet is 90 times that of Earth at sea level (90 atmosphere). (NASA/Mattias Malmer)

➌ The Magellan planetary explorer mapped the surface of Venus using radar. This radar map shows the volcano Sapas Mons. Old lava flows can be clearly seen on the radar image because – as on Earth – they have a rough surface. (NASA/JPL)

Fallen Planet

William Herschel Discovers the Planet Uranus

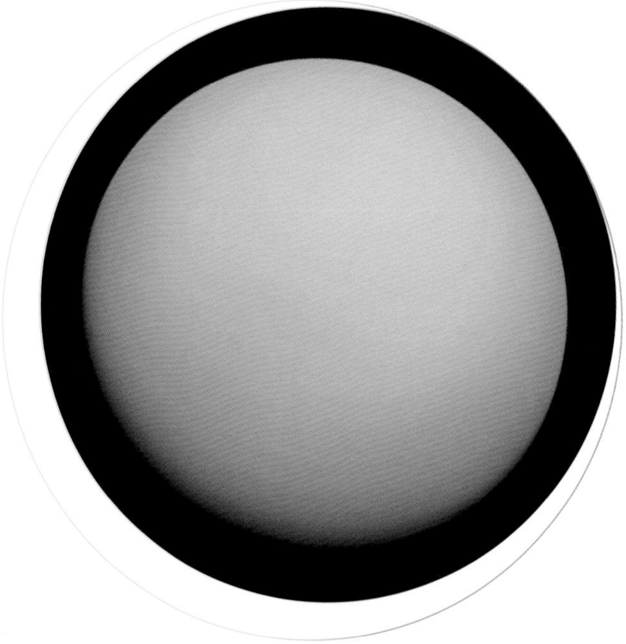

William Herschel is a composer and musician, not an astronomer. In 1756, aged 18, together with his brother Jacob, he moves from his birthplace in Hanover to England, where he plays oboe, violin and organ. He writes more than 20 symphonies and is leader of the orchestra in his new home city of Bath. In 1772 he is joined by his younger sister Caroline.

How does a respected musician become one of the most renowned astronomers in history? Well, discover a new planet and the rest follows automatically. Herschel becomes interested in astronomy in 1773, builds telescopes and spends his nights observing the Moon and the stars.

On Tuesday March 13, 1781, between 10 and 11 in the evening, Herschel is standing in the garden of his house on New King Street with his self-built 15-centimeter telescope. In the constellation of Gemini he sees a star that is larger than the rest: it is not a minuscule point of light, but a sharply defined disk. A day later the mysterious object has moved a little across the sky.

Herschel thinks that he has discovered a new comet at a great distance from the Sun, but there is no sign of a tail. He watches the comet for a few weeks and corresponds with Nevil Maskelyne, the Astronomer Royal, about it. He suggests that it might be a new planet, far beyond the orbit of Saturn. German astronomer Johann Bode agrees.

Herschel himself is not convinced that he has discovered a new planet until 1783. King George III is so impressed with Herschel's discovery that he awards him an annual grant of £200. As a token of his gratitude Herschel wants to call the new planet *Georgium Sidus* ('George's Star') and the English Nautical Almanac Office uses the name until 1850, but by then the name Uranus – proposed by Bode – has become widely accepted.

It is nearly impossible to detect any details on the small, faraway planetary disk, but on January 11, 1787, Herschel discovers two moons around the planet, which are later given the names Oberon and Titania. (In the summer of 1789 he also finds two new moons around Saturn: Enceladus and Mimas.) It is immediately clear that there is something strange about Uranus. The orbits of its moons are perpendicular to Uranus' own orbital plane. If you assume that they move in the planet's equatorial plane, Uranus must be a fallen planet, whose axis lies practically in its orbital plane.

Although three more moons are discovered around Uranus in 1851 (Ariel and Umbriel) and 1948 (Miranda), the distant planet remains a mystery until January 24, 1986, when it receives a visit from the American space probe Voyager 2.

Herschel is the first man in history to discover a planet. John Flamsteed could have beaten him to it nearly a hundred years previously. In 1690 he observed Uranus, but didn't think it was anything special. He recorded it as just another star in the constellation Taurus.

Uranus was studied close up for the first time in January 1986, by the American planetary explorer Voyager 2. The blue-green color of the atmosphere indicates the presence of methane gas. There are almost no clouds to be seen; they are located too deep in the atmosphere. (NASA/JPL)

The dark ring system, a few of the many moons, and several clouds are visible on this infrared photo of Uranus, made by the Hubble Space Telescope in 1998. This picture clearly shows that the axis of the planet more or less coincides with its orbital plane. (NASA/JPL/STScI)

1781

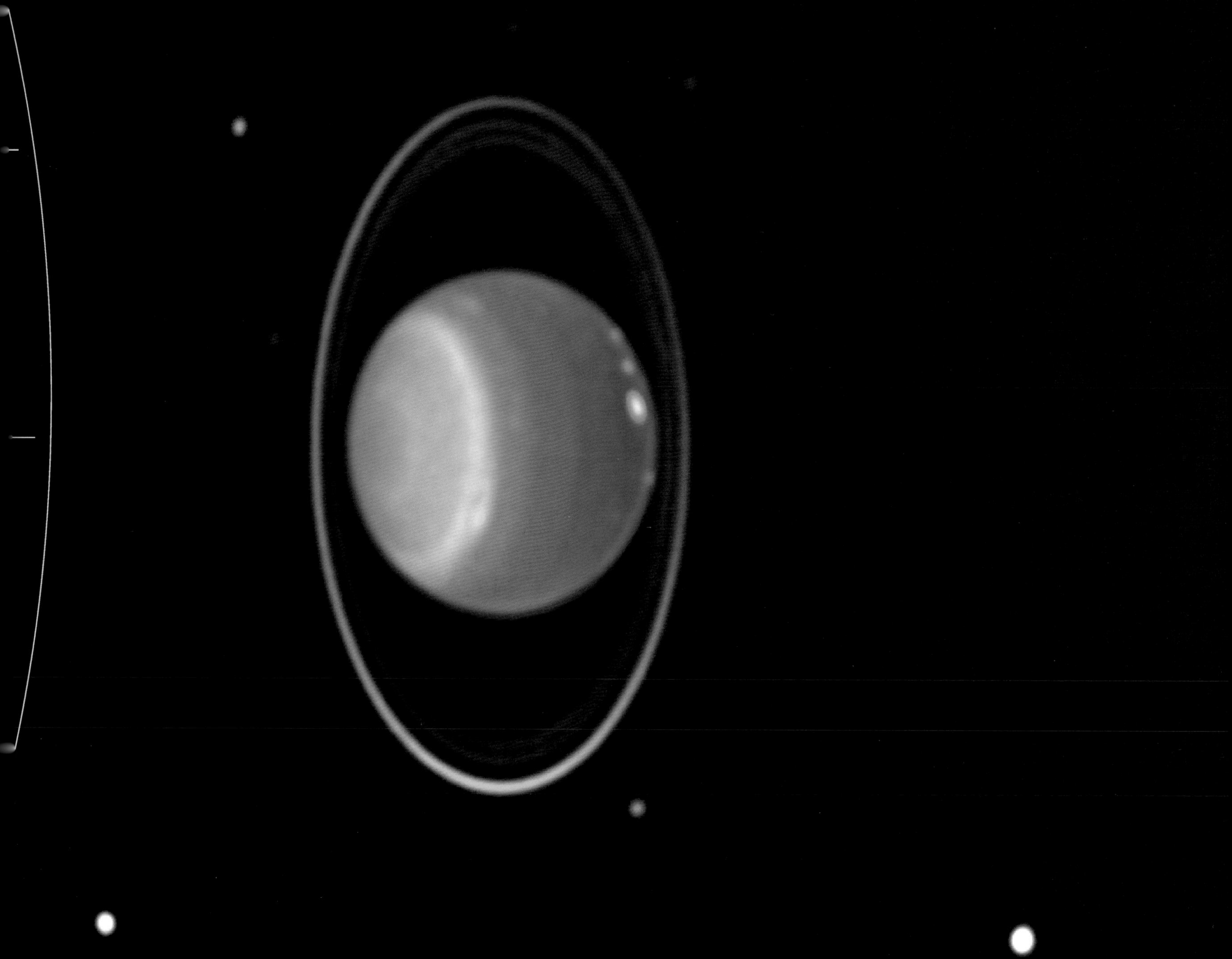

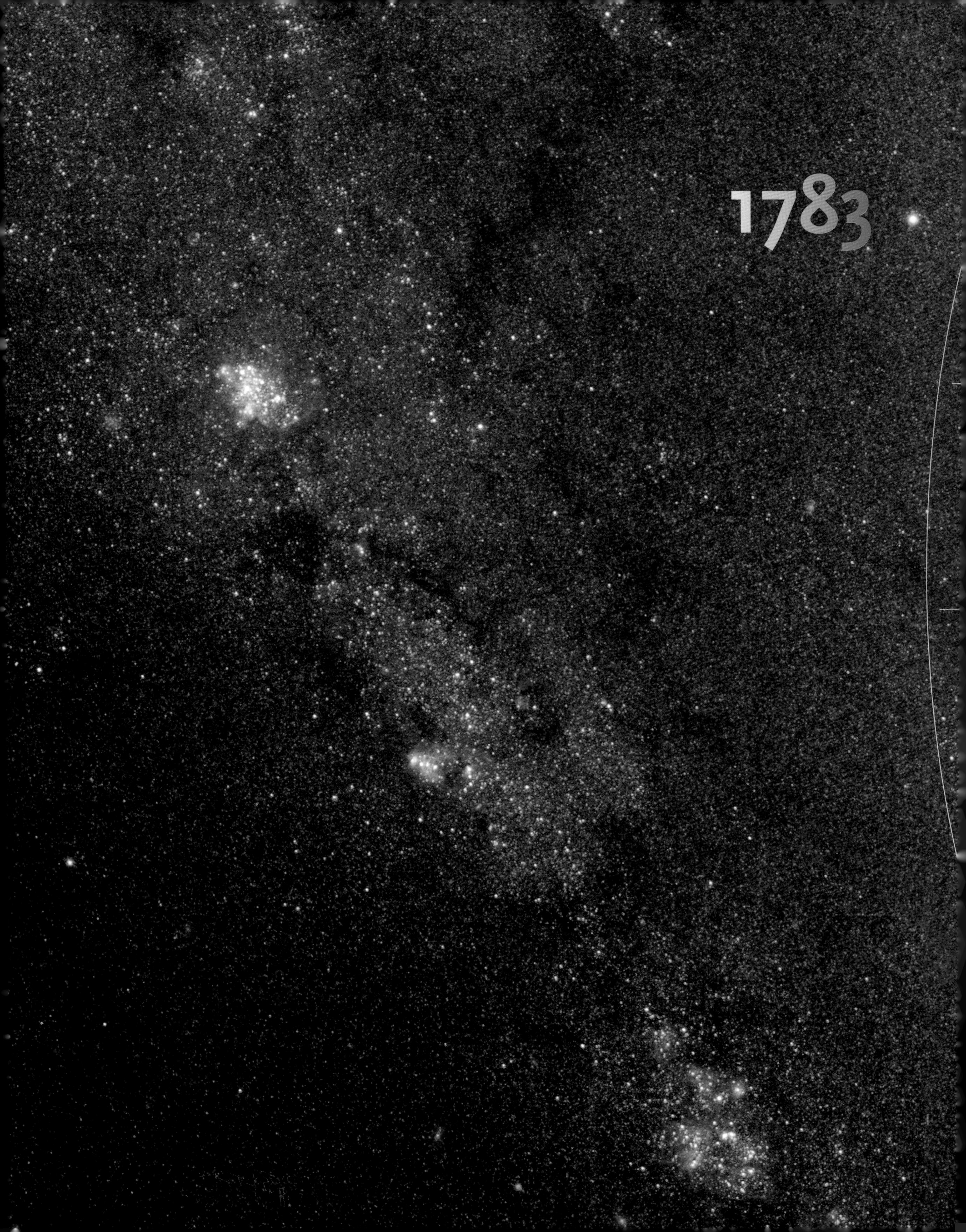
1783

Route du soleil

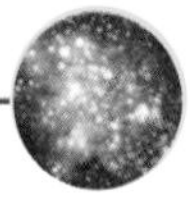

William Herschel Discovers the Motion of the Sun Through Space

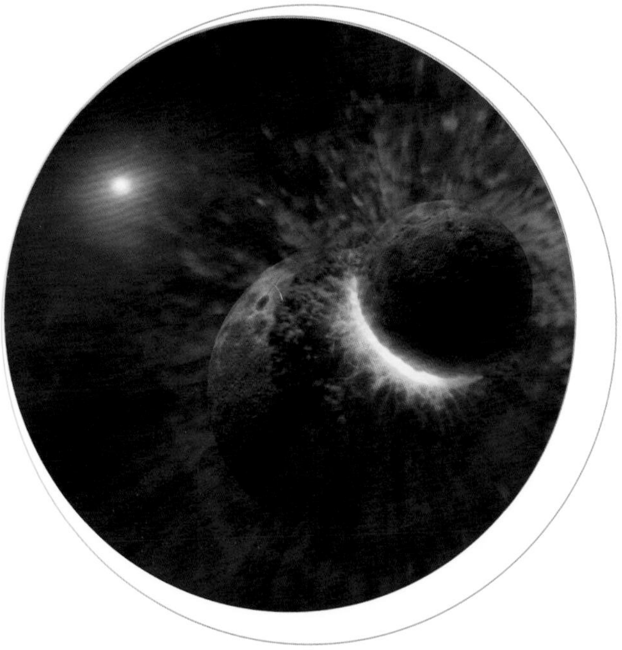

Star Trek fans need no introduction to the solar apex. When the *USS Enterprise* goes into warp speed, you see the stars in the Milky Way shoot through the sky at high speed. They seem to be racing away from the point towards which the spaceship is moving. The same effect can be seen from the cab of an express train. The point towards which the observer is moving, and from which the surrounding environment seems to be fleeing, is known as the apex.

William Herschel introduces the term in 1783, when he discovers the apex of the Sun – the point towards which the Sun is moving in the sky. Apparently it is not only the Earth that moves through space, the Sun does, too, and Herschel's discovery shows that the Sun is not the fixed center of the cosmos.

Herschel elaborates on Edmund Halley's discovery of the proper motion of the stars. In 1718 Halley discovers that Arcturus, Sirius and Aldebaran are not in the same position in the sky as they were in the time of the Greek astronomer Hipparchus. But Herschel's positional measurements are much more accurate than Halley's; he has the best telescopes available in his time.

Herschel does not compare his measurements with those of Hipparchus, which are not precise enough. Instead he uses the star catalog compiled by John Flamsteed at the start of the eighteenth century. Most of the fainter stars are in the same position as they were in Flamsteed's time. (They are probably so far away that, even after more than one and a half centuries, their proper motion is immeasurably small.) But many brighter stars prove to have moved a little since Flamsteed took his measurements.

Herschel concludes that, if all the other stars move through space, the Sun certainly does, too. The Sun's proper motion should be discernible from the motions of stars in the immediate vicinity. Herschel may never have heard of Star Trek, but he realizes that the stars will have to be moving systematically away from the solar apex.

We now know the direction in which the Sun is moving with reasonable accuracy, thanks to precision measurements on hundreds of thousands of stars. It is heading towards a point in the sky in the constellation Hercules, not far from the bright star Vega, in Lyra. But Herschel has much less data to work with. He measures the proper motion of no more than 14 stars and concludes that the solar apex is near the star Lambda Herculis – surprisingly close to its actual position.

In 1783 Herschel publishes his results in the *Philosophical Transactions of the Royal Society*. Astronomers have still not succeeded in determining the distance to a star, so they therefore know little about the size of the universe. But Herschel's discovery makes it clear that our solar system is not the fixed center of that universe. Just like all the other stars, the Sun is a small pinhead of light, journeying through the vastness of the cosmos.

The dust disk around the young, bright star Vega was probably created by a collision between two embryonic planets, a million years ago at the most. Vega is the star that the Sun is more or less moving towards as it rotates around the center of the Milky Way. (NASA/JPL/Tim Pyle)

Bright star formation areas mark one of the spiral arms of the M81 galaxy in Ursa Major. Our own Milky Way looks roughly the same from a great distance. The Sun and all the other stars orbit the core. (NASA/ESA/Hubble Heritage Team)

Deadly Beauty

William Herschel Discovers the Central Star in a Planetary Nebula

In 1781 the French astronomer Charles Messier publishes his *Catalogue des Nébuleuses & des Amas d'Étoiles (Catalog of Nebulae and Star Clusters)*. It contains four planetary nebulae, including the Dumbbell Nebula in the constellation Vulpecula and the Ring Nebula in Lyra. But Messier has no notion of the true nature of these faint objects. He thinks that the almost circular nebular spots are star clusters whose stars cannot be seen individually.

In the summer of 1782, a year after he discovers the planet Uranus, William Herschel begins his own large-scale project to map the night sky in which he records as many star clusters, nebular spots and binary stars as possible. One of his first discoveries, on September 7, is a small, faint nebula in the constellation Aquarius. It is almost circular, quite sharply defined, and slightly green in color – just like the Dumbbell Nebula and the Ring Nebula. It looks a little like the blue-green planetary disk of Uranus.

When Herschel creates a classification system for nebulae in the 1780's he calls these faint, symmetrical spots of light 'planetary nebulae'. Of course he knows that the nebulae in reality have nothing to do with planets, but the name ultimately sticks, and Herschel still has no idea what kinds of objects they really are.

That changes on November 13, 1790. In the constellation of Taurus, Herschel discovers a planetary nebula that now has the catalog number NGC 1514. It is as round, sharply defined and green as the others, but it has a clearly visible star in the middle. Herschel concludes that the nebula is not a far off collection of faint stars, but a cloud of gas or dust that has something to do with the star itself.

Later, with larger and more powerful telescopes, it becomes clear that *all* planetary nebulae have central stars. Astronomers now know that planetary nebulae are the final breaths of dying stars – swollen giants blasting their outer layers into space. While the nebula expands and dissolves, the giant star shrinks to become a hot, white dwarf. The ultraviolet radiation of the dwarf heats the rarefied gas, lighting it up in spectacular colors.

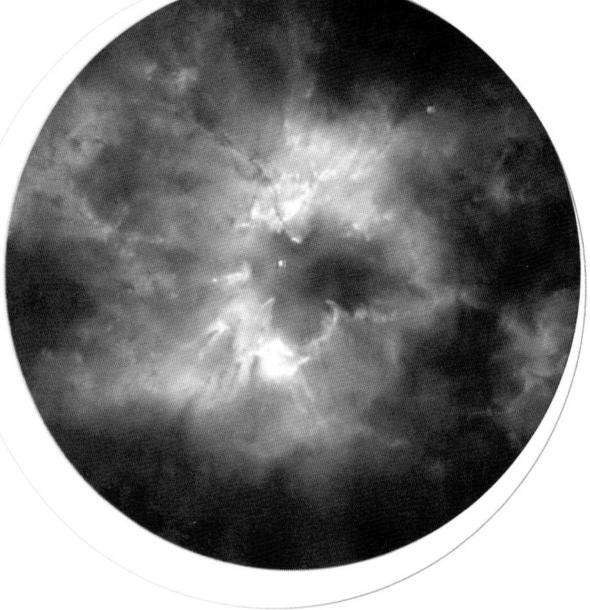

In a few billion years our own Sun will swell up to become a red giant surrounded by an expanding planetary nebula. Surprisingly enough the structure of the nebula will possibly be influenced by the presence of the giant planets and the flattened belts of asteroids and ice dwarfs that orbit the Sun.

Many planetary nebulae are bipolar: most of the gas ends up in two large lobes on opposite sides of the central star. That is probably because the dying star is part of a binary star system or because, like the Sun, it has a planetary system. In hindsight, Herschel's name 'planetary nebula' was perhaps not so wide off the mark after all!

The hot, white dwarf star in the center of the planetary nebula NGC 2440 has a surface temperature of about 200,000 degrees Celsius. The high-energy radiation of the star heats up the nebular material expelled into space at an earlier stage, when the star was still a red giant.
(NASA/ESA/Hubble Heritage Team)

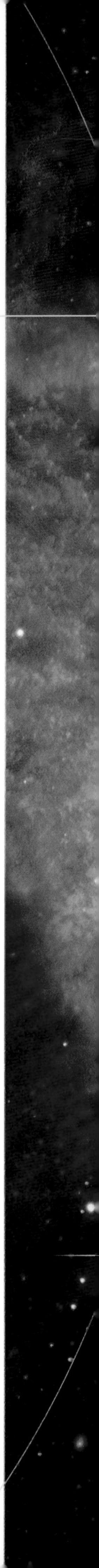

The Helix Nebula, in the constellation of Aquarius, is one of the closest planetary nebulae. This colorful image was composed from photographs taken by the Hubble Space Telescope and the infrared Spitzer Space Telescope.
(NASA/JPL/ESA)

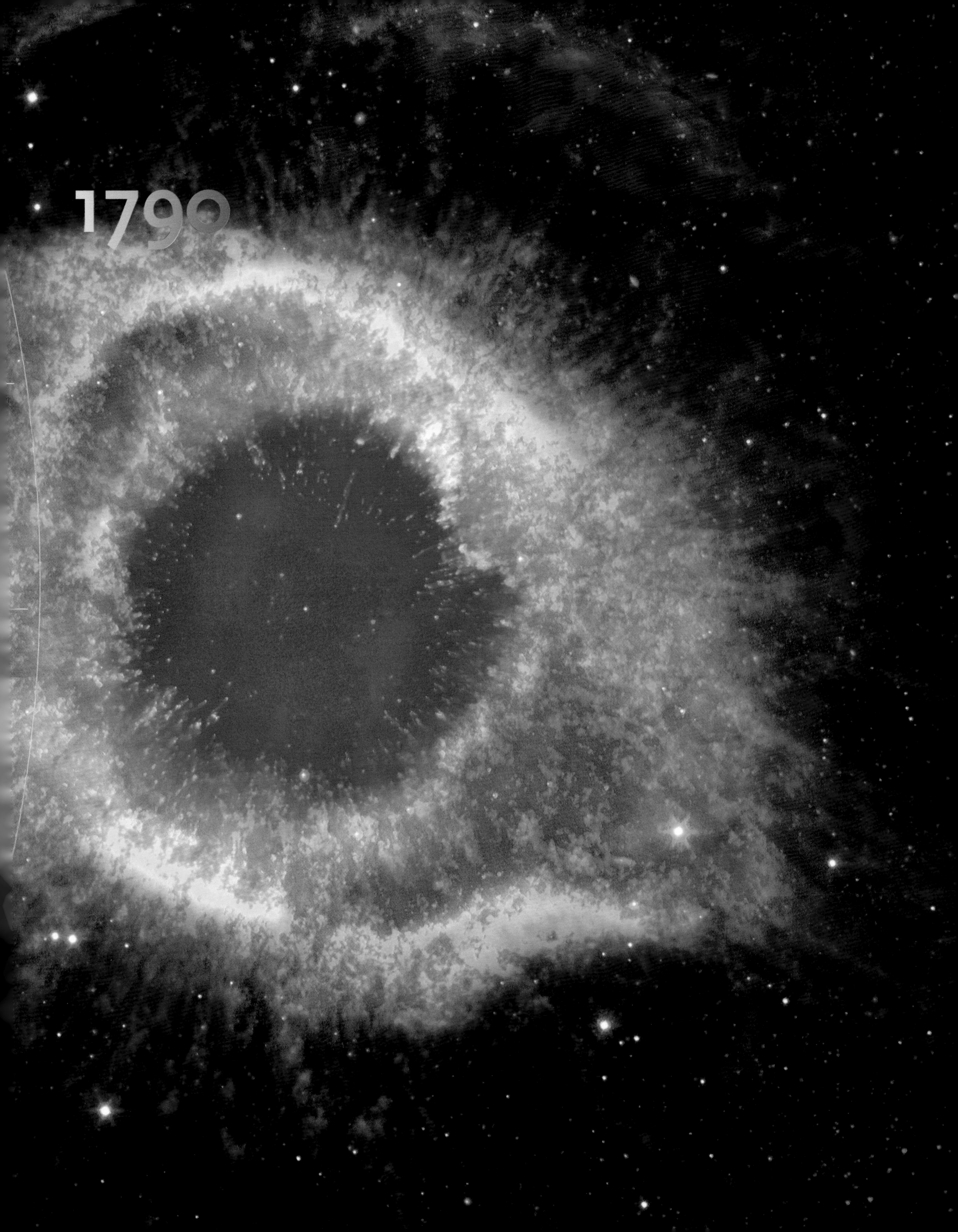
1790

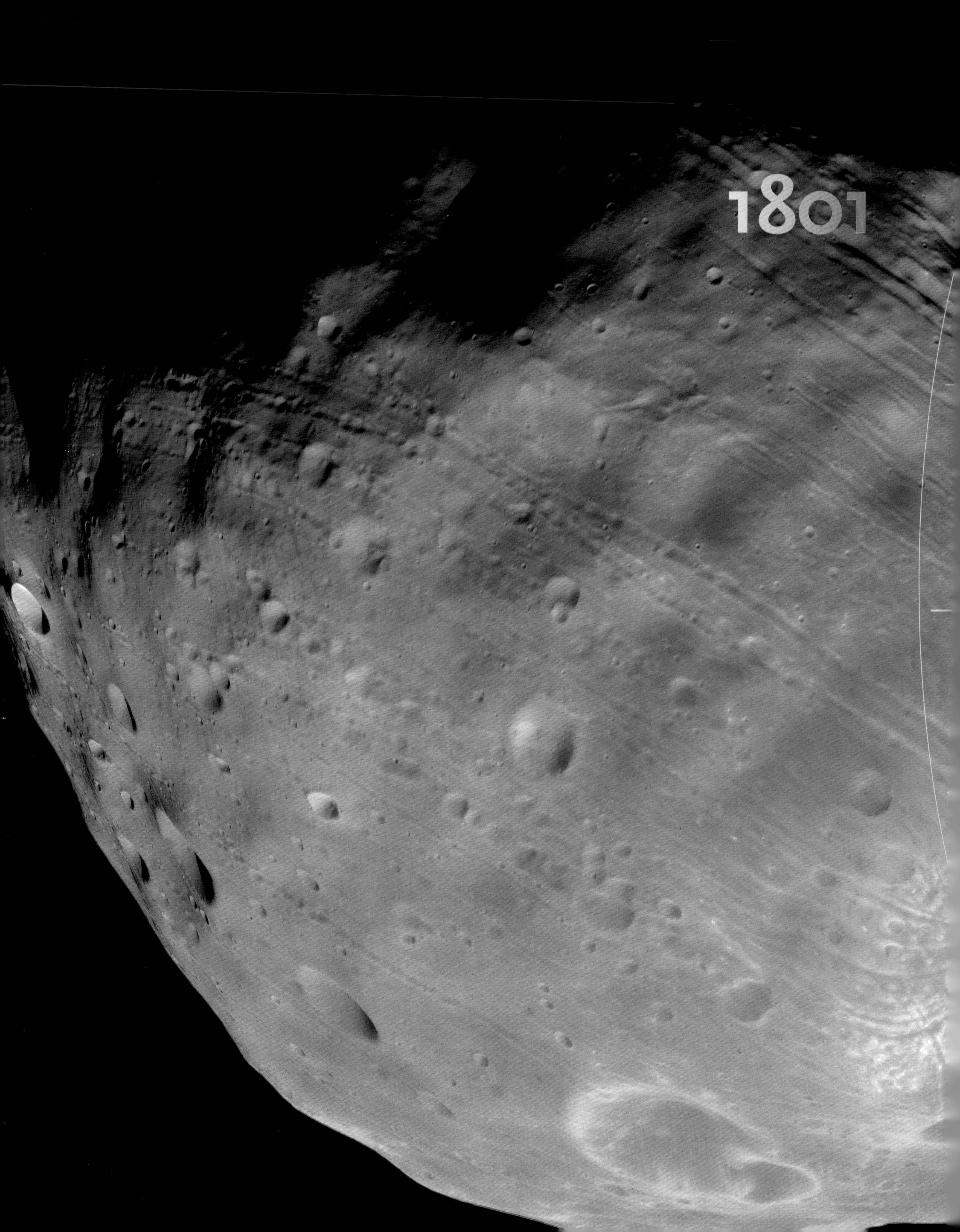
1801

Celestial Vermin

Giuseppe Piazzi Discovers Ceres, the First Asteroid

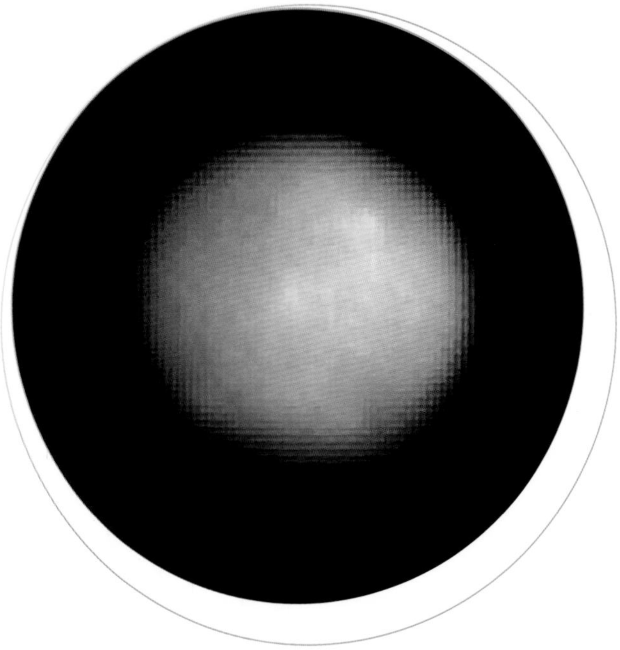

After William Herschel's accidental discovery of Uranus in 1781, other astronomers are also inspired to hunt for new planets. With a planet having been found beyond the orbit of Saturn, it was high time for the 'missing' planet between the orbits of Mars and Jupiter to be tracked down.

Johannes Kepler already expressed his surprise at the relatively large gap between the orbits of the two planets. In 1772, Johann Bode publishes a mathematical rule that describes the distances between the planets relatively accurately, but only if there is 'something' between Mars and Jupiter.

In September 1800, six European astronomers decide to organize a hunting party for the unknown planet. They call in the help of a few dozen colleagues to comb the night sky for a moving speck of light, each of them focusing on a different section.

The Sicilian astronomer Giuseppe Piazzi is among those invited to join this *'Celestial Police'*. But, for one reason or another, the letter from Germany does not reach him, so when he discovers the long-sought planet on the first day of the nineteenth century, it is pure coincidence.

Piazzi is working on an accurate star catalog, and on January 1, 1801, he measures the positions of a large number of stars in the constellation Taurus. He repeats the measurements the following day and sees that one of the stars has shifted position slightly. Has he discovered a new comet? Or has he – like Herschel – found an as yet unknown planet?

Piazzi calls the new object Ceres Ferdinandae, after the patroness of Sicily and King Ferdinand IV. But in the early spring, Taurus is not visible and calculating the orbit of the new object proves difficult; Ceres disappears from view. It is not sighted again until December 1801, thanks to a revolutionary calculation method devised by mathematician Carl Friedrich Gauss.

Ceres is heralded as the eighth planet. But when three more small objects are discovered between the orbits of Mars and Jupiter in 1802, 1804 and 1807 (Pallas, Juno and Vesta), some astronomers are of the opinion that the four bodies do not really merit the name 'planet'. After all, they do not show a planetary disk, like Uranus. Yet, until the mid-nineteenth century, many astronomy books continue to speak of 11 planets.

That changes in the years after 1845, as the number of new objects increases rapidly. By 1868 a hundred of these 'asteroids' have been discovered. Astronomers describe the small bodies as the 'vermin of the skies'. In 2008, the counter has passed the hundred thousand mark. With a diameter of around 950 kilometers, Ceres is simply the largest, and was therefore the first to be found.

Giuseppe Piazzi continues to be convinced until his death in 1826 that he has discovered a new planet. The Sicilian achieves immortality in 1923, when the thousandth asteroid, discovered by Karl Reinmuth, is called Piazzia.

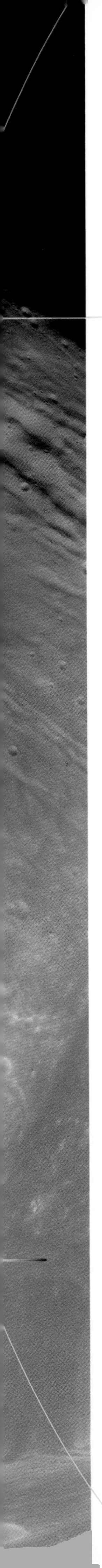

The dwarf planet Ceres, with a diameter of around 950 kilometers, is the largest object in the asteroid belt. We do not know much about it; this image, made with the Hubble Space Telescope, is the best available to astronomers so far. (NASA/ESA/J. Parker)

The Martian moon Phobos is almost certainly a captured asteroid. The surface of the irregularly shaped lump of rock, 27 × 21 × 18 kilometers, is pockmarked with countless impact craters, fault lines and crater chains. (NASA/JPL/University of Arizona)

Rocks from Space

Jean-Baptiste Biot Discovers the Extraterrestrial Origins of Meteorites

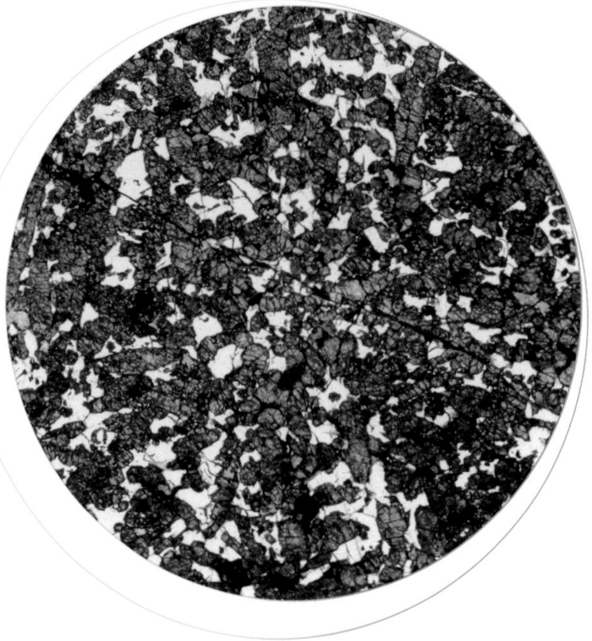

Our planet is continually bombarded from space. Space dust drifts down and settles on the surface of the Earth, falling grains of sand and pebbles produce 'shooting stars' in the atmosphere, and every now and then a piece of space rock will land on the ground. Yet for centuries serious scientists consign stories of the extraterrestrial origins of such meteorites to the realms of myth. The rocks must come from the condensation of clouds or the impact of lightning. Rocks from outer space? That has to be superstition.

The German physicist and musician Ernst Chladni thinks differently. From his colleague Peter Pallas he hears about a 700-kilogram iron meteorite found near the Russian town of Krasnoyarsk. Pallas discovers unknown minerals in the meteorite, and Chladni becomes convinced that meteorites come from outer space. He outlines his hypothesis in a book in 1794, but almost no one takes it seriously.

Not long after, on December 13, 1795, the people of Wold Cottage, England, see a rock weighing 25 kilogram fall from the sky in broad daylight. Chemist Edward Howard studies the rock and arrives at the same conclusion as Chladni. Still most astronomers do not want to believe it.

It is not until 1803 that they are forced to think again. On April 26, in l'Aigle, in the French *departement* of Orne about 70 kilometers to the west of Paris, at least 3,000 rocks fall from the sky. The shower is seen by many dozens of witnesses, and the French Academy of Sciences therefore decides it is time to find out once and for all where these meteorites come from.

Interior minister (and chemist) Jean-Antoine Chaptal sends the 29-year-old Jean-Baptiste Biot to investigate. Biot has studied mathematics, physics and astronomy, and is one of the Academy's promising young members. But there is another reason for choosing Biot: Chaptal attaches great importance to the hypothesis of his German colleague Chladni, and he knows that Biot also believes that meteorites have extraterrestrial origins – he thinks that they come from volcanic activity on the Moon.

Biot travels to l'Aigle to study the mysterious rocks and interview eye witnesses. On July 17, 1803, he presents a soundly argued report at a meeting of the Academy in Paris. In Biot's view, the exceptional nature of the meteorites, the fact that similar rocks do not occur naturally in the affected area, and that the shower occurred in broad daylight, all leave no room for doubt: meteorites are rocks from the cosmos.

We now know that most meteorites come from the asteroid belt between the orbits of Mars and Jupiter. Studying meteorites provides us with valuable information on the origins of the solar system.

Thin section of a stony meteorite from Mars. The meteorite (NWA 3171) was found in the northwest of Africa. The brown and white minerals are pyroxene and maskelynite. The section in the photograph is around one and a half centimeters in size. (Anthony Irving/Scott Kuehner)

The chemical and mineralogical composition of meteorites provides information on the gas and dust cloud from which the solar system evolved. New stars and planetary systems are also formed in the Corona Australis star forming region, shown here on an infrared photograph taken with the Spitzer Space Telescope. (NASA/JPL/CfA)

1803

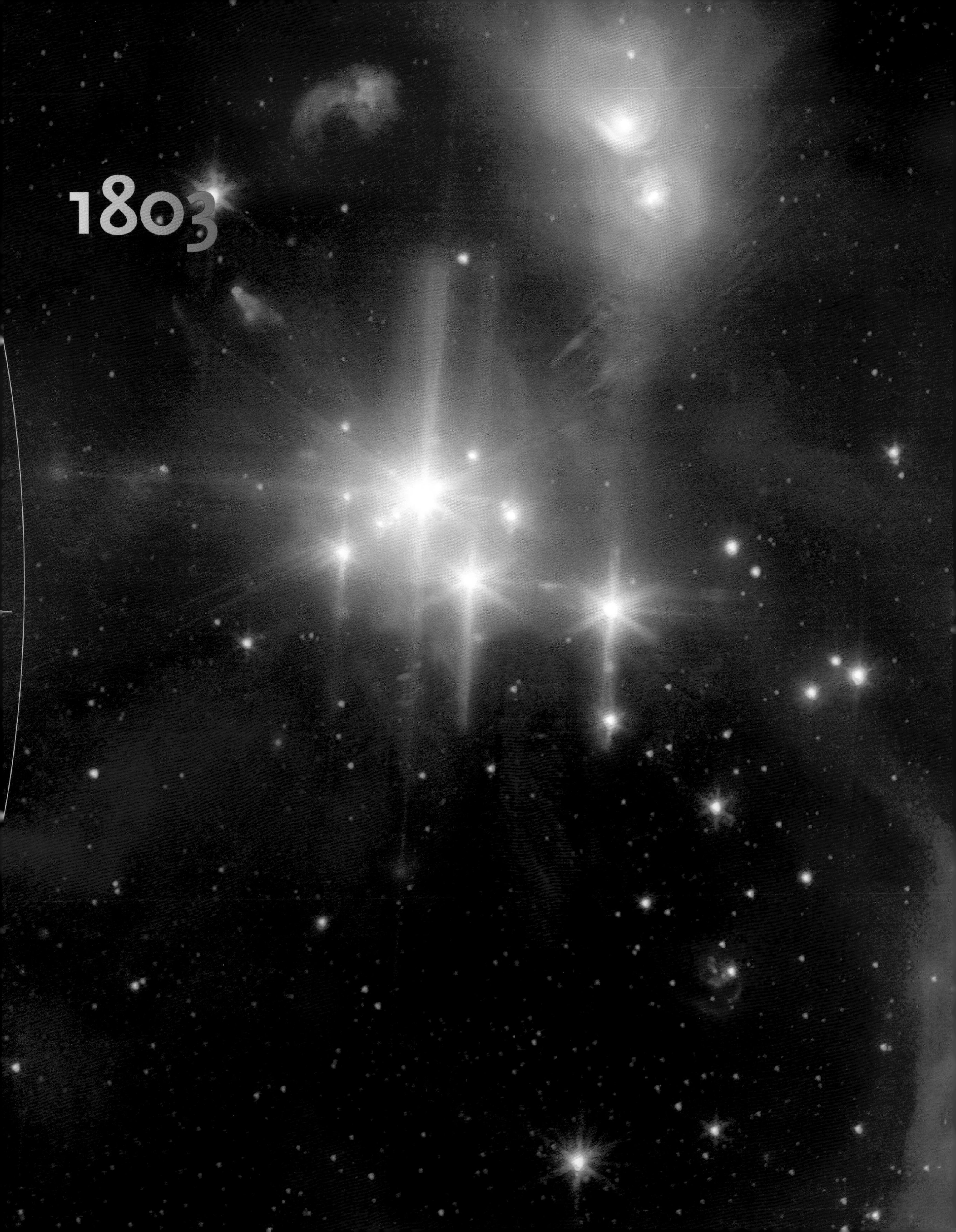

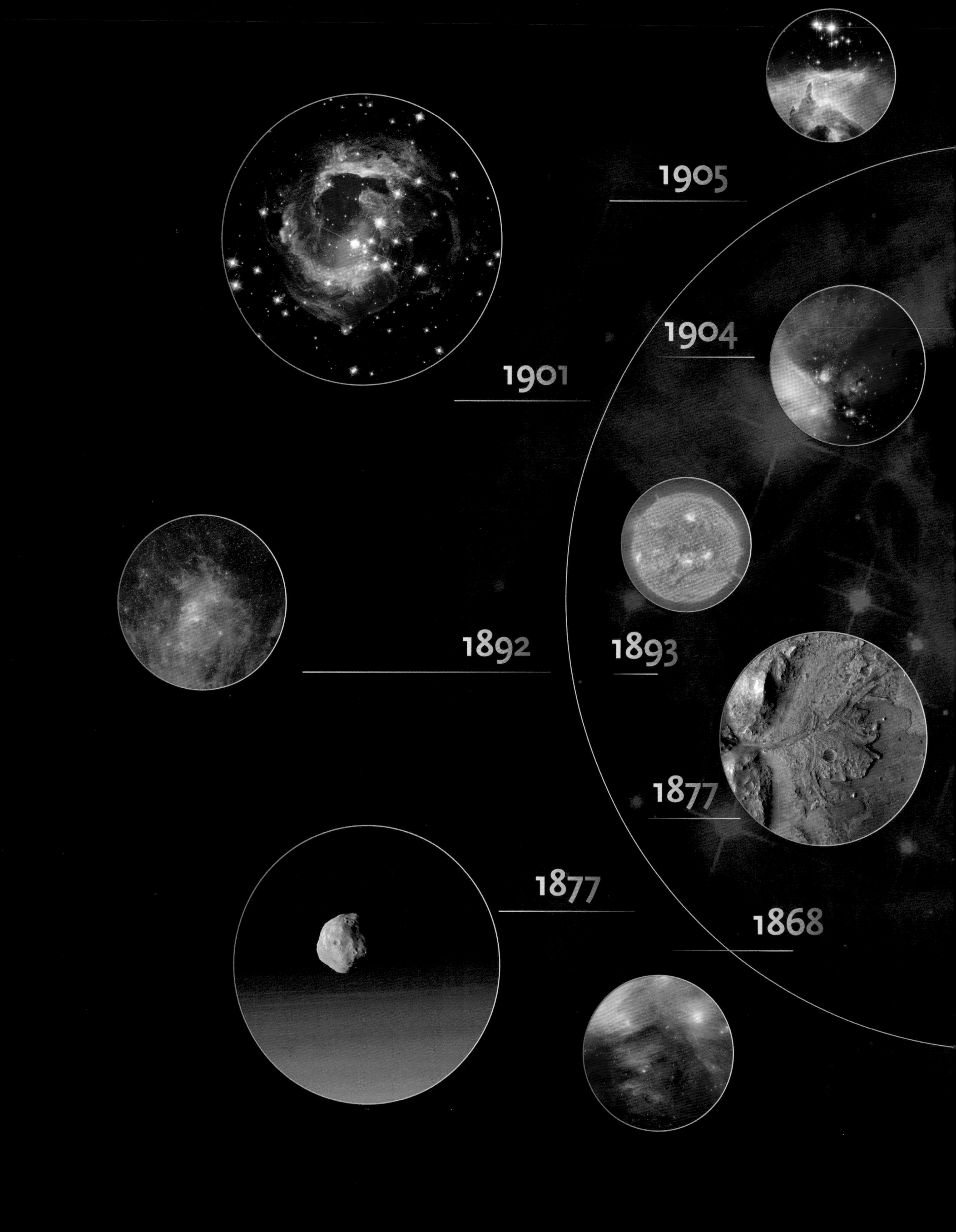
1905
1904
1901
1892
1893
1877
1877
1868

1808 - 1908

Astronomy is an exceptional science. Astronomers cannot take the objects of their research back into the laboratory and cannot experiment with them or conduct physical measurements. They have to extract all their information from the light of unreachable celestial bodies – light that has taken hundreds, or even thousands, of years to reach the Earth.

Paving the Way for Major

While in the eighteenth century astronomers concentrated mainly on determining the direction from which light was coming and its intensity – the exact position of an object in space and how bright it is – in the nineteenth century the focus of attention shifts to the composition of the light that reaches the Earth. With prisms, diffraction gratings and spectroscopes, they unravel and analyze the light from stars, planets and nebular spots. By determining how much radiation is emitted at various wavelengths, astronomers can obtain information about the chemical composition and velocity of distant celestial objects.

But that is not all. In the nineteenth century, discoveries are made whose full significance is not immediately clear, but which pave the way for major theoretical breakthroughs. Astronomers determine the

Theoretical Breakthroughs

physical properties of stars, measure the composition of the Sun, and discover spiral nebulae, white dwarfs and stellar explosions. Around 1900, astronomy is pregnant with new insights into the birth and life of stars and the place of the Earth and humankind in the evolving universe.

These revolutionary developments are made possible by the construction of a new generation of giant telescopes.

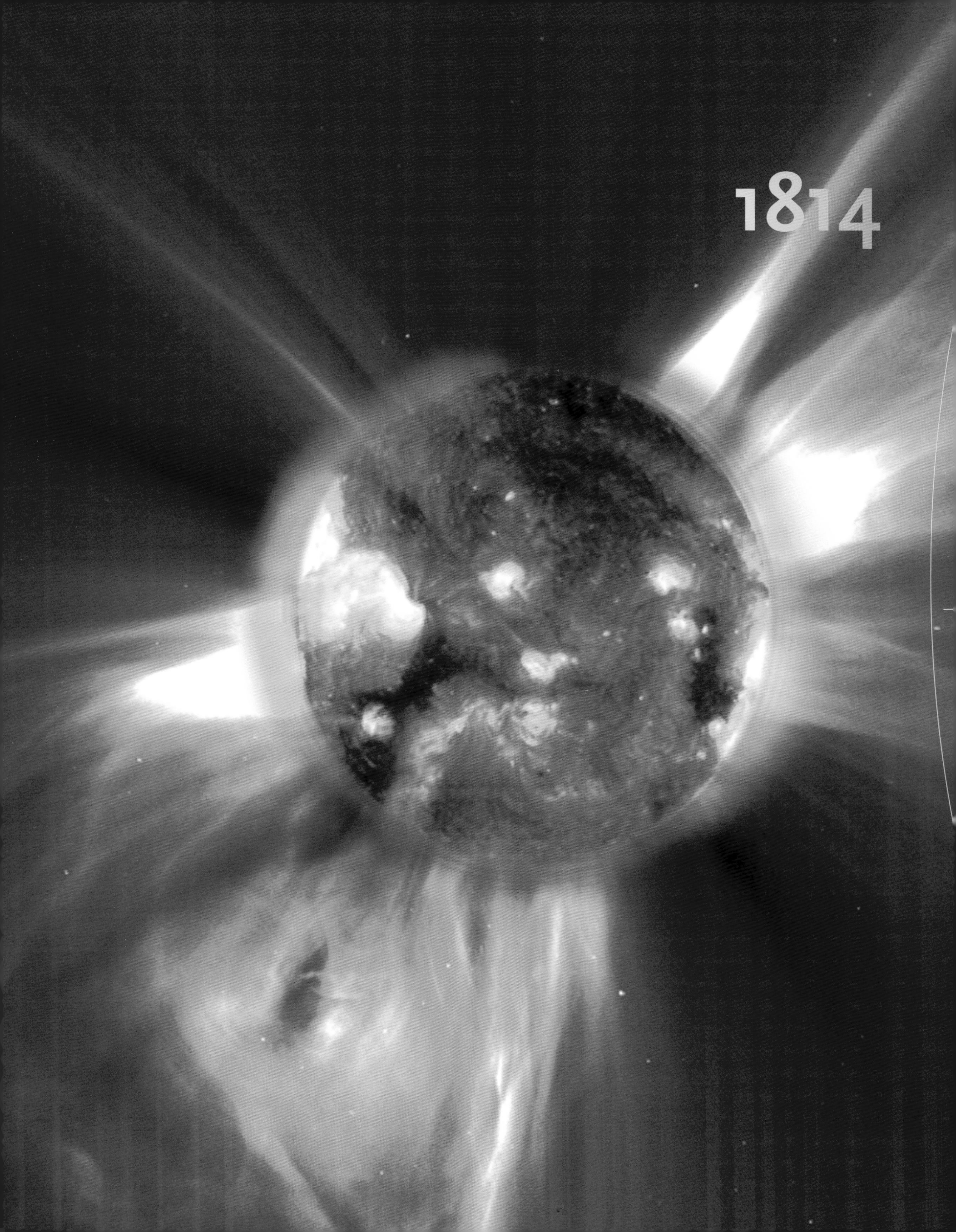

1814

Deceptive Lines

Joseph Fraunhofer Discovers the Solar Spectrum

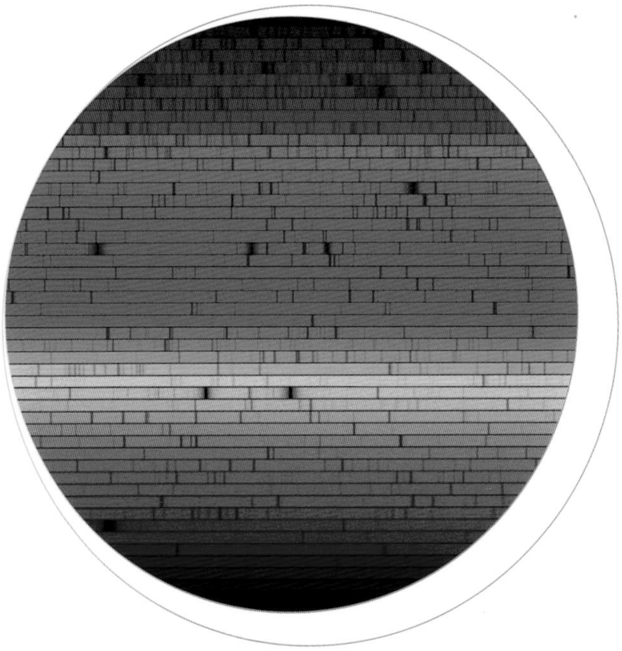

Joseph Fraunhofer is 14 years old in 1801 when he is buried under the rubble after the workshop of his employer, glass-maker Philipp Weichelsberger, collapses in southern Germany. He survives the disaster thanks to a quick rescue operation led by Prince Maximilian IV, and the Prince takes the young boy under his wing. Joseph is an orphan, his employer is harsh, and allows him no time to read or study, but this changes.

With the Prince's help, Joseph finds a job at the Optical Institute in Benediktbeuern, a former monastery in Bavaria where the thirteenth-century manuscript of the *Carmina Burana* was found a few years before. He specializes in glass-making, and within a short time has become the best glass-maker in the world. Fraunhofer develops incredibly accurate methods of measuring the refractive index of glass and builds achromatic telescope lenses of extremely high quality. Together with the French opticist Pierre-Louis Guinand he builds a 24-centimeter telescope for the observatory in Dorpat, which is for a long time the best telescope in the world.

In 1814, Fraunhofer invents the spectroscope. Isaac Newton and William Herschel have earlier succeeded in splitting white sunlight into the colors of the rainbow, and in 1802, William Wollaston discovers dark lines in the color bands. However, Fraunhofer's spectroscope is of such excellent quality that it allows the solar spectrum to be studied in detail for the very first time.

In total, Fraunhofer identifies more than 500 dark, narrow lines in the solar spectrum and measures the wavelength of more than 300 as accurately as possible. It is as though very specific colors are missing in the sunlight. Fraunhofer even succeeds in studying the spectrum of the bright star Sirius, and concludes that its visible lines are not the same as those in the solar spectrum. Yet he is still unable to explain the true nature of these 'Fraunhofer lines'.

In 1818, Fraunhofer becomes the director of the Optical Institute and a few years later he is awarded an honorary doctorate at the University of Erlangen, becomes a noble (*Ritter von Fraunhofer*) and is made an honorary citizen of Munich. In 1826 he dies of tuberculosis, at the age of 39.

The dark lines in the solar spectrum are not explained until 1859 when Gustav Kirchhoff and Robert Bunsen discover that relatively cool gas absorbs light at certain wavelengths. That means that every element in the outer layers of the Sun leaves its own characteristic spectral 'fingerprint' behind in the sunlight. The spectrum therefore acts as a kind of cosmic barcode revealing the chemical composition of the light source.

Spectroscopy lies at the basis of modern astrophysics – the physics of the stars. After the telescope, the spectroscope is undeniably the most important instrument in the history of astronomy.

The spectrum of the Sun, from violet to red, is cut into pieces which are shown here from bottom to top. The countless dark Fraunhofer lines arise because atoms and molecules in the outer layers of the Sun absorb light at certain wavelengths. (AURA)

The American/European space probe SOHO holds the Sun under constant surveillance, including at ultraviolet wavelengths. Bottom left in this composite photograph a coronal mass ejection can be seen, a high-energy explosion of electrically charged particles. (NASA/ESA)

Distant Suns

Friedrich Bessel Discovers the Parallax of a Star

Nicolaus Copernicus knew it long before the telescope had been invented: if the Earth revolves around the Sun, the stars must be very far away, otherwise you would see them shift slightly in the course of a year. Since the early eighteenth century astronomers had been searching for the 'stellar parallax', which would finally enable them to calculate the distance to the stars. Yet although they made all kinds of other interesting discoveries – like the proper motion of the stars and the aberration of starlight – the stellar parallax remained beyond their reach.

Until 1838 when the German astronomer Friedrich Bessel publishes his calculations for the distance of the star 61 Cygni, in the constellation Cygnus, in the *Astronomische Nachrichten*. Finally astronomers now have some conception of the scale of the universe.

Bessel is crazy about numbers and about the stars. His calculations of the orbit of Halley's Comet attract the attention of astronomer Heinrich Olbers from Bremen. In 1806, he helps young Friedrich to get a job at Johann Schröter's observatory in Lilienthal. Four years later, when he has just turned 26, Bessel is appointed director of Königsberg Observatory in East Prussia (now Kaliningrad).

In 1834 Bessel devotes himself to the parallax problem, focusing on 61 Cygni. The star's proper motion of more than five arcseconds per year is substantial enough to suggest that it is close and should have a relatively large parallax.

By the autumn of 1838 Bessel has collected hundreds of accurate positional measurements for 61 Cygni. Most are made using a heliometer built by Joseph Fraunhofer. Indeed, the star proves to trace a small ellipse in the sky each year – a reflection of the Earth's orbit. Bessel determines its parallax at 0.314 arcsec, placing it at a distance of more than a hundred trillion kilometers from the Earth.

Bessel is very nearly pipped at the post. At the observatory in Dorpat (now Tartu in Estonia), Wilhelm Struve is also searching for the stellar parallax. Struve conducts measurements of the bright star Vega in the constellation Lyra, using Fraunhofer's large telescope. And as early as 1833, Scottish astronomer Thomas Henderson is already making positional measurements of Alpha Centauri from the southern hemisphere. Henderson, however, has little confidence in his own measurements and does not analyze them until many years later. Both Struve and Henderson publish their results in 1839.

With increasingly better telescopes and more sensitive measuring instruments, astronomers finally succeed in calculating the parallax of thousands of stars and so determine their distances from the Earth. The most accurate measurements made so far are conducted in the early 1990s by the European Hipparcos satellite. In the future, the Gaia satellite should be able to measure stellar parallaxes up to distances of 30,000 lightyears.

◐ A solar eclipse that is total in Africa will be seen in Southern Europe as a partial eclipse. That is also the consequence of parallax: from different places on Earth, we see the Moon in different positions in the sky. The greater the distance to an object, the smaller the parallax. (Helder da Rocha)

➲ The individual stars in the globular star cluster NGC 6397 are all at practically the same distance, and therefore have the same (extremely small) parallax. A star that happens to be in the foreground would stand out immediately, as it would have a much greater parallax. (NASA/ESA/H. Richer)

1838

1842

Moving Waves

Christian Doppler Discovers the Doppler Effect

Stars are suns at unimaginable distances from the Earth, which are moving through the universe at high speed. But why are some stars orange, while others are red, or blue? The difference in color is especially striking in certain binary stars. In 1842, the Austrian astronomer and mathematician Christian Doppler, professor at the University of Prague, thinks he has found an explanation.

According to Doppler, light is a wave phenomenon and the color of the light we observe is determined by the frequency of the light waves. The frequency must depend on the relative speed between the source of the light and the observer. If they are approaching each other, more light waves will arrive per second. The observed frequency will then be higher and the light will get bluer. If the distance between the two is increasing, the observed frequency is lower and the observer will see a redder color.

On May 25, 1842, Doppler gives a lecture at the *Königlich-Böhmische Gesellschaft der Wissenschaften* in Prague in which he explains his theory. A year later, the theory is published. (Doppler also mentions in passing that the same effect should be observable with sound waves.)

In the Netherlands, Christophorus Buys Ballot – who later founds the KNMI, the royal Dutch meteorological institute – does not believe Doppler's explanation. In 1844, he admits in his doctoral thesis that the effect probably exists, but it does not explain the color differences between binary stars.

In 1845, Buys Ballot is the first to demonstrate the Doppler effect with sound waves. He arranges for a locomotive to ride back and forth on the newly opened railway line between Utrecht and Maarssen, with a horn player on board. Another horn player is positioned alongside the track. They play the same tone yet, to a practiced ear, it is clear that the frequency changes as the train moves.

The Doppler effect does not explain the colors of the stars because the differences are caused by different surface temperatures. The reason the color differences are especially noticeable in binary stars is primarily because of the contrast between the two adjacent bodies. Yet the Doppler effect remains one of the most important instruments in modern astronomy; the rotations and motions of celestial bodies can be calculated from the minute shifts in frequency of the light they emit.

Precision measurements of the Doppler effect in the light from stars make it possible to determine their radial velocity (towards or away from us). By combining that with a star's proper motion and its distance – based on the parallax – its actual true motion can be calculated.

In 1995, astronomers succeed for the first time in using the Doppler effect to demonstrate the existence of planets orbiting other stars, and the latest spectrographs allow speeds of only a few meters per second to be detected.

The striking red and blue colors of some stars are not caused by the Doppler effect, as Christian Doppler first thought, but indicate different surface temperatures: red stars are cool, while blue ones are extremely hot. (NASA/ESA/Davide De Martin/Edward Olszewski)

Doppler measurements of individual stars provide information on the rotation of the Andromeda Galaxy. As a whole, the galaxy displays a blue shift, caused by the fact that it is moving at a speed of 35 kilometers per second towards the Milky Way. (R. Gendler)

Inconstant Sun

Heinrich Schwabe Discovers the Solar Cycle

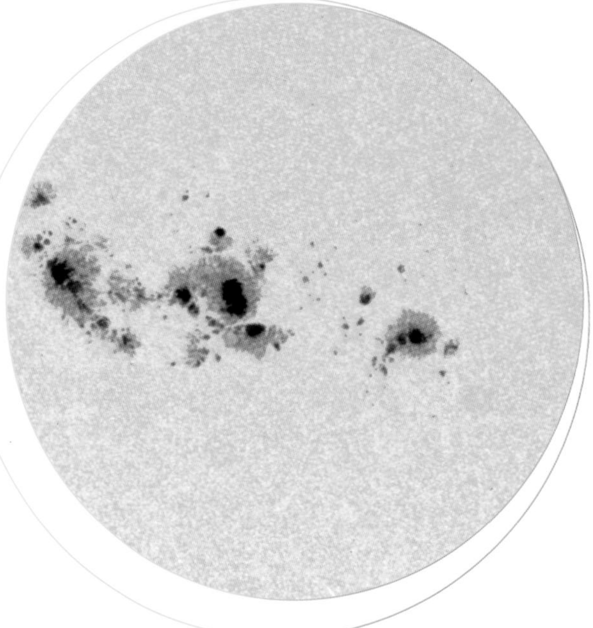

German astronomer Heinrich Schwabe wins his first telescope in a lottery in 1825. Schwabe is an apothecary, but is also very interested in astronomy. He is fascinated by the idea that there may be another planet, inside the orbit of Mercury. This hypothetical planet, known as Vulcan, is of course not visible in the bright glow of the Sun, but could perhaps betray its existence when it moved across the face of the Sun, as seen from the Earth.

From October 11, 1825, Schwabe aims his telescope at the Sun on every clear day, in the hope of catching sight of Vulcan. He soon orders a much better telescope from Joseph Fraunhofer, and in 1829 he sells the family apothecary shop to devote himself entirely to astronomy. He keeps up his daily observations of the Sun for 42 years, but he never finds a new planet. He does discover the renowned 11-year sunspot cycle, though.

Sunspots have been observed since the discovery of the telescope, but little is known about them in Schwabe's time. Every day Schwabe notes the number of sunspots, individually and in groups, in minute detail. The Sun has never been studied so closely and for such a long time without interruption. He soon notices that the number of visible sunspots can vary enormously. In 1828 and 1829 not a day goes by, literally, without him seeing spots, while in 1833 the Sun is 'spotless' for no less than 139 days.

In the second half of the 1830s, a large number of sunspots are once again visible, but in 1843 the situation seems to repeat itself. That year, in the German astronomy journal *Astronomische Nachrichten*, Schwabe publishes his suspicions that the sunspots follow a 10-year cycle.

At first almost no one pays any attention to Schwabe's discovery. But that is to change in 1851, after his remarkable discovery is mentioned in the comprehensive, five-volume work *Kosmos* by Alexander von Humboldt. Around that time the next period of minimal sunspot activity is due and Schwabe's suspicions are confirmed: the Sun proves to have an activity cycle of an average of 11 years.

In 1908, George Ellery Hale discovers that the cycle is actually 22 years long. Sunspots prove to have very strong magnetic fields, and the magnetic polarity in any 11-year cycle is exactly opposite to that in the following cycle.

During a period of maximum activity, the Sun not only displays more dark spots than average, but also more bright flares and energetic eruptions. That means that the total energy production is higher during a solar maximum than during a minimum, despite the dark spots. Although there is widespread agreement that solar activity has an influence on the Earth's climate, there is considerable disagreement about the Sun's role in the current period of global warming.

◐ In July 2002, around a year after the most recent activity maximum, a gigantic group of sunspots was visible on the Sun. During periods of high solar activity powerful solar flares also occur. These are mostly only visible with X-ray telescopes. (NASA/Lockheed Martin)

➲ The structure of the magnetic field above a sunspot on the rim of the Sun becomes visible when hot gas flows away along solar field lines. This X-ray photo was made by the Japanese Hinode satellite. (NASA/JAXA)

1843

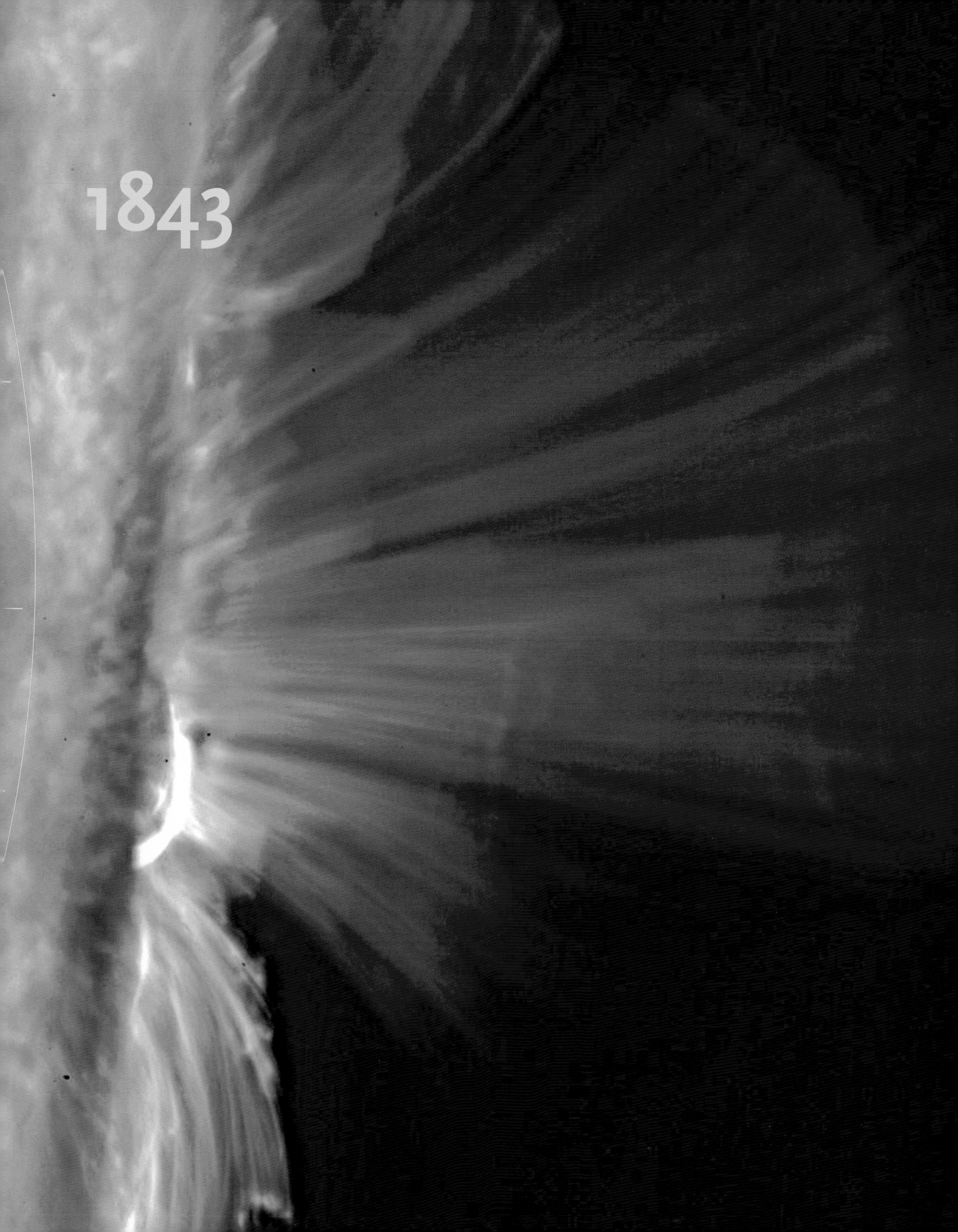

1845

Swirling Veils

William Parsons Discovers Spiral Nebulae

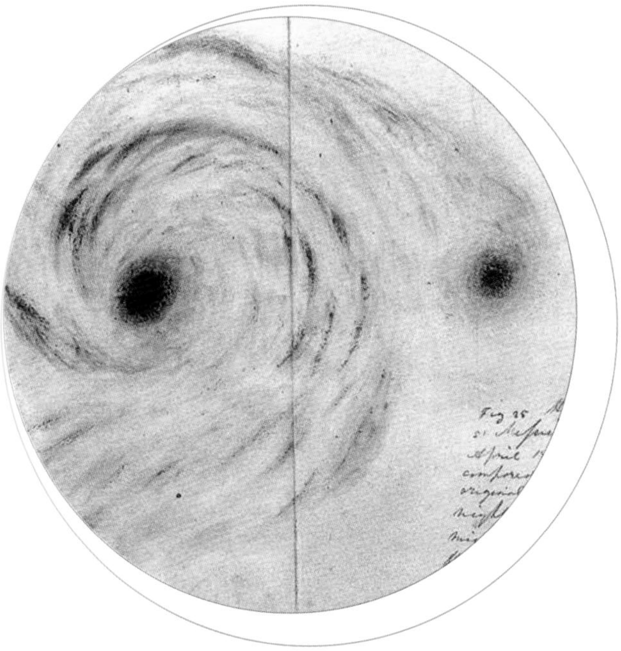

The telescopes that William Herschel builds at the end of the eighteenth century are gigantic; the largest has a mirror with a diameter of no less than 120 centimeters. But William Parsons goes a step further, and in the mid-nineteenth century he completes the construction of the largest telescope of all time: a colossus with a mirror 183 centimeters in diameter and an optical tube almost 20 meters long. With the 'Leviathan of Parsonstown', as it is called, he discovers in 1845 that some nebulae in the night sky have a spiral form.

Parsons, the third Earl of Rosse, is a noble and a large landowner in central Ireland. He studies at Oxford and is a member of the House of Lords for most of his adult life. In 1842, he starts building a telescope on his estate at Birr Castle. The work stops in 1845 because of the Irish potato famine, but when the telescope is completed in 1847, the first observations have already been conducted. The metal mirror weighs three and half tons and the telescope has to be trained at the heavens using ropes and winches.

Lord Rosse, as Parsons is often called, invites professional astronomers to come and use his telescope, but spends as much time at the eyepiece himself as the Irish climate permits. He studies the countless nebular spots that have previously been charted by Herschel and catalogued by the Frenchman Charles Messier. Some of the nebulae prove to be clouds of countless faint stars, while others retain their misty appearance even when magnified to the maximum.

In 1845, still before his giant telescope is finished, Parsons aims it at Messier 51, a nebula in the constellation Canes Venatici, just under the handle of the Big Dipper. It has a clear spiral structure – something that no one has ever noticed before. Could it be a rotating cloud from which a new star will be born, or a revolving group of stars too far away to be seen individually?

The Whirlpool Nebula, as Parsons calls it, proves not to be the only spiral nebula and within a few years many more are discovered. Parsons' observations attract the attention of Danish astronomer Johann Dreyer, and at Birr Castle he starts work on his *New General Catalogue* (NGC), an extensive catalog of nebulae and star clusters. It is not until 1924 that Edwin Hubble succeeds in observing the individual stars in spiral nebulae, and they prove to be full-fledged galaxies, like our own Milky Way.

Parsons' youngest son Charles, the inventor of the steam turbine, is one of the founders of the optics firm Grubb-Parsons in 1935. The company builds the 2.5-meter Isaac Newton Telescope, which is now at the Roque de los Muchachos Observatory on La Palma, in the Canary Islands.

In 1845, while sitting at the eyepiece of his 183-centimeter reflecting telescope, William Parsons made this sketch of a nebular spot in the constellation of Canes Venatici. Parsons called it the Whirlpool Nebula, because of its striking spiral structure. The nebula's companion is also clearly visible.

An extremely detailed image of the Whirlpool Nebula (M51) by the Hubble Space Telescope. It is a spiral galaxy at a distance of 23 million lightyears that we can see almost directly from above. Most dust clouds and star forming regions are located in its spiral arms. (NASA/ESA/Hubble Heritage Team)

Newton's Triumph

Johann Galle Discovers the Planet Neptune

Newton's law of gravity makes it possible to calculate the orbit of the newly discovered planet Uranus accurately, although this is no easy task because the gravitational pull of Jupiter and Saturn has to be taken into account. In the early nineteenth century, it becomes clear, though, that something is not right; Uranus is increasingly further away from its predicted position, as though it is being slowed down by the gravitational pull of an unknown celestial body.

In 1832, English astronomer and later Astronomer Royal George Airy calls the Uranus mystery one of the greatest unsolved puzzles in astronomy. Nine years later, 22-year-old mathematician John Adams decides to solve the problem. If there is another planet beyond the orbit of Uranus, which is responsible for the observed orbital deviations, it should be possible to calculate the position of the unknown planet by analyzing the deviations in great detail.

Adams spends many years on the problem, but his results are not conclusive and he receives little support from his astronomy colleagues at Cambridge. But in 1845, the French astronomer and mathematician Urbain Le Verrier also turns his attention to the mystery. In August 1846, he is convinced that the unknown planet must be in the constellation Aquarius, in an orbit far beyond that of Uranus; that is the only explanation for Uranus' orbital deviations.

At the Paris Observatory, however, Le Verrier's prediction is not taken seriously. On September 18, he writes a letter to Johann Galle in Berlin asking if the Fraunhofer telescope at the Berlin Observatory could perhaps take a look at the section of the night sky that Le Verrier is interested in.

Galle receives the letter on Wednesday, September 23. That same evening, assisted by Heinrich d'Arrest, he searches the night sky. Only a few degrees away from the predicted position he discovers the planet Neptune, and the discovery is nothing short of a triumph for Newton's theory of gravity.

Now that the gravitational disturbance of Neptune can be taken into account, Uranus' orbit becomes much more predictable. Yet at the end of the nineteenth century it once again seems to display very small orbital deviations. These abnormalities give rise to Percival Lowell's hunt for Planet X, which leads to the discovery of Pluto in 1930.

In 1847, William Lassell discovers the large Neptunian moon Triton – which, strangely enough, moves around its parent planet in opposition to 'normal' direction – and in 1948 Gerard Kuiper finds the smaller satellite Nereid. But Neptune reveals no more of its secrets until August 1989, 143 years after its discovery, when the planet is visited by an unmanned space probe. Voyager 2 flies past Neptune and Triton, and takes breathtaking photographs of a deep-blue planet with white wisps of cloud, of small dark moons and incomplete ring arcs, and of a large icy moon with active nitrogen geysers.

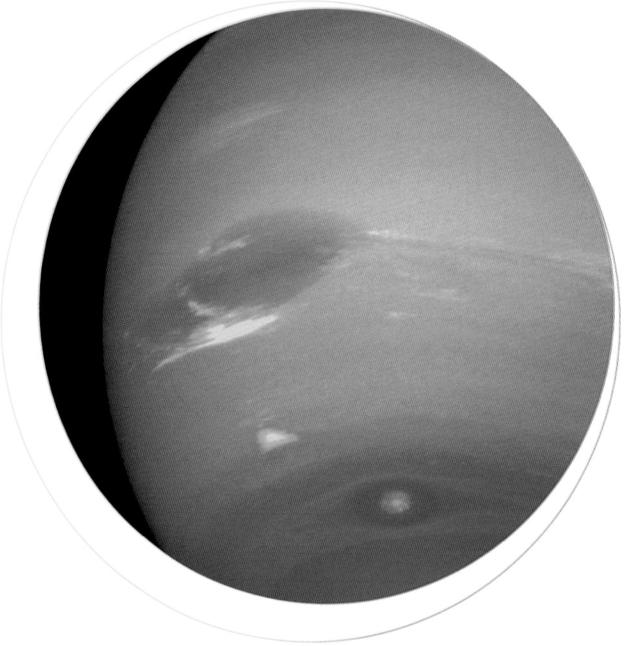

The distant planet Neptune was studied from close up for the first time in August 1989, by the American planetary explorer Voyager 2. This photo shows various cloud patterns, including the Great Dark Spot. (NASA/JPL)

Shortly after its closest approach to Neptune, Voyager 2 took this photo of the distant planet's night side. Its southern pole is still just illuminated by the Sun. The space probe, like its predecessor Voyager 1, has since left the solar system on its way to the stars. (NASA/JPL)

1846

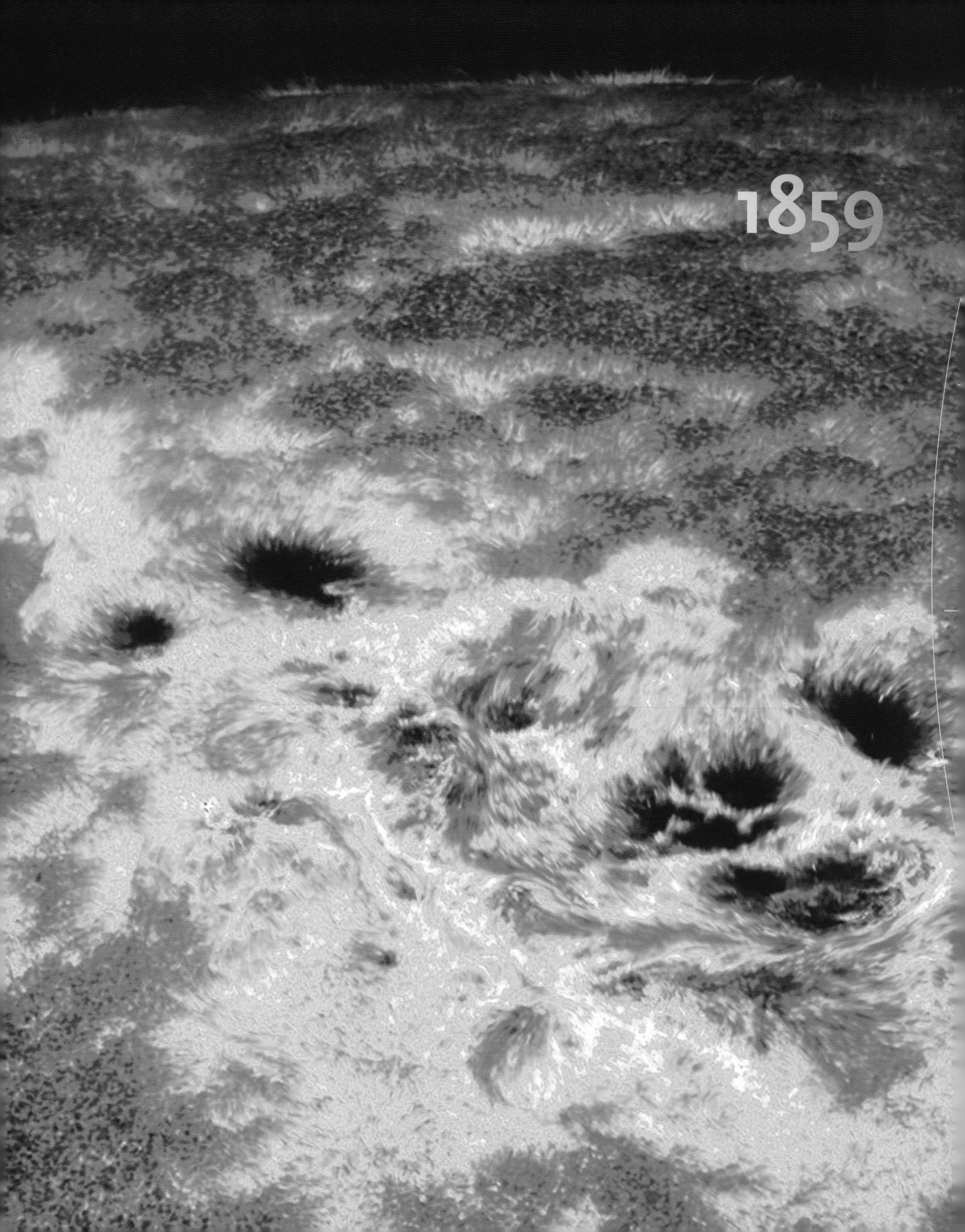
1859

Enervating Explosions

Richard Carrington Discovers Solar Flares

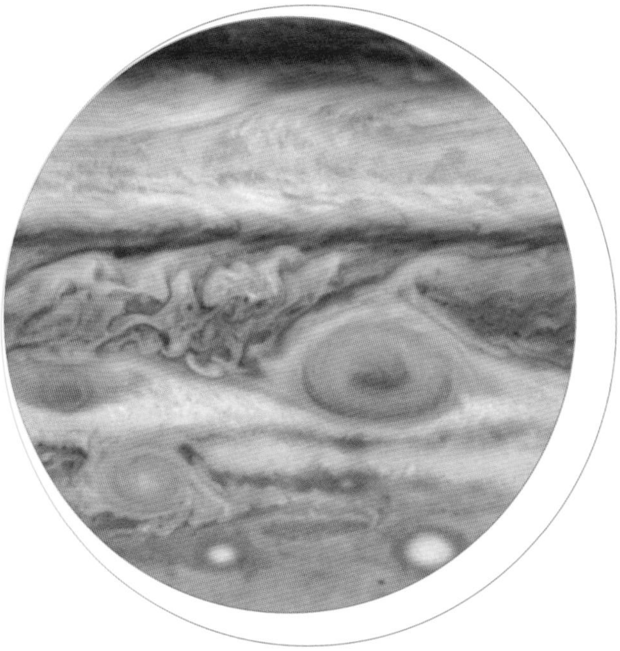

'One swallow doesn't make a summer', warns Richard Carrington during his presentation at the Royal Astronomical Society on November 11, 1859. The fact that shortly after he has observed an eruption on the Sun there are unusual auroras and a geomagnetic storm can indeed just be coincidence. Yet everything seems to suggest that solar activity does affect events on Earth.

Amateur astronomer and brewer Carrington is just 33 when he takes his regular morning look at the Sun on Thursday September 1, 1859. With his telescope he projects the image of the Sun onto a screen. Meticulously, he takes note of the form and position of the sunspots. From observations like these Carrington later deduces that, at the beginning of each new activity cycle, sunspots are formed at a great distance on both sides of the solar equator. Later in the cycle they appear at much lower latitudes. He also discovers the differential rotation of the Sun: at its equator the gaseous body makes one full rotation in 25.38 days, while at the poles it has a rotation period of 36 days.

Around 11:20 in the morning, two extremely bright spots suddenly appear in a large group of sunspots, only to disappear a few minutes later. Later it is discovered that amateur astronomer Richard Hodgson also saw the eruption; it is as though an even brighter star is burning on the bright surface of the Sun.

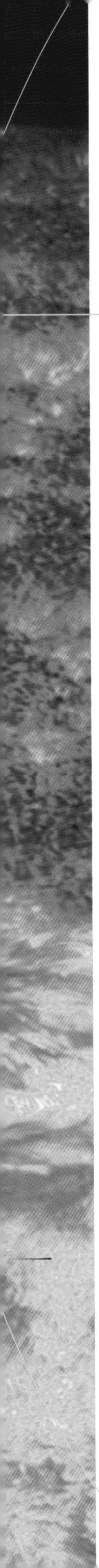

Eighteen hours later magnetometers register enormous disturbances in the Earth's magnetic field. On Friday September 2, spectacular auroras can be seen everywhere on the Earth, even in the Tropics. At the same time telegraph communications in Europe and the United States are almost brought to a standstill as a result of induction currents in the telegraph cables. Such geomagnetic disturbances have been recorded before, but the link to solar activity has never been so clear.

Optical solar flares, like those seen by Carrington and Hodgson, are extremely rare. Mostly these high-energy eruptions are only observable at X-ray wavelengths. Magnetic discharges blast enormous quantities of electrically charged particles into space. Such a coronal mass ejection speeds through space at almost 10 million kilometers an hour, arriving at the Earth within a day. Once there, the charged particles cause auroras and disturb the magnetic field of our planet.

On March 13, 1989, 130 years after Carrington's discovery, there is another optical solar flare of unprecedented power. Radio communications break down, satellites are disrupted and, in Quebec, more than six million people have to go without power for an entire day.

On November 4, 2003, there is yet another gigantic solar flare. It occurs on the edge of the visible solar disk, so the resulting particle cloud presents no threat to the Earth, but races through the solar system in another direction.

➋ Like the Sun, the giant planet Jupiter undergoes differential rotation: the rotation period of the equator (9 h, 50 min and 30 s) is shorter than that of the poles (9 h, 55 min and 41 s). Also, in the various cloud layers different wind speeds occur. (Imke de Pater)

➌ In early November 2003, the Dutch Open Telescope on La Palma started photographing an active area on the Sun in the light of ionized calcium atoms. The area also produced a very powerful solar flare, but not as extreme as that of September 2, 1859. (University of Utrecht)

Stellar Runt

Alvan Graham Clark Discovers Sirius' White-Dwarf Companion

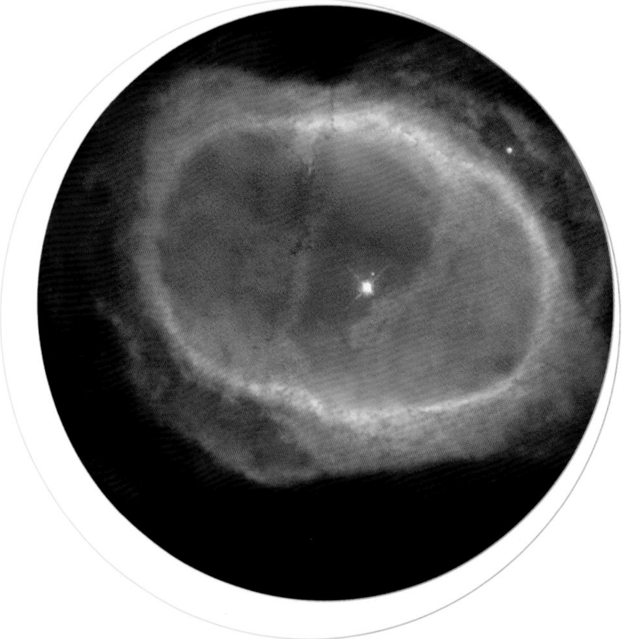

Sirius, the main star in the constellation Canis Major, is the brightest star in the night sky. That is not so surprising; at a distance of less than nine lightyears from the Earth, Sirius is one of the closest stars to us in the universe. Because of its proximity, the star also has a very noticeable proper motion, as Edmund Halley discovered in 1718.

Precision measurements by Friedrich Bessel show that Sirius' proper motion is a little irregular. You would expect the star to move in a straight line and at a constant speed, but Sirius turns out to move with a slight wobble. In 1844, Bessel concludes that the star is accompanied by an invisible companion. He also observes that Procyon, the main star in the constellation Canis Minor, oscillates a little as it moves, meaning that it too must have a 'dark companion'.

It is 18 years before Sirius' companion is discovered, and Bessel does not live to see it. The 29-year-old astronomer and telescope-builder, Alvan Graham Clark, first sees the tiny, faint star in 1862, and Procyon's companion is not identified until 1896.

Clark is the youngest son of the famous American telescope-builder Alvan Clark. The company Alvan Clark & Sons builds large refracting telescopes and counts American universities among its clientele. At the end of January 1862, Clark Junior is making test observations with the 45-centimeter Dearborn Telescope, being constructed for the Northwestern University in Evanston, Illinois. On January 31, close to the bright star Sirius, he sees an unimpressive speck of light, which is given the name Sirius B.

In 1915, at Mount Wilson Observatory in California, Walter Adams succeeds in identifying the spectrum of Sirius B. It turns out to be a very hot, white star. That it is so faint, despite its high surface temperature, can mean only one thing: Sirius B must be very small. In 1922, Dutch astronomer Willem Luyten calls these small, hot stars 'white dwarfs'. Luyten later himself discovers many new examples.

Sirius B is the first star to be predicted on the basis of its gravitational influence, but it is not the first white dwarf. In 1910, astronomers discover that the companion of the star 40 Eridani – first observed by William Herschel in 1783 – is also faint yet extremely hot; the same is true for Procyon B. (White dwarfs are the compact remains of Sun-like stars.) Nuclear reactions no longer occur in their interiors and, over a period of several billion years, they will cool off to become dark stellar cinders.

Detailed orbital measurements show that Sirius B is about as massive as our Sun, and its surface temperature is 25,000 degrees Celsius. The observed luminosity suggests a diameter of 12,000 kilometers – a little smaller than that of the Earth. White dwarfs have a very high density: one cubic centimeter of white-dwarf matter weighs about a ton.

White dwarfs are small, compact hot stars. Their mass is comparable to that of the Sun, but they are hardly larger than the Earth. When a red giant changes into a white dwarf, it blasts a shell of gas into space which is visible as a planetary nebula. (NASA/ESA/Hubble Heritage Team)

The white dwarf star Sirius B is the small, faint, inconspicuous companion to Sirius A, the brightest star in the night sky. On this Hubble photograph, Sirius B is just visible, close to one of the 'rays' of the overexposed main star Sirius A. The two stars are 8.6 lightyears away from the Earth. (NASA/ESA)

1862

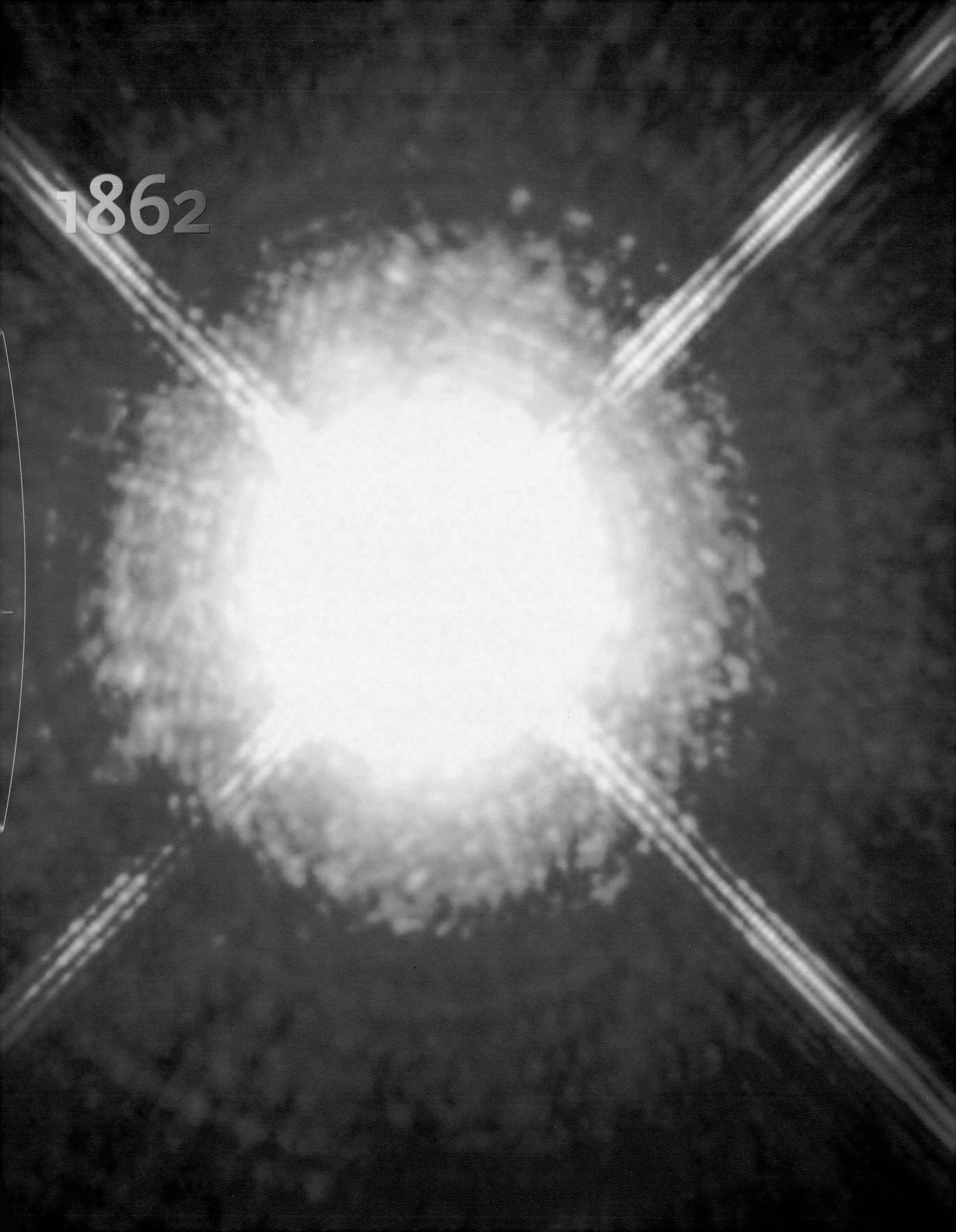

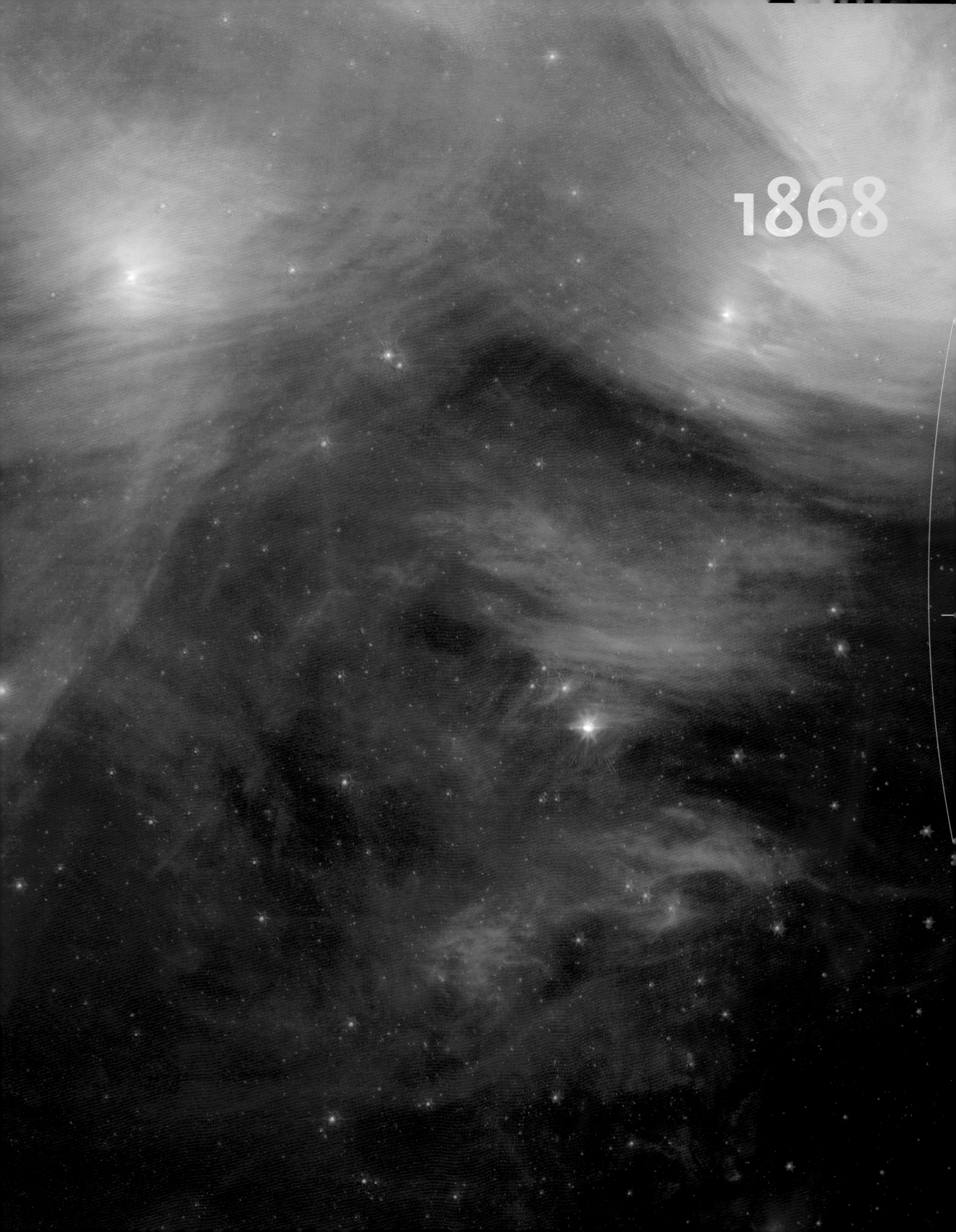

1868

Elementary Puzzles

Pierre Janssen and Norman Lockyer Discover Helium in the Sun

About one-quarter of the universe consists of the light noble gas helium. Helium is the second most common element in the cosmos, after hydrogen, yet it was not discovered until 1862. The reason: helium is very rare on Earth, since it is extremely light and does not bond with other atoms.

The nineteenth century sees great leaps forward in the field of spectroscopy. Volatile gases leave characteristic fingerprints in the solar spectrum, in the form of dark absorption lines or, if the gases are extremely hot, bright emission lines. Laboratory experiments make it possible to identify all these lines in the spectrum. For example, the two conspicuous lines in the orange part of the solar spectrum (designated D1 and D2 by Joseph Fraunhofer) are caused by sodium gas.

In the summer of 1868, the French astronomer Pierre Janssen – later director of Meudon Observatory– travels to Guntur, on the east coast of India, to see the total solar eclipse on August 18. Janssen is especially interested in observing prominences – gas explosions on the Sun which are sometimes visible for a short time on the edge of the solar disk during an eclipse.

A day after the eclipse Janssen realizes that the emission spectrum of the prominences is so bright that it should also be visible when there is no eclipse. He discovers a mysterious yellow line in the spectrum, apparently caused by a yet unknown chemical element.

The Englishman Norman Lockyer, founder of the scientific weekly *Nature* at the end of 1889, discovers exactly the same spectral line in October 1868. He originally calls the line D3, because it is so close to the two sodium lines. But Lockyer too concludes that there is an element that occurs in the Sun that is unknown on Earth. He calls it helium, after the Greek word for sun.

Janssen's report of his observations, which he sends in September, takes a few months to arrive in Paris, and in the meantime the French Academy of Sciences has heard of Lockyer's work. Usually, both astronomers are credited with the discovery of helium.

It is not until March 26, 1895 that William Ramsay succeeds in isolating helium from terrestrial minerals. In 1907, Ernest Rutherford shows that the mysterious alpha particles released during radioactive processes are in fact positively charged helium nuclei. And a year later in Leiden, Heike Kamerlingh Onnes is the first to cool helium to 1 degree above absolute zero, at which point the gas becomes liquid.

Janssen's expedition to observe the eclipse of 1868 is perhaps his most interesting, but not the most spectacular. In December 1870, while Paris is occupied by German soldiers, he flees the city by air balloon to observe a total solar eclipse in Algeria. Unfortunately, during the eclipse, the sky is cloudy all day.

Just below the surface of the Moon, there are helium atoms that originate from the Sun, and have been carried along with the solar wind. They include light helium-3 atoms, which can in theory be used as fuel for nuclear fusion. (NASA)

Three-quarters of the universe consists of the lightest element, hydrogen, and almost a quarter of helium. Yet helium is very rare on Earth. This infrared image of the Pleiades was made by the Spitzer Space Telescope. (NASA/JPL)

Fear and Dread

Asaph Hall Discovers the Moons of Mars

In January 1610, Galileo Galilei discovers the four large moons of Jupiter. Astronomer, mystic and numerologist Johannes Kepler immediately draws a remarkable conclusion: if the Earth has one moon, and Jupiter four, then Mars must have two. In *Gulliver's Travels* (1726) Jonathan Swift describes how the astronomers of Laputa have indeed discovered two small moons orbiting Mars. And in 1750, Voltaire also writes of two Martian moons in *Micromégas*.

In the summer of 1877, the American astronomer Asaph Hall starts searching specifically for the elusive moons of Mars. At the United States Naval Observatory in Washington, Hall has access to the best and largest telescope of his time, with a lens diameter of 66 centimeters. On August 12, despite less than favorable weather conditions, he observes a small, faint speck of light close to Mars. A few days later it proves to be not a regular star, but an object moving through the sky along with the planet. And on August 18 Hall sees another moon, at an even shorter distance from Mars.

Hall is an expert at determining astronomical orbits. The distances and orbital periods of the two small moons – which are astoundingly close to those described by Swift in *Gulliver's Travels* – allow him to calculate the mass of Mars for the first time. It also becomes clear that the closer of the two moons has an exceptional orbit: it is less than 10,000 kilometers from the planet and completes an orbit in under eight hours – shorter than the rotation period of Mars. That means that from the surface of the planet you would see the moon rise in the west and set in the east, instead of vice versa.

The small moons are called Phobos (Fear) and Deimos (Dread), after the horses that pulled the chariot of Ares, the Greek god of war. They are irregular in shape and are probably reasonably porous lumps of rock. Phobos is about 27 by 21 by 18 kilometers, and Deimos is half that size. Phobos sports an enormous crater, the result of an impact that must have nearly shattered the little moon. The crater is named Stickney, after Asaph Hall's wife.

In the mid-twentieth century astronomers discover that Phobos' orbital period is becoming shorter as the moon seems to be continually moving into a lower orbit. Russian astronomer Iosif Shklovsky claims that this can only be explained if Phobos has so little mass that it is slowed down by the extremely thin outer layers of the Martian atmosphere. This gives rise to the theory that Phobos is hollow, and there are even suggestions that it is an artificial object, built by intelligent inhabitants of Mars.

Phobos' shrinking orbit – currently about one and a half meters per century – is however the result of Mars' strong tidal forces. In about 50 million years, the moon will be so close to the planet that the same tidal forces will rip it apart. Mars will then be surrounded by a ring of dust and rubble, just like Jupiter.

◐ At the end of the 1970s, the Mars explorer Viking 2 flew past the small Martian moon Deimos at a distance of 30 kilometers. This photo, which covers an area of less than a square kilometer, shows details a few meters in size. The small craters on Deimos are covered with a layer of dust several meters thick. (NASA/JPL)

➲ In this photo, taken by the European Mars Express space probe at the beginning of 2007, the Martian moon Phobos floats above the thin atmosphere of Mars. In about 50 million years, Phobos will be torn apart by the planet's tidal forces. (ESA/DLR/G. Neukum)

1877

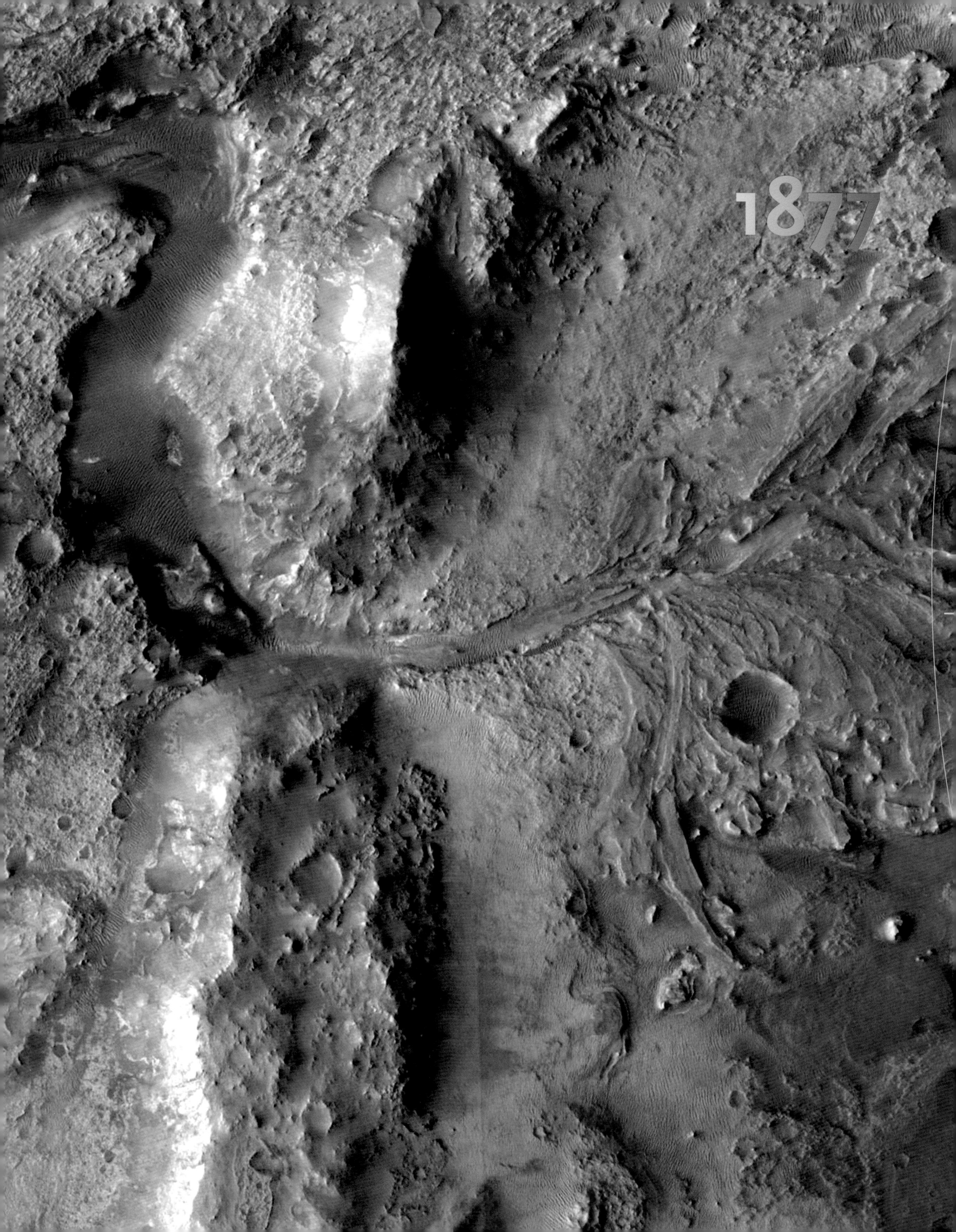
1877

Extraterrestrial Waterways

Giovanni Schiaparelli Discovers Canals on Mars

Every 26 months, Mars is exactly opposite the Sun in the sky. During this opposition, the planet is visible in the sky for the whole night, and is closest to the Earth. The orbit of Mars, however, is not a complete circle, so some oppositions are more favorable than others, and in September 1877 the planet is only 56 million kilometers away.

Giovanni Schiaparelli, director of Brera Observatory in Milan and the uncle of renowned couturier Elsa Schiaparelli, takes the opportunity to observe Mars closely with the observatory's 22-centimeter telescope. For many weeks he stays away from alcohol and coffee so as not to affect his eyesight, and on the basis of his observations he produces detailed maps of Mars, on which he gives dark and light areas poetic names like Syrtis Major, Elysium and Zephyria.

Schiaparelli also discovers several dozen straight, dark lines on the surface of Mars which he calls *canali*, the Italian word for channel or strait. In his publications, he compares the *canali* to the English Channel between France and England and the Mozambique Channel off the coast of Africa. But he does not rule out the possibility that the channels are artificial, and moreover the Italian word *canali* is incorrectly translated into English as *canals*.

In the 1890s, in the United States, aristocrat, diplomat and amateur astronomer Percival Lowell becomes fascinated by the canals of Mars. In 1894, in Flagstaff, Arizona, he opens the Lowell Observatory especially to unravel the mystery. Lowell is convinced that they are part of a massive irrigation network on the red planet, constructed by intelligent Martians. According to Lowell, the dark lines are not the canals themselves, but wide strips of vegetation along their banks, and where the canals meet there are fertile oases.

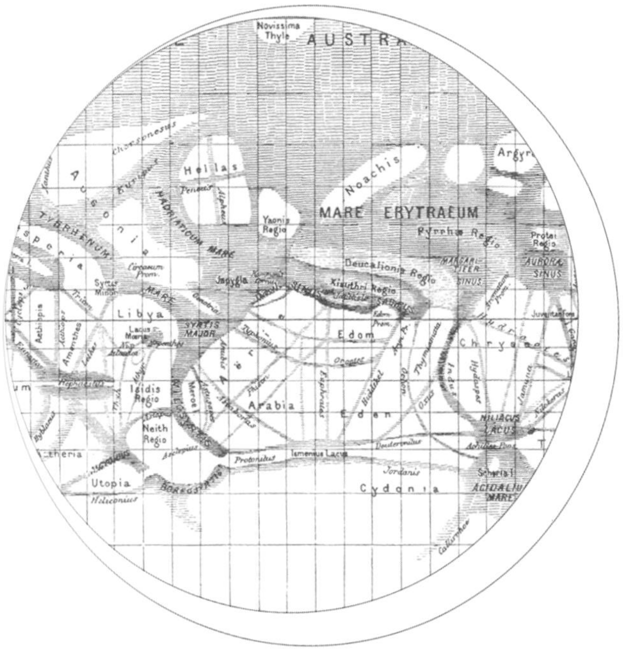

➋ Map of Mars by the Italian astronomer Giovanni Schiaparelli. Schiaparelli thought that the dark areas on Mars were seas and oceans. His observations of dead straight, dark Martian 'canals', many of which he gave wonderful names, proved, however, to be optical illusions.

➌ Martian canals, new-style: delta-like flow patterns and clay sediments have been found in the Jezero crater. On this false-color photo, the sediments are shown in green. In the distant past, the crater must have been filled with water. (NASA/JPL/JHU-APL/MSSS/Brown University)

Lowell's ideas on intelligent life on Mars are not taken very seriously by most professional astronomers of his time, but the wider public laps them up. As early as 1895, Lowell is giving countless lectures, publishing popular-scientific articles and publishes a bestseller entitled *Mars*. The book is the inspiration for the famous science fiction novel *The War of the Worlds* by H.G. Wells, published in 1898. In the early years of the twentieth century, Lowell publishes two more much-discussed books on the canals and the inhabitants of Mars.

Both Schiaparelli and Lowell remain convinced until their deaths that the canals of Mars really exist, although they differ as to their true nature. It is not until later in the first half of the twentieth century that it becomes clear that the canals are an optical illusion and wishful thinking. There is no running water on Mars because the temperature and air pressure on the surface are too low. Surprisingly, however, later space probes do detect dried up river beds and other traces of flowing water on the red planet.

Celestial Fireworks

Edward Emerson Barnard Discovers that Novae Are Exploding Stars

In 1572, Tycho Brahe sees a new star – a *Stella Nova* – flare up in the night sky for the first time; Johannes Kepler observes something similar in 1604. In both cases they are bright stars, which are very conspicuous, but later when more accurate charts of the night sky are available and astronomers can observe the firmament more closely, much fainter novae are discovered. In most cases they are only visible for a few months before they are slowly extinguished, and therefore practically nothing is known of their true nature.

In 1892, Edward Emerson Barnard comes to the conclusion that novae are exploding stars. Barnard is a meticulous observer. As an amateur astronomer he attends a meeting of the American Association for the Advancement of Science in his birthplace, Nashville, in August 1877. Barnard is 19, and has just bought a 12.5-centimeter telescope. Famed astronomer Simon Newcomb advises him to hunt for comets, and between 1881 and 1892 he discovers no less than fourteen.

1892 is a good year for Barnard: he has been working at Lick Observatory on Mount Hamilton in California for five years, and has access to one of the best telescopes in the world. On September 9, while using the telescope, he discovers a small moon orbiting close to Jupiter. The moon, named Amalthea, is the first new Jupiter satellite to be found since 1610, and the last planetary moon that will be discovered visually. At the end of 1892, Barnard also discovers craters on the planet Mars.

On August 19, 1892, Barnard aims the telescope at the constellation Auriga. A nova was discovered there at the beginning of the year, which was at its brightest in early February and then became steadily fainter. In the summer, however, it flares up again – reason enough to keep a close eye on the new star.

Barnard observes Nova Aurigae on almost every clear night. He soon sees that the star is surrounded by a sharply defined, round nebula. If the nebula had been there earlier in the year, it would certainly not have escaped Barnard's notice, therefore it must have developed only recently.

At the end of 1892, Barnard publishes his observations in *Astronomische Nachrichten*. In his article he concludes that 'a star has transformed into a nebula in a period of only four months'. Novae are therefore not new stars which are born from an existing nebula, but existing stars that produce nebula-like phenomena. Given the scale of the nebula around Nova Aurigae, it must be the result of a high-energy explosion of some kind.

We now know that a nova is a thermonuclear explosion on the surface of a white dwarf that is part of a binary system. Gas from the companion flows towards the white dwarf and accumulates on the surface until the temperature and density are so high that spontaneous nuclear reactions occur. Because the white dwarf does not actually explode, a nova can generate multiple bright explosions.

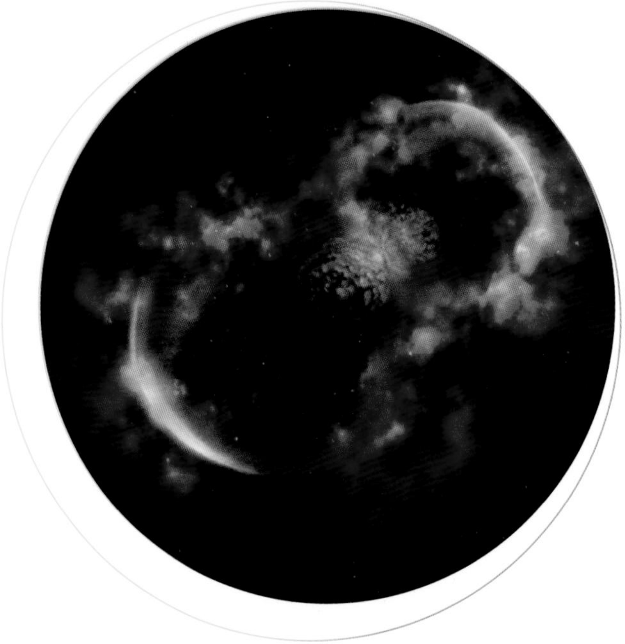

◐ The star Eta Carinae, in the southern sky, experienced a powerful explosion in 1843. This illustration shows the small Homunculus Nebula that was created by the explosion, together with a much larger nebula formed by a previous explosion about a thousand years ago. (Gemini Observatory/Lynette Cook)

➲ The Bubble Nebula, in the constellation of Cassiopeia, has a diameter of about six lightyears. It was formed long ago by a nova-like explosion of the massive star BD+602522, which is close to the center of the gas shell. (WIYN/NOAO/AURA/NSF)

1892

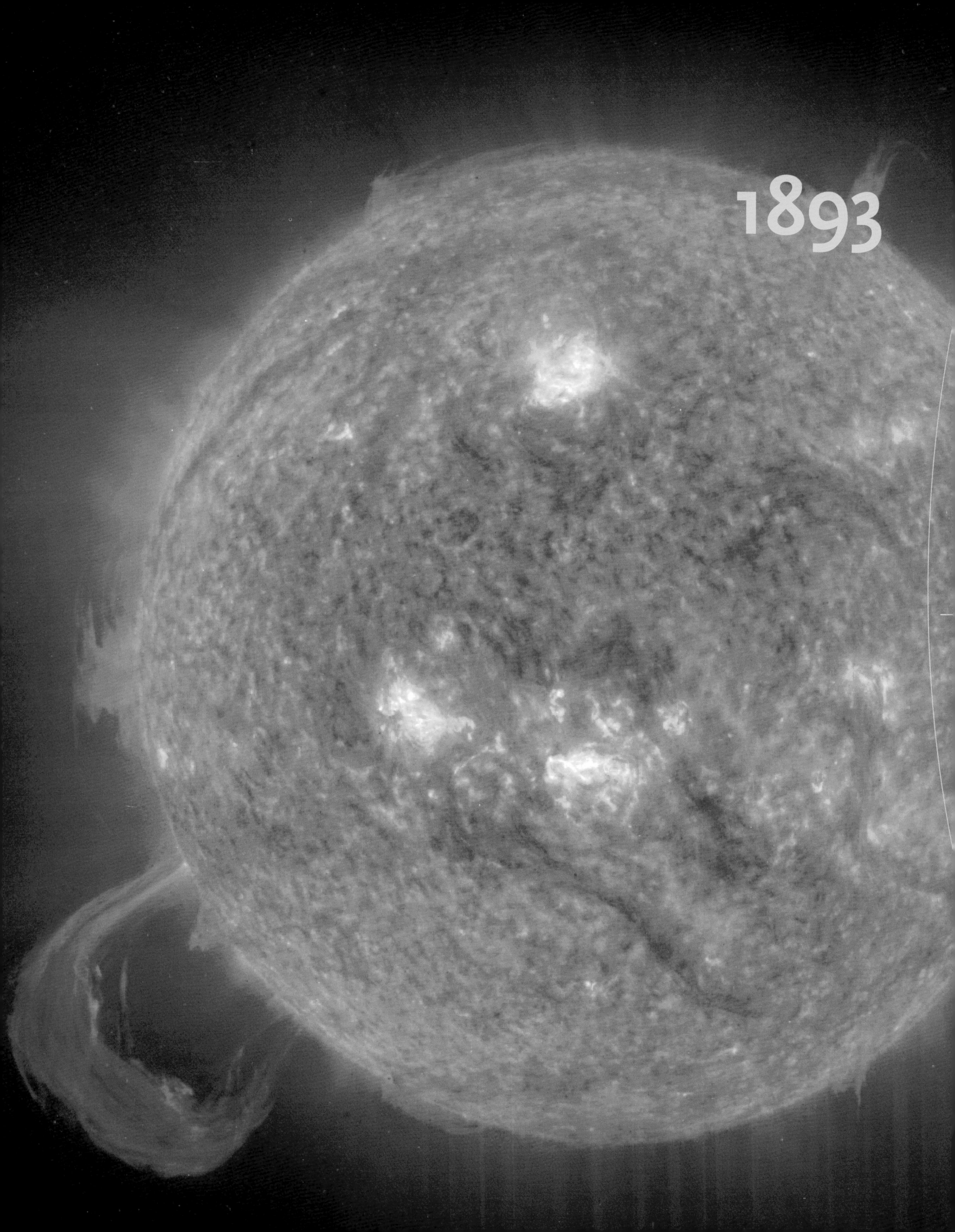
1893

Minimum Temperature

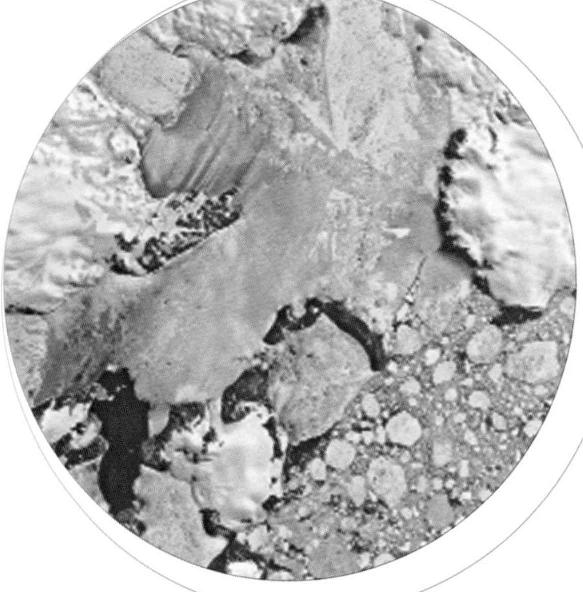

Edward Maunder Discovers an Extended Solar Minimum

By the end of the nineteenth century, astronomers are fully familiar with the solar cycle. There is an activity minimum every 11 years, when fewer dark spots are visible than normal. The Sun's corona – which is only visible during a total solar eclipse – also looks different than during the intervening activity maxima. It is generally assumed that the solar cycle has always existed, even though there are no reliable observations of sunspots before the discovery of the telescope.

The German researcher Gustav Spörer discovers that there is a link between solar activity and the relative quantity of carbon-14 in tree rings. Rings dating from the time of a solar minimum contain proportionately more of these heavy radioactive carbon atoms. Spörer does not understand the cause of this correlation but he does discover that, between 1420 and 1550, there must have been an extended activity minimum, as tree rings from that period all have a high carbon-14 content.

Spörer's discovery arouses the interest of Edward Maunder, who has been associated with the Royal Greenwich Observatory since 1873. Maunder conducts a great deal of research into sunspots and wonders whether the 11-year cycle has also experienced irregularities in the more recent past. In 1890, together with his second wife, mathematician Annie Russell, he embarks on a study of all sunspot observations in the observatory's archives.

In 1893, Maunder publishes his remarkable findings: between 1645 and 1710 there was indeed an 11-year activity cycle but, generally speaking, solar activity stayed at a very low level for more than half a century. In fact, there had been a single extended solar minimum. Maunder's results attract little attention, however, and many astronomers believe that the apparent anomaly in the second half of the seventeenth century is the consequence of inaccurate observations.

It is not until 1976 that American solar physicist Jack Eddy of the High Altitude Observatory in Boulder, Colorado, makes the link between the extended solar activity minimum and the Little Ice Age, a period when the winters in Europe were extremely severe. Eddy calls the period without sunspot activity between 1645 and 1715 the Maunder Minimum, and suggests that there is a causal link between solar activity and the climate on Earth.

The temperature on Earth proves also to have been lower than average during the Spörer Minimum (1420–1550). The cause of these extended activity minima (which may themselves recur in cycles) is unknown. The next Maunder Minimum, with an accompanying Little Ice Age, can in theory occur at any moment, and bring the current period of global warming to a temporary halt.

➋ Satellite image of glaciers on Antarctica. If solar activity stays at a minimum for an extended period, the average temperature on Earth falls. In the distant past, the planet has at times been almost completely frozen, possibly as the result of erratic behavior of the Sun. (NASA)

➌ The gigantic prominence on this ultraviolet image of the Sun – an explosion of hot gas – was photographed in September 1999. It reached an altitude of a few hundred thousand kilometers and a temperature of 60,000 degrees Celsius. During the Maunder Minimum such explosions occurred very rarely. (NASA/ESA)

Luminous Thunder

Max Wolf Discovers the First Light Echo, of Nova Persei

It's not often that you look up at the sky at night and discover a new star. But on February 21, 1901, that is what happens to the Scottish preacher and amateur astronomer Thomas Anderson. He knows the night sky like the back of his hand and realizes that the magnitude 3 star he has observed in the constellation of Perseus does not belong there. Anderson discovers the first nova of the twentieth century, which is just seven weeks old.

A nova is a gigantic thermonuclear explosion on the surface of a white dwarf. The dwarf attracts gas from a companion star over a period of time, and at some point, the accumulated layer of gas is so thick and hot that it explodes like a hydrogen bomb, expelling enormous quantities of material into space. The faint white dwarf becomes a million times brighter over a period of only a few days, and then gradually becomes fainter again in the months that follow.

But Nova Persei has a surprise in store. In the summer of 1901, Max Wolf, of the Heidelberg Observatory in Germany, observes faint wisps of nebular gas around the exploded star. Surely it cannot be the stellar gas expelled during the explosion? At a distance of several hundred lightyears the expanding shell cannot already be visible six months after the explosion. Furthermore, some of the arc-shaped wisps of gas change position within a few weeks!

Astronomers have never witnessed such extremely rapid motion before. The measured dimensions of the shell and changes of position suggest that the gas is moving at almost the speed of light, and there is no way at all of explaining such high velocities.

Are astronomers perhaps not seeing real movements, but a kind of echo effect? This ingenious solution is first proposed by the Dutch astronomer Jacobus Kapteyn. There is an afterglow from the nova explosion as the light is reflected in existing dust clouds, in the same way that a short clap of thunder at the end of a valley becomes a sustained rumble as it echoes off the mountainsides.

In 1939, French astronomer Paul Couderc shows that light echoes are usually generated by dust that is closer to us than the exploding star, just as the mountainsides that echo a

thunderclap are closer to the observer than the actual lightning strike. The phenomenon is, however quite rare; light echoes are observed only in the case of a handful of supernovae, most spectacularly V838 Monocerotis, a mysterious star that explodes in 2002.

Yet light echoes offer more than just pretty pictures. As Kapteyn already realizes in 1901, they literally shed light on the nature of interstellar material. Careful analysis of the light echoes of an exploding star can provide valuable information on the three-dimensional structure of the illuminated dust clouds. Nature's light show is always worth taking a good look at.

➊ Slowly expanding gas shells are still visible around variable star GK Persei, which experienced a nova explosion in 1901. Shortly after the explosion the gas shells caused light echoes, so that they appeared to move faster than light. (USNO/Al Kelly)

➲ In 2005, the Hubble Space Telescope took this photograph of spectacular light echoes around the exploded star V838 Monocerotis. The stellar explosion occurred at the beginning of 2002, but the light echoes could still be seen years later. (NASA/ESA/Hubble Heritage Team)

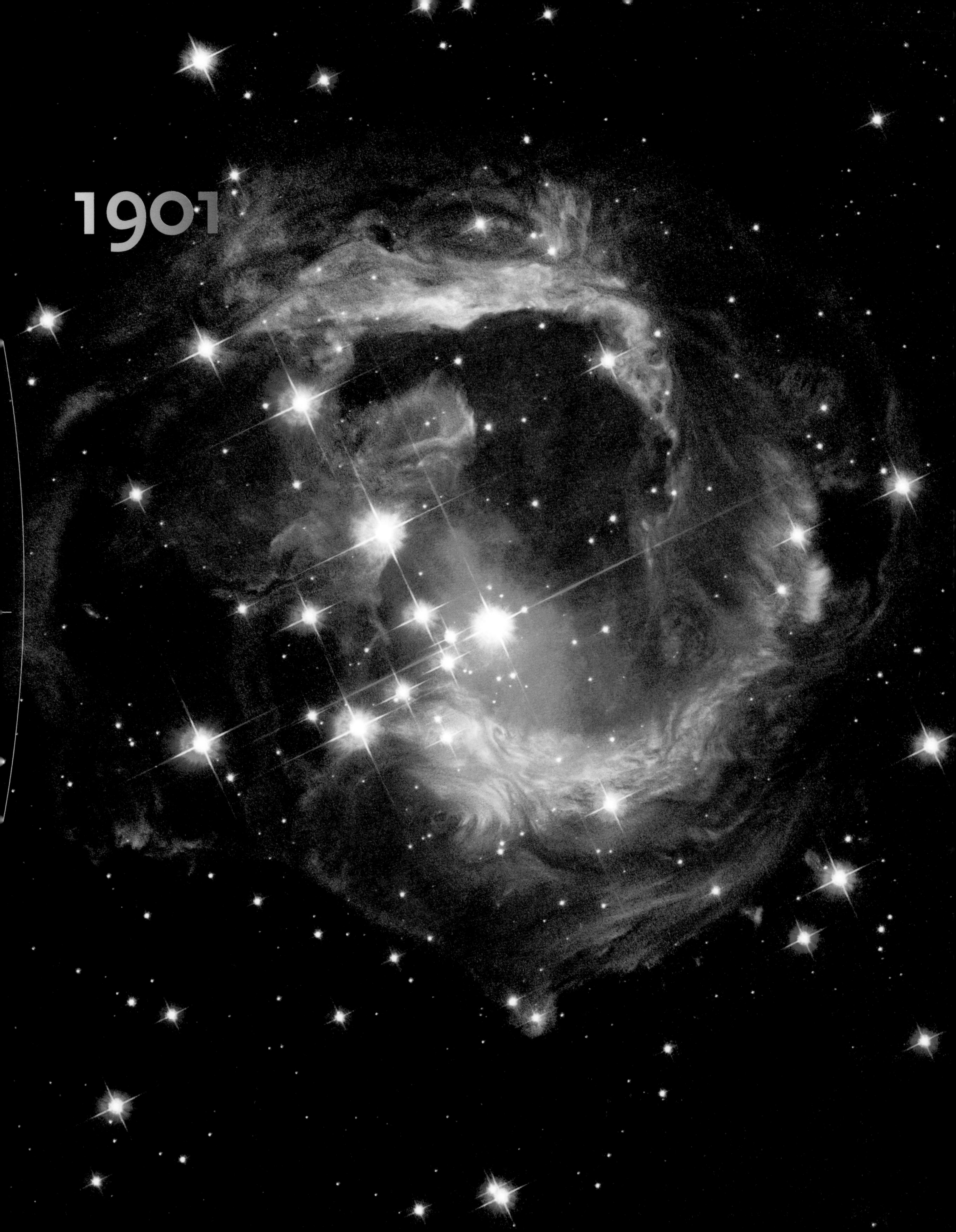

1901

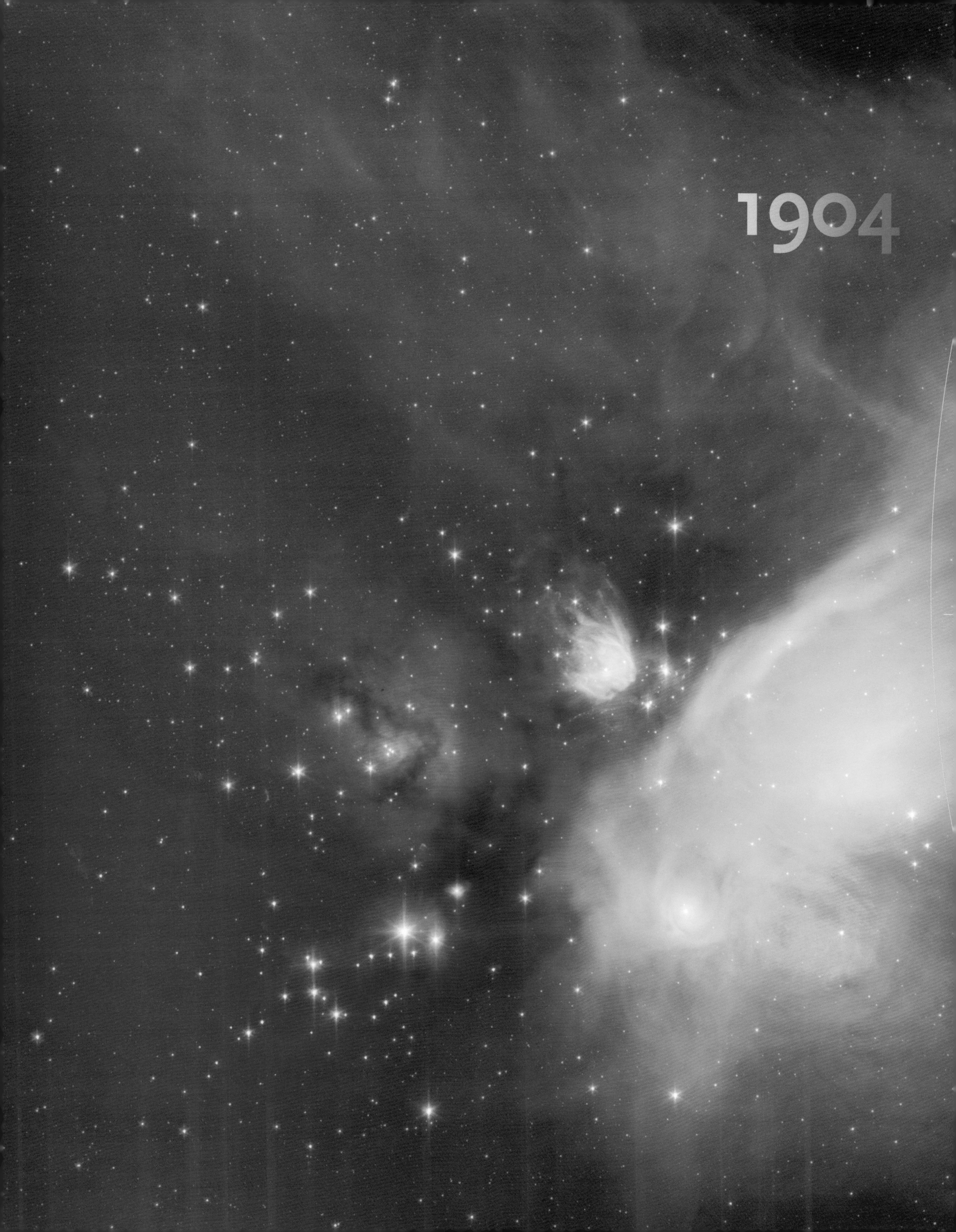
1904

Almost Empty

Johannes Hartmann Discovers Interstellar Material

The universe contains an enormous diversity of celestial objects. We have known about the Sun, the Moon, the planets and the stars since ancient times. Comets, too, proved to be objects far from Earth, and since the discovery of the telescope, we can add planetary moons, asteroids, star clusters and nebulae to the list. Generally speaking, however, the space between all these objects is considered to be empty like a vacuum.

In 1904, Johannes Hartmann discovers that there is also material in the 'empty' space between the stars. This interstellar material may be extremely rarefied, but because the universe is so vast, its total mass is larger than that of all the stars together.

Hartmann works at the observatory in Potsdam, where he devotes a lot of time and energy to developing optical tests for telescope lenses. Thanks to the tests and the improvements they bring about, the observatory's 80-centimeter telescope is one of the best of its time. Hartmann conducts extensive research into stellar spectra and is an authority on astrophysics, which studies the temperature, composition and other physical properties of stars.

❷ The three stars in the belt of Orion are more than a thousand lightyears away. Beneath the left-hand star (Alnitak), the dark Horsehead Nebula is visible. In the light of the right-hand star (Mintaka), in 1904, Johannes Hartmann discovered absorption lines of calcium – proof of the existence of interstellar matter. (Digitized Sky Survey/Davide De Martin)

❸ The star formation area Rho Ophiuchi consists of interstellar gas and dust clouds, from which new stars form. The newborn stars are visible in this infrared photograph by the Spitzer Space Telescope. (NASA/JPL/CfA)

In 1904, Hartmann photographs the spectrum of Mintaka, the right-most of the stars in the belt of the constellation Orion. Spectroscopic research has shown that Mintaka is a binary star, and the variable orbital velocity of the star produces minuscule periodical shifts in the dark lines of its spectrum. However, Hartmann also discovers an absorption line that does not take part in the shifts: the 'K line' of calcium atoms. The calcium atoms are apparently not in the star's atmosphere, but somewhere between the star and the telescope.

Since the absorption line is not on the exact wavelength observed during laboratory measurements, it cannot have arisen in the Earth's atmosphere either; the wavelength difference means that the calcium atoms must have a certain velocity with respect to the Earth. Hartmann concludes that somewhere in space there must be a tenuous cloud of matter that absorbs a little of the light from Mintaka. Later, astronomers also discover sodium absorption lines in the star's spectrum, and find the spectrographic fingerprint of interstellar material in the spectra of other stars.

The interstellar material in the Milky Way comprises approximately one million particles per cubic meter. That may sound like a lot, but it is considerably less than in the best vacuum that can be created in a terrestrial laboratory. Ninety-nine percent of interstellar material consists of gas atoms (mainly hydrogen and helium); only 1% is made up of microscopic dust particles.

Color Coding

Ejnar Hertzsprung Discovers the Link Between the Color and Luminosity of Stars

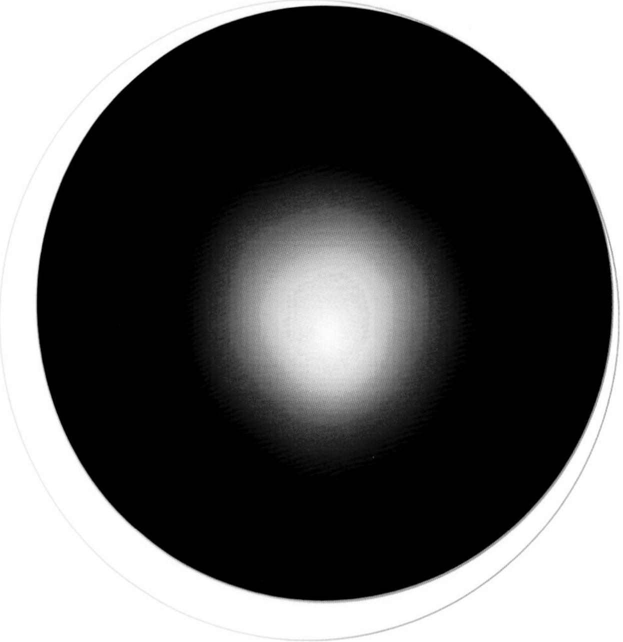

If you want recognition for your work as a scientist, it is not always enough to make a revolutionary discovery – good public relations are often just as important. Danish astronomer Ejnar Hertzsprung experiences this first hand in 1905 when he publishes his discovery of the link between the color and luminosity of stars in an obscure German magazine. Because the article practically goes unnoticed, the honor of making the discovery almost goes to an American.

Hertzsprung is actually a chemist specializing in photographic techniques and emulsions. At first, astronomy is just a hobby. In 1905, he discovers that orange and red stars are divided into two groups and that these stars are relatively cool, with surface temperatures of 3,000–4,000 degrees Celsius. Most of them radiate less light than the Sun. This means that they are relatively small, yet there are other orange and red stars that generate enormous quantities of energy. With such low surface temperatures, these stars must be gigantic.

Hertzsprung publishes his discovery in the *Zeitschrift für Wissenschaftliche Photographie*. His article suggests that stars pass through various stages of evolution. Because giant stars are relatively rare, he concludes that stars only have such enormous dimensions for a short time. He develops these ideas further in a second article published, in 1907.

In the United States in 1910, Henry Norris Russell comes to the same conclusions, independently of Hertzsprung. Around 1911, both Hertzsprung and Russell publish diagrams showing the relation between the color of a star and its luminosity. The diagrams clearly show the various 'population groups' of stars. Most stars, including the Sun, are on the 'main sequence' – a diagonal band in the diagram running from bright, hot, blue-white stars to faint, cool, red dwarf stars. In addition, there are also red giants, blue and red supergiants, and white dwarfs.

The diagram clearly offers the key to a better understanding of the way stars evolve. The diagram is called the Russell diagram by American astronomers. Hertzsprung, who is appointed vice-director of Leiden Observatory in 1919, does not appear overly concerned, but for Dutch astronomer Willem Luyten, who obtains his Ph.D. under Hertzsprung before emigrating to the United States in 1921, it is a sore point. In his columns for *The New York Times*, he repeatedly points out that it was Hertzsprung who first discovered the link between the color of stars and their luminosity. Thanks to his lobbying, the name Hertzsprung–Russell diagram eventually becomes widely used.

It is not until much later that astronomers obtain a better picture of how stars evolve. They spend most of their life on the main sequence, where their original mass determines their temperature and luminosity.

Betelgeuse, the star in the top left-hand corner of the constellation of Orion, is a red supergiant. The color indicates a relatively low surface temperature of 3,300 degrees Celsius. Betelgeuse was the first star, after the Sun, whose surface was photographed. (NASA/ESA/STScI)

Blue stars, like the young supergiants in the star cluster Pismis 24, have surface temperatures of tens of thousands of degrees and emit large quantities of high-energy ultraviolet radiation that heats up the surrounding gas and dust clouds. (NASA/ESA/J.M. Apellániz/Davide De Martin)

1905

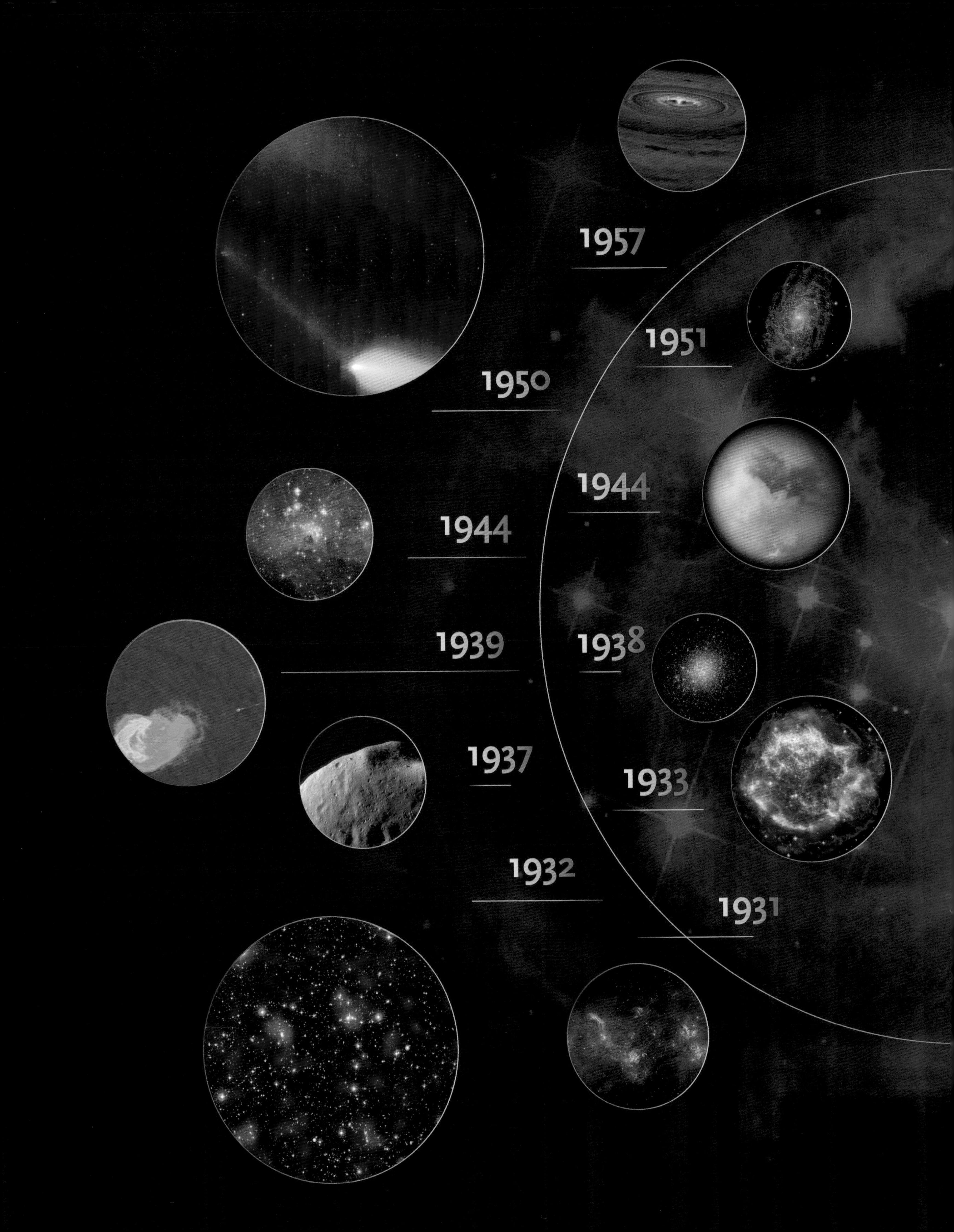
1957
1951
1950
1944
1944
1939
1938
1937
1933
1932
1931

1908 - 1958

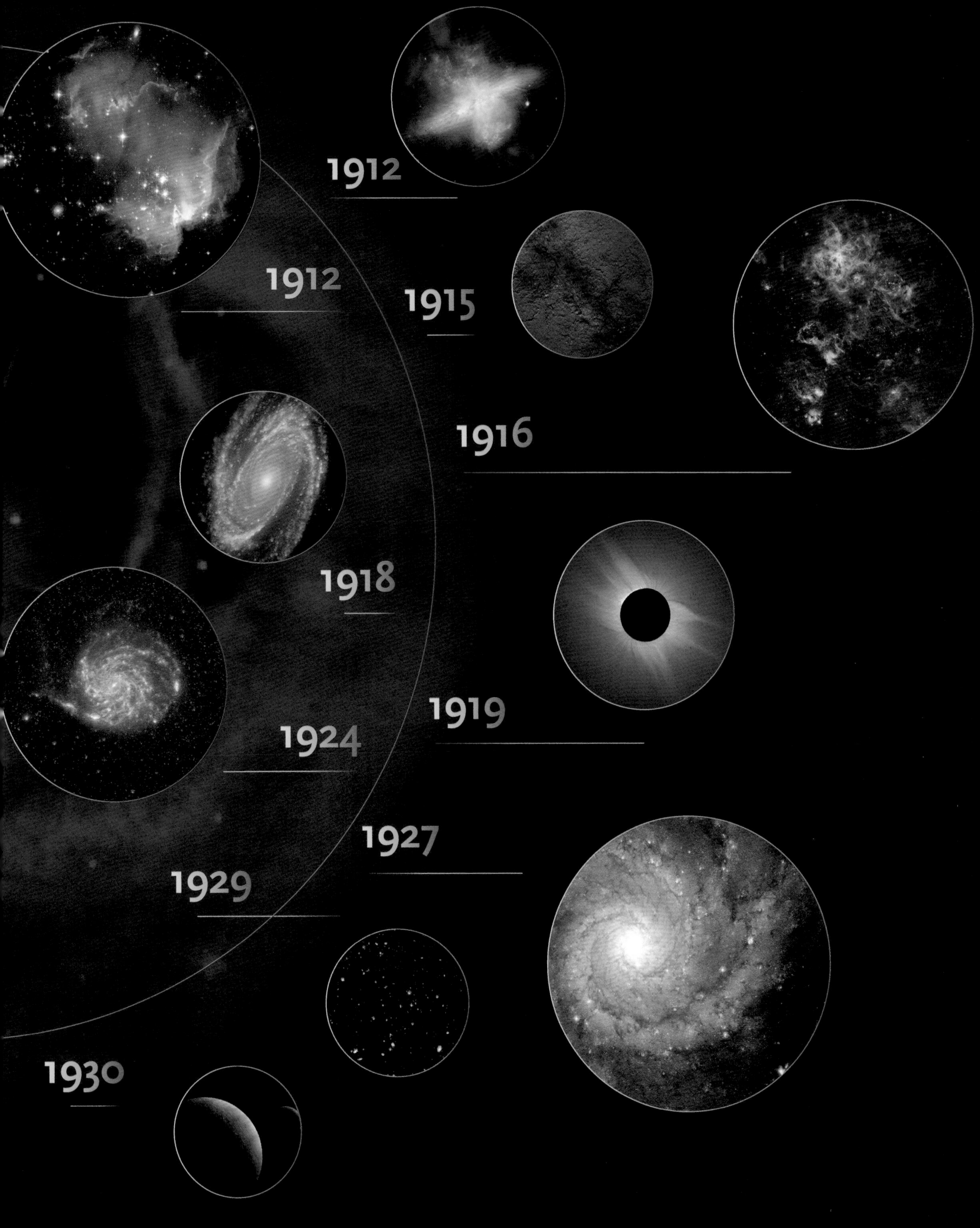

The discovery that the universe is expanding is without a doubt one of the greatest triumphs of astronomy. Not only do biological species evolve, as Charles Darwin described in the nineteenth century, and not only is the Earth subject to change, as geologists have discovered, but the universe as a whole has its own evolutionary process of birth, growth and development.

A Speck of Cosmic Dust

The astronomical breakthroughs in the first half of the last century are made possible by the large reflecting telescopes found in the United States, but more particularly by the acumen and creativity of the scientists who analyze and interpret the observations made using these instruments. Astronomy also owes a great deal to revolutionary developments in physics; the theory of relativity and quantum mechanics lay the basis for modern cosmology, and nuclear physics makes it possible to understand how stars work.

Astronomers spread their wings in time and space, unveiling a universe in which our Milky Way is little more than a backwater, and in which the Earth and humankind are relative latecomers to the cosmic stage. They also devise new ways of unraveling the secrets of the cosmos, by studying not only visible light from the universe, but also weak radio waves, which lift a completely new corner of the veil.

in an Evolving Cosmos

Homo sapiens may be insignificant in the great expanse of the universe but, in the middle of the twentieth century, astronomy unexpectedly acquires a human face when astrophysicists discover the origin of the elements in nature. No one can deny any longer that we are an irrevocable part of the evolving universe, literally composed of stardust.

1912

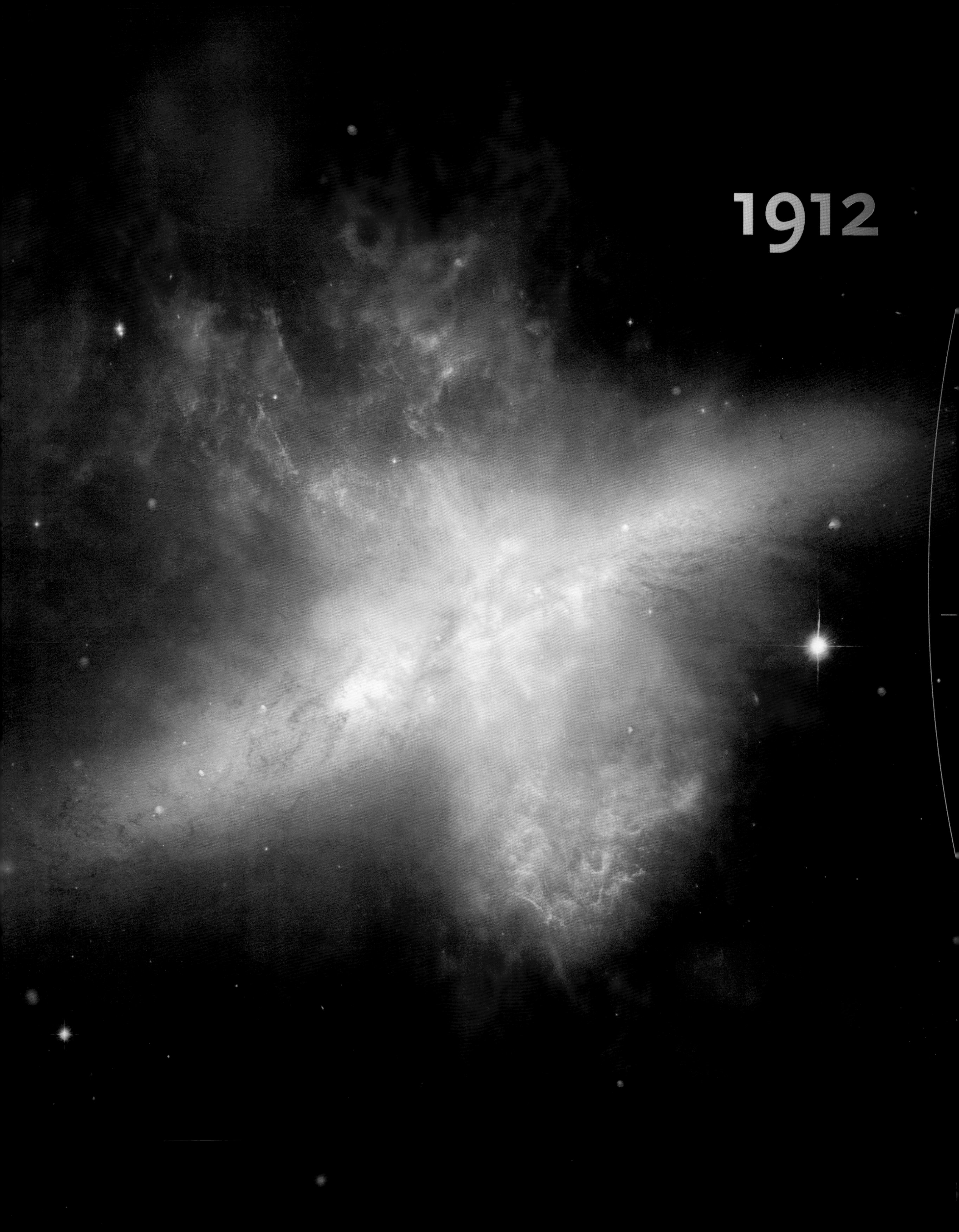

Heavenly Messengers

Victor Hess Discovers Cosmic Rays

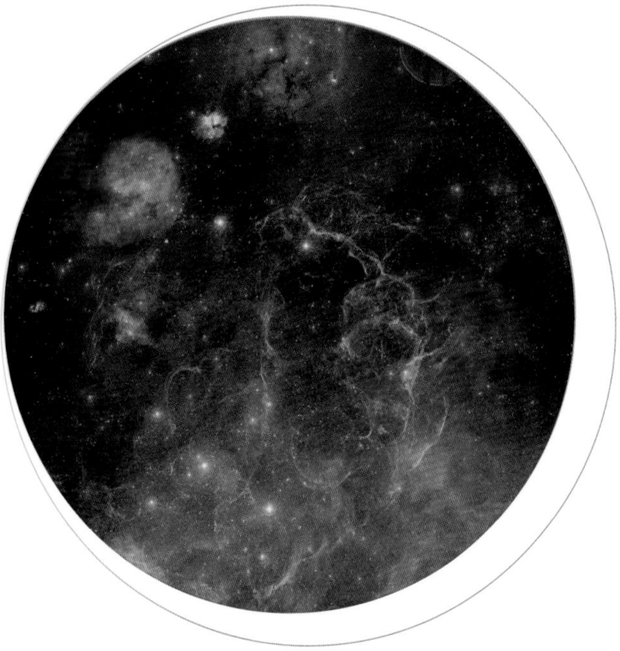

At the beginning of the twentieth century, while astronomers are racking their brains about the scale of the universe, the true nature of spiral nebulae, the canals on Mars and the hunt for a planet beyond the orbit of Neptune, terrestrial physicists are completely preoccupied with radioactivity. Radioactive elements produce high-energy alpha, beta and gamma rays that can ionize neutral atoms. A new element discovered by Marie and Pierre Curie in 1898 is even given the name 'radium' because of its radioactive properties.

Austrian physicist Victor Hess has been working at the Institute of Radium Research at the Viennese Academy of Sciences since 1910. He is mainly interested in the origin of the ionized radiation which, according to measurements of positively charged ions and negatively charged electrons, is present everywhere in the Earth's atmosphere; the whole planet appears to be a source of ionized radiation.

German physicist and Jesuit priest Theodor Wulf, who teaches physics at the Jesuit University in Valkenburg, develops a sensitive electrometer to measure the level of ionization in the atmosphere. In 1910, he conducts measurements at the bottom and top of the Eiffel Tower, and Wulf concludes that the level of radiation actually increases the further he moves away from the Earth's surface.

Many physicists do not take Wulf's results seriously, and Victor Hess decides to solve the puzzle once and for all. Since there are no buildings higher than the Eiffel Tower, he resorts to hot-air balloons. In 1911 and 1912, Hess undertakes nerve-racking balloon flights up to heights of 5,300 meter, by day and night. His sensitive instruments leave no room for doubt: a few kilometers above the ground the amount of ionized radiation is several times larger than on the Earth's surface.

Hess concludes that the radiation must be of extraterrestrial origin, suddenly providing the new area of research into radioactivity with an astronomical component. His conclusions are confirmed in 1925 by the American physicist Robert Millikan, who is the first to use the term 'cosmic rays'.

It is not discovered until the end of the 1920s that the phenomenon is not caused by high-energy electromagnetic radiation, but by very fast-moving electrically charged particles, mainly protons (hydrogen nuclei). These protons come predominantly from the Sun, supernova explosions in the Milky Way and probably the active nuclei of distant galaxies.

In 1936, Victor Hess is awarded the Nobel Prize for Physics, together with the American, Carl Anderson, the discoverer of anti-matter. Systematic research into cosmic rays does not, however, gather pace until the end of the twentieth century.

The wispy trails of gas in this photo are the remnants of a star that exploded around 11,000 years ago in the constellation of Vela. During such supernova explosions, electrically charged particles can be whipped up to enormous speeds and levels of energy. (Digitized Sky Survey/Davide De Martin)

The explosive galaxy M82 in Ursa Major has a supermassive black hole at its core. In the vicinity of such black holes electrons and atomic nuclei reach extremely high speeds and reach Earth tens of millions of years later in the form of cosmic rays. (NASA/ESA/JPL/CXC)

Cosmic Yardstick

Henrietta Leavitt Discovers the Period-Luminosity Law of Cepheids

They are known as 'Pickering's harem', the female 'computers' at the Harvard College Observatory in Cambridge, Massachusetts. Edward Charles Pickering is appointed director of the observatory in 1877, when it is already about 40 years old. He hires women at 25 cents an hour to conduct accurate measurements of photographic plates of the night sky and process the results. One of the most brilliant women in Pickering's harem is Henrietta Leavitt.

In 1895, Leavitt is already doing voluntary work at Harvard. In 1902, after a long illness which has left her nearly completely deaf, Pickering gives her a permanent job where she eventually becomes head of stellar photometry at the observatory. For many years she mainly occupies herself with measuring the brightness of stars on photographic plates of the southern hemisphere.

Over a period of many years, Leavitt discovers nearly 2,000 variable stars in the Magellanic Clouds, two large nebula-like objects only clearly visible from the southern hemisphere. There are so many images of the Magellanic Clouds available that she can determine the brightness variations of most of the stars.

Several dozen variable stars, primarily in the Small Magellanic Cloud, prove with great regularity to become brighter and then fainter again over a period of a few days or weeks. Leavitt discovers that there is something strange about these Cepheids, named after Delta Cephei, whose variable character was first discovered by John Goodricke in 1784.

There appears to be a correlation between the average luminosity of a Cepheid and its pulsation period. Bright Cepheids take about a month to become brighter and then fainter again; faint Cepheids complete the same cycle in only a few days. Leavitt describes her discovery as early as 1908 in the *Harvard College Observatory Circulars*, and in 1912 she publishes her 'period-luminosity law'.

A year later, Ejnar Hertzsprung succeeds in testing Leavitt's law by determining the distance of a number of Cepheids in our own Milky Way, thereby enabling their real luminosity to be calculated. In this way, he is the first to be able to deduce the distance to the Small Magellanic Cloud.

Cepheids are yellow giant stars passing through a short, unstable stage in their evolution. They swell up and then shrink again, causing their temperature and luminosity to vary continually. Thanks to Leavitt's discovery, they can be used as cosmic milestones; their pulsation period, and therefore their luminosity, is easily measured. Comparing that with their observed brightness in the sky enables their distance to be calculated. In 1924, Edwin Hubble uses this method to determine the distance of the Andromeda nebula.

The Large Magellanic Cloud is one of the companion galaxies of our Milky Way. It is more than 170,000 lightyears away. The distances to the Magellanic Clouds were determined for the first time in the early twentieth century, using the Cepheids method. (ESO/Yuri Beletsky)

The Small Magellanic Cloud, the distance of which was first measured in 1913, contains a large number of star clusters and star forming regions. On this photograph, taken with the Hubble Space Telescope, a remote background galaxy is also visible, left of center. (NASA/ESA/Hubble Heritage Team)

1912

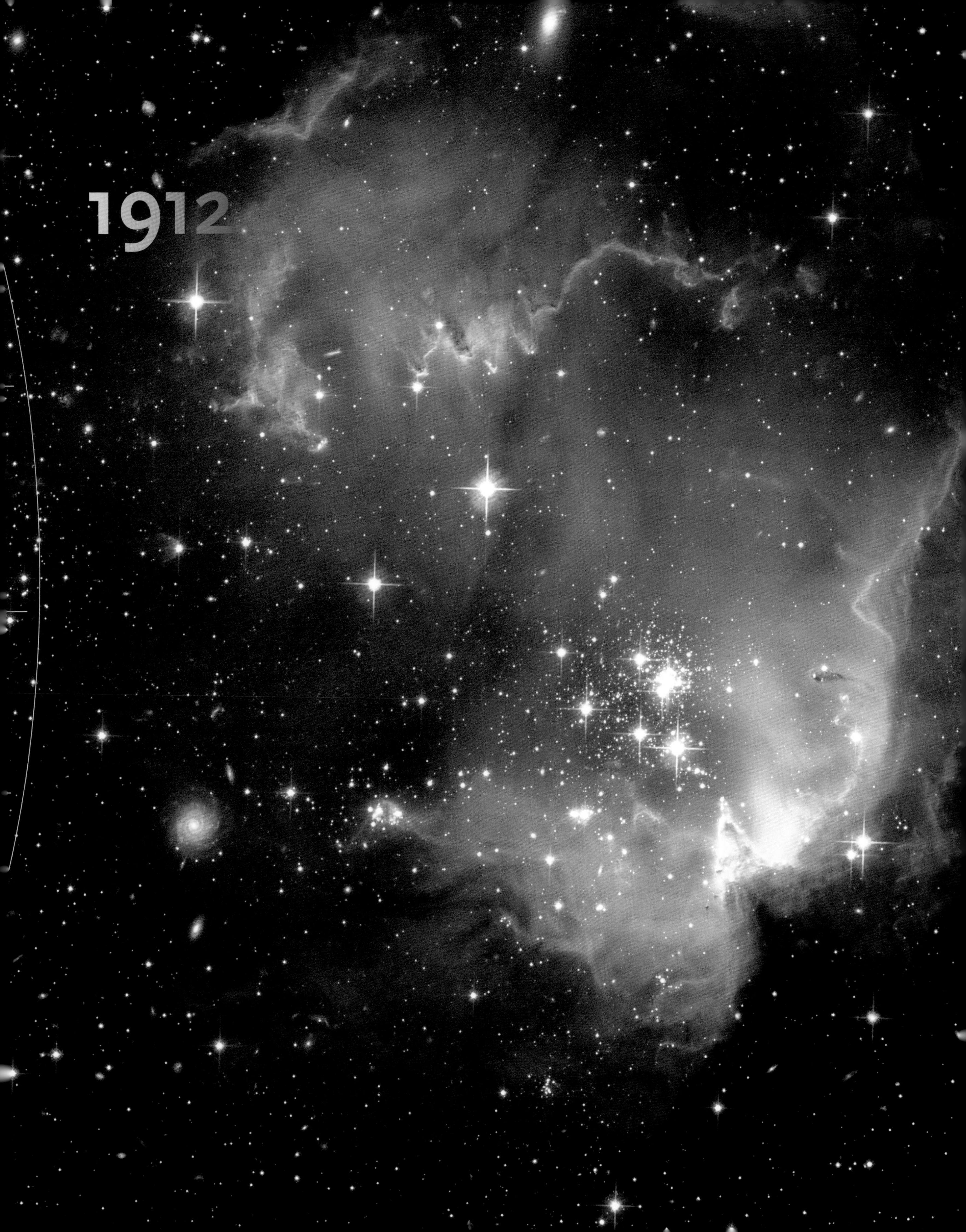

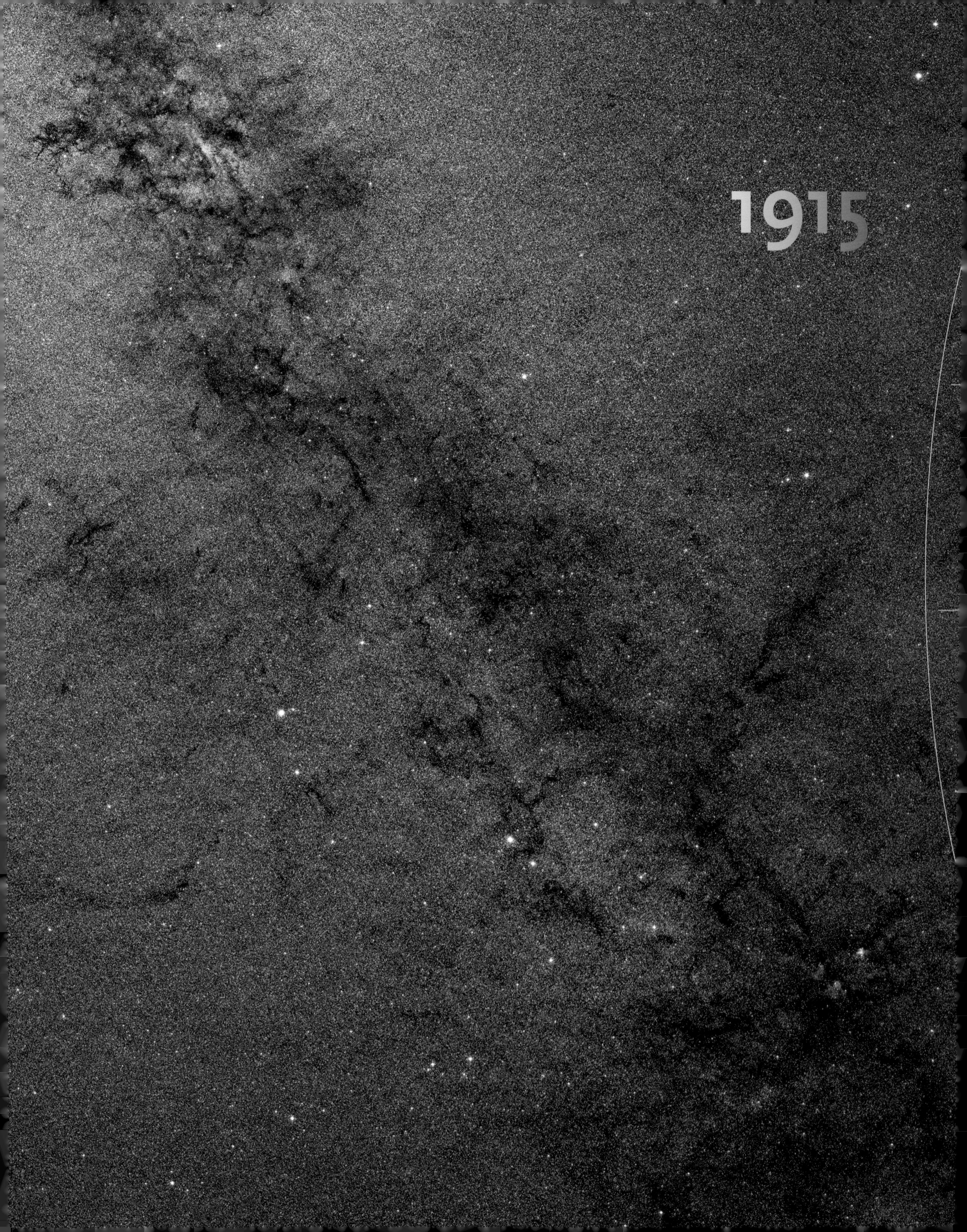
1915

Nearest Neighbor

Robert Innes Discovers the Nearest Star, Proxima Centauri

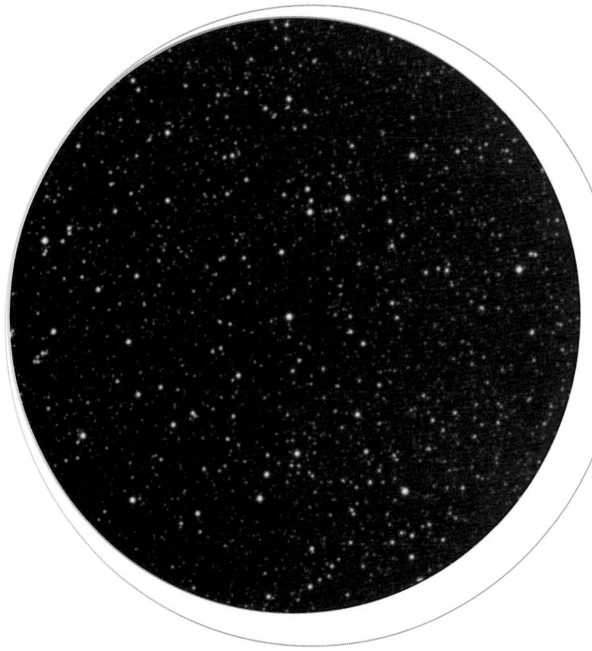

Strange as it may seem, the nearest star to the Sun was not discovered until more than 300 years after the telescope was invented. Proxima Centauri is an extremely faint dwarf star which, despite being only 4.22 lightyears away, can only be seen with a powerful telescope.

Since Thomas Henderson's parallax measurements in the 1930s, the bright binary star Alpha Centauri has been considered the nearest star to the Sun. At a short distance of approximately 4.3 lightyears, or more than 40 trillion kilometers, Alpha Centauri has a relatively large proper motion in the sky of 3.7 arcseconds per year.

Robert Innes, a Sydney-based Scottish wine merchant who has discovered countless new binary star systems in the southern night sky during his spare time, specializes in the proper motion of stars. In 1896, David Gill, director of the Royal Observatory at the Cape of Good Hope, offers him a job, and in 1903 Innes becomes director of the Meteorological Observatory in Johannesburg, which under his leadership is transformed into the Transvaal, and later the Union, Observatory.

In 1915, using the Observatory's 22-centimeter telescope, Innes discovers a faint star a few degrees away from Alpha Centauri with exactly the same proper motion. It must be at more or less the same distance as Alpha Centauri, and move with the binary star through the universe. Additionally because it is so faint and yet so close, it must emit very little light.

Despite its relative proximity – it is only 4.22 lightyears away – the red dwarf star Proxima Centauri is hardly visible. It is exactly in the center of this photograph, taken with the 5-m Hale Telescope at Palomar Observatory.
(Digitized Sky Survey)

Tens of thousands of stars are visible on this infrared photo of the Milky Way, in the direction of the constellation of Sagittarius. The center of the Milky Way, partly concealed by dust clouds, is more than 20,000 lightyears away – 5,000 times further than the nearest star. (2MASS)

Innes is convinced that the dwarf star is a little closer to the Sun than Alpha Centauri and gives it the name Proxima, meaning 'nearby'. His assumption appears to be confirmed in 1917 by parallax measurements conducted by the Dutch astronomer, Joan Vôute, at the Cape Observatory. Later, precision measurements by the European satellite Hipparcos in the 1990s determine the distance to Proxima Centauri at 4.22 lightyears.

In the sky, Proxima Centauri is a hundred times fainter than the faintest star that can be seen with the naked eye. It emits 18,000 times less light than the Sun, has about 12% of the Sun's mass, and is 7 times smaller. In fact, Proxima's diameter is only one and a half times that of the giant planet Jupiter, although it weighs 130 times as much.

Proxima is about 0.21 lightyears away from Alpha Centauri. It probably moves in a very wide orbit around the bright binary star, making it in effect a triple system, and Proxima's orbital period is measured somewhere between 500,000 and 2 million years.

It is not inconceivable that there is another faint dwarf star even closer than Proxima Centauri; new red dwarfs are regularly discovered in the vicinity of the Sun.

Sprinter in the Night Sky

Edward Emerson Barnard Discovers the Star with the Largest Proper Motion

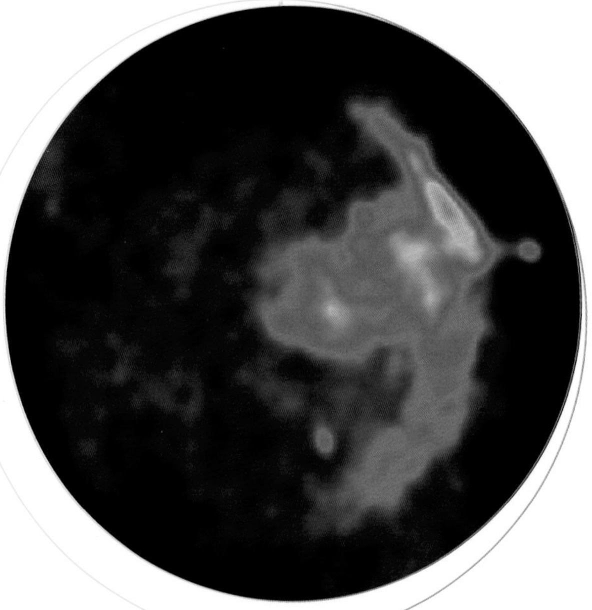

A bird flying past at a short distance moves across the sky faster than a fighter jet at high altitude. The same applies to stars. Although most stars are moving through the universe at high speed, if they are at great distances, they hardly seem to move. Closer stars, therefore, move through the sky more quickly.

In 1916, while studying long-exposure images, Edward Emerson Barnard discovers a small, faint star in the constellation Ophiuchus, with an unprecedented proper motion of 10.3 arcseconds per year. In 175 years, the star moves half a degree, equal to the apparent diameter of the Full Moon! It is immediately obvious that this must be one of the nearest stars in the universe.

Parallax measurements show that Barnard's Star, as the new discovery is called, is at a distance of only 5.97 lightyears. After the binary star Alpha Centauri and the red dwarf Proxima (which is almost certainly in orbit around Alpha Centauri), Barnard's Star is the nearest star to the Sun. Like Proxima it is a red dwarf, but it is a little bigger, brighter and more massive.

In 1969, the Dutch-American astronomer Peter van de Kamp publishes his 'discovery' of a planet in orbit around Barnard's Star. Using the 60-centimeter telescope at Sproul Observatory in Swarthmore, Pennsylvania, Van de Kamp takes photographs of the star over a period of several decades. The positional measurements he takes from the more than 2,000 photographic plates show that, during its hasty journey through the sky, Barnard's Star wobbles very slightly. He deduces from this that the star is accompanied by a planet 60–70% more massive than Jupiter with an orbital period of around 25 years.

In 1982, Van de Kamp changes his mind: Barnard's Star is not accompanied by one but *two* planets with orbital periods of 12 and 20 years. The planets are 50% and 70% the mass of Jupiter. In the 1990s, however, precision measurements with more powerful instruments, including the Hubble Space Telescope, prove that the planets do not exist. Van de Kamp was probably misled by subtle systematic errors in his positional measurements, which may have been caused when astronomers removed the lens of the Sproul telescope for cleaning in the 1960s.

Barnard's Star has a spatial velocity of 166 kilometers per second, and its distance from the Sun decreases by 140 kilometers every second. In a little less than 10,000 years, in the year 11800, it will pass by the Sun at a distance of only 3.8 lightyears. But even then, the dwarf star will be too faint to see with the naked eye.

Radio image of the remnant of a supernova 15,000 lightyears away, in the constellation of Sagittarius. The pulsar created by the explosion races through the universe at a speed of more than 550 kilometers per second and is now outside the supernova shell, just to the right of the brightest part. (NRAO)

The proper motion of a star tends to be greater the closer the star is to Earth. Despite this, astronomers have succeeded in measuring the tiny proper motion of stars in the Large Magellanic Cloud: 0.002 arcseconds per year. (ESO)

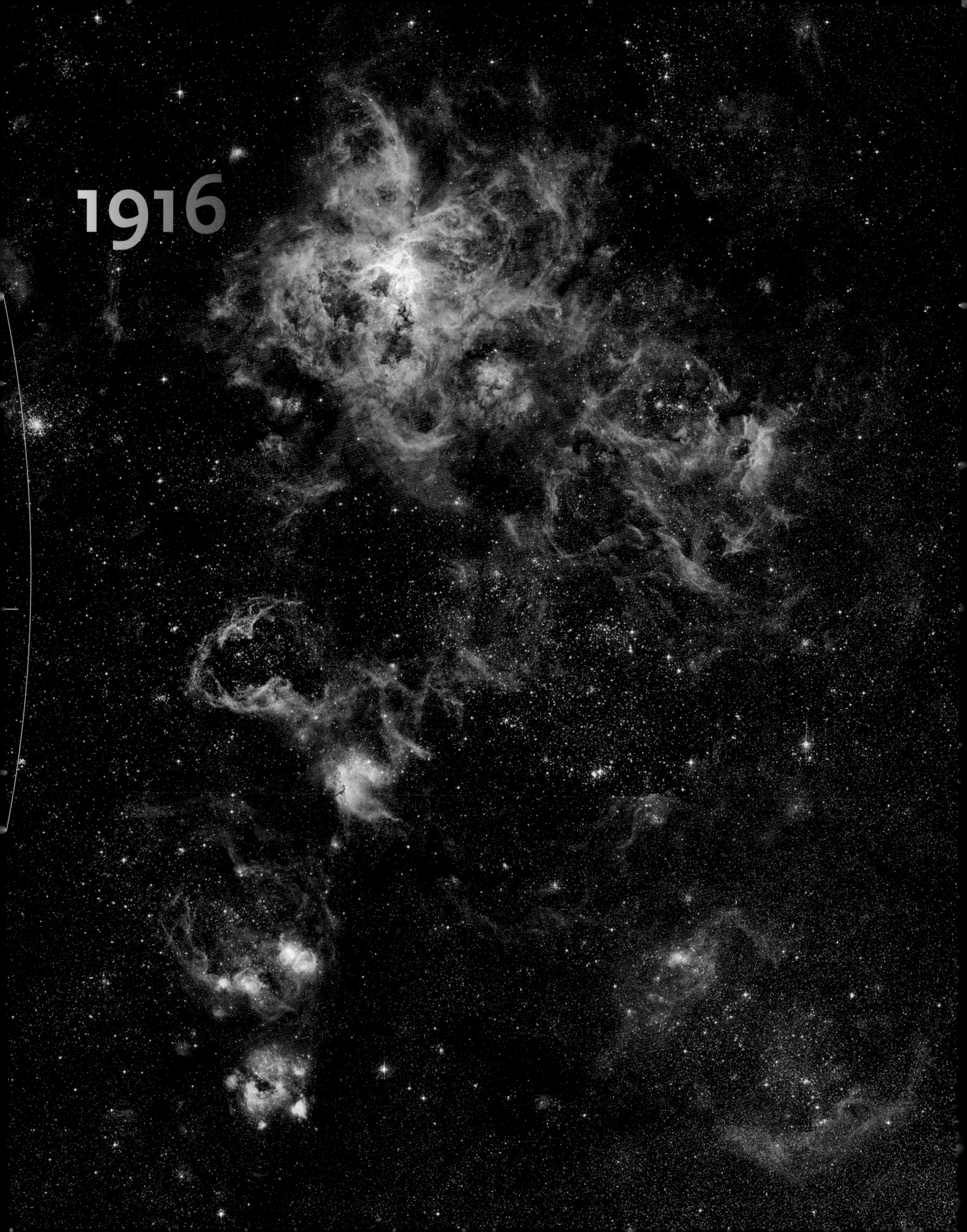
1916

1918

Galactic Dimensions

Harlow Shapley Discovers the Size of the Milky Way

Star counts by William Herschel at the end of the eighteenth century show that the Milky Way is an extremely flattened collection of countless stars. In 1906, the Dutch astronomer Jacobus Kapteyn starts an international observation campaign to determine the dimensions of the Milky Way, leading him to conclude that it is a flat, round disk with a diameter of around 40,000 lightyears. The Sun is nearly in the middle, only 2,000 lightyears from the actual center.

Yet there is something strange going on. Stars and star clusters are reasonably evenly distributed throughout the Milky Way band. Spiral nebulae, however, are spread evenly across the heavens, while globular star clusters tend to prefer one-half of the night sky, centered on the constellation of Sagittarius.

Harlow Shapley charts the three-dimensional distribution of globular star clusters and comes to the conclusion that the Milky Way is much larger than the small 'Kapteyn system'. Shapley actually wants to study journalism – he has some experience as a crime reporter – but on a whim chooses astronomy instead, studying under Henry Russell in Princeton, among others. In 1914, he is given a job at Mount Wilson Observatory.

Globular star clusters are enormous collections of hundreds of thousands of stars. These include Cepheids – variable stars that expand and contract again with great regularity. With the aid of the period-luminosity law, discovered by Henrietta Leavitt in 1912, Shapley can determine the distance to the Cepheids; the duration of their pulsation period shows their real luminosity and, by comparing this with their observed brightness, it is possible to calculate the distance. This also enables the distances of the globular clusters and their distribution in space to be determined.

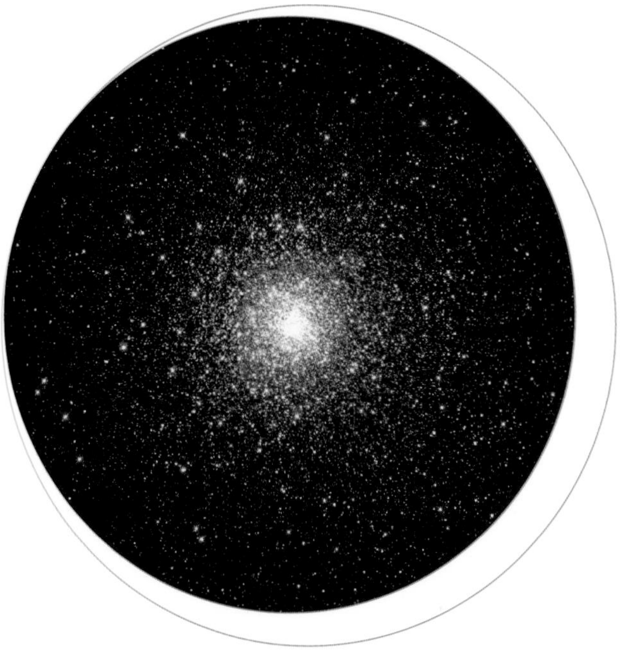

Shapley publishes his results in 1918. The globular clusters prove to be spread out over a large, more or less spherical halo, with its center in the constellation Sagittarius, at a distance of about 50,000 lightyears. According to Shapley that is the real midpoint of the Milky Way. The system is therefore much larger than Kapteyn thinks, and the Sun is not at all at the center. Like Kapteyn, however, Shapley does not take account of the absorbing effect of interstellar dust, and that means his estimate of the size of the Milky Way later has to be adjusted downwards.

On April 26, 1920, Shapley conducts a large public debate in Washington with Heber Curtis, director of the Allegheny Observatory. Curtis believes that spiral nebulae, like the Andromeda Nebula, are distant galaxies similar to our own Milky Way. Shapley thinks that, like the globular star clusters, they are small nebular spots that are part of the colossal Milky Way system. It becomes clear a few years later that is mistaken on that count.

Seen from the Earth, most globular star clusters are located more or less in the direction of the center of the Milky Way. Harlow Shapley used their distances and positions in the sky to deduce the dimensions of the Milky Way. This Hubble photo shows the globular cluster M80, at a distance of 28,000 lightyears. (Hubble Heritage Team)

M81 is a spiral galaxy in the constellation of Ursa Major. This photograph combines infrared, optical and ultraviolet observations. If it were an image of our own Milky Way, the Sun would be somewhere on the inner edge of one of the outer spiral arms. (NASA/ESA/JPL/CfA)

Einstein Proved Right

Arthur Eddington Discovers the Deflection of Starlight

Albert Einstein publishes his general theory of relativity in 1915, during the First World War. Communication between German and British scientists is almost at a standstill, but news of the revolutionary theory of gravity eventually reaches London and Cambridge, through Willem de Sitter, a physicist from Leiden. There it is primarily the mathematician and astronomer Arthur Stanley Eddington who shows an interest in the theory and publicizes it.

Eddington is already a respected scientist, a professor at Cambridge and director of the Observatory there. He is a brilliant astrophysicist, who makes his name in 1926 with his book *The Internal Constitution of Stars*. Although little is still known about where stars obtain their energy, Eddington's theories allow astronomers to calculate the pressure, temperature and gas density of any point in the interior of a star.

According to Einstein's theory of relativity, space in the immediate vicinity of massive celestial bodies is very slightly curved, causing light rays to be deflected. Together with Astronomer Royal Frank Watson Dyson, Eddington decides to test this proposition. If Einstein is right, the observed positions of stars in the sky will be slightly influenced by the presence of the Sun, and therefore it should be possible to measure that during a solar eclipse.

The eclipse of May 29, 1919 is perfect for the great Einstein test because it lasts a long time and the Sun is close to the Hyades, an open star cluster in the constellation Taurus. That means there are lots of relatively bright stars available on which to conduct measurements. Eddington leads an observation expedition to the island of Principe, in the Gulf of Guinea, while Andrew Crommelin of the Royal Greenwich Observatory travels to Sobral, in Brazil.

The expeditions do not go off without some hitches-Eddington suffers from bad weather, Crommelin from technical problems. Yet they do manage to take a series of photographs of the stars in the vicinity of the eclipsed Sun. During that summer, Eddington conducts detailed measurements of the images, and on November 6, 1919, he announces the results at a crowded meeting of the Royal Society in London stating that the observations of the eclipse are completely as predicted by Einstein's theory.

Although Einstein is later proved correct time and again, physicists and astronomers are initially skeptical, especially because of the disappointing quality of the observations. Eddington's astounding conclusion is, however, embraced by the press and the general public. So soon after the horrors of the First World War, people clearly need some good news, and also the fact that the theory of a German pacifist is confirmed by the observations of a British scientist captures the public imagination. Einstein owes much of his later widespread cult status to Eddington's eclipse expedition.

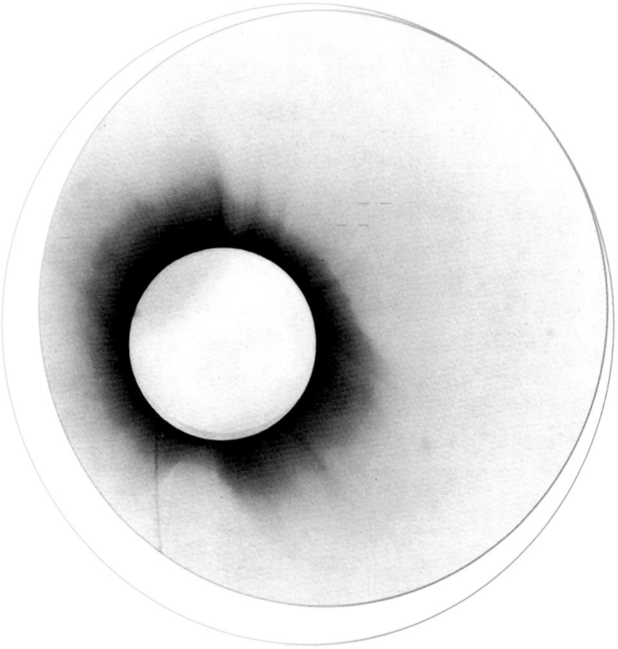

Negative photograph of the solar eclipse of May 29, 1919, taken in Brazil by Andrew Crommelin. The short horizontal lines mark stars whose positions have been affected very slightly by their light being bent by the Sun's gravitational field.

A total solar eclipse is one of the most spectacular natural phenomena. When the bright disk of the Sun is covered by the Moon, the thin, silver-white solar corona becomes visible. This impressive composite photograph shows the total eclipse of March 29, 2006 in Libya. (Arnaud van Kranenburg)

1919

1924

Island Universes

Edwin Hubble Discovers the True Nature of Spiral Galaxies

Athlete, boxer, lawyer, cosmologist: Edwin Powell Hubble is all of these. At the age of 16 he sets the high-jump record in the state of Illinois. He studies law and Spanish at Oxford. After returning to the United States and being awarded his Ph.D. in astronomy at Yerkes Observatory in 1917, he is invited by George Ellery Hale to come and work at Mount Wilson Observatory. But Hubble first goes to serve his country, and in a telegram to Hale he writes: 'Regret cannot accept your invitation. Am off to war'.

In the summer of 1919, Hubble returns from France, where he has risen to the rank of major. Later the same year he finally joins the staff at Mount Wilson and soon becomes one of the most frequent users of the brand-new 2.5-meter Hooker Telescope, the largest in the world at the time. With this fantastic instrument it must be possible to unravel the secrets of spiral nebulae!

Spiral nebulae, like the Andromeda Nebula, can be found pretty much everywhere in the sky, but their distances are unknown. No one knows if they are part of our own Milky Way or – as some astronomers believe – if they are extremely far away and form autonomous 'island universes'. Hubble is the first to discover their true nature.

In 1922 and 1923, he uses the Hooker Telescope to make countless photographs of the Andromeda Nebula and the Triangulum Nebula, another striking spiral nebula in the northern sky. Individual stars are visible on the extremely detailed images, and by comparing different photographs, Hubble discovers that the brightness of some of the stars varies over time. In early 1924, it becomes clear that one of the variable stars is a Cepheid.

Cepheids have an easily recognizable brightness variation, and their pulsation period is a direct measure of their average luminosity. The Cepheid in the Andromeda Nebula has a period of a few weeks, which means that its average luminosity must be very high. By comparing that calculated luminosity with the low observed brightness of the star in the sky, Hubble determines the distance to the Andromeda Nebula at around one million lightyears. He also discovers Cepheids in other spiral nebulae. The conclusion is inescapable: these nebulae are located far beyond the Milky Way.

On February 19, 1924, Hubble describes his discovery in a letter to his ex-colleague and arch-rival Harlow Shapley, who has recently been appointed director of Harvard College Observatory, and who initially does not believe in the extragalactic nature of the spiral nebulae. Hubble's publication on his spectacular findings follows and appears on January 1, 1925.

During the Second World War, new insights lead to Hubble's distance scale being adjusted. The Andromeda Nebula proves to be approximately two, not one, million lightyears away.

Ultraviolet image of the Andromeda Galaxy, the nearest large neighbor to our own Milky Way. Edwin Hubble showed in 1924 that spiral galaxies like Andromeda are not part of the Milky Way but are separate galaxies. (NASA/JPL)

The spiral galaxy M101, in the constellation of Ursa Major, is also known as the Pinwheel Galaxy. It is 27 million lightyears from the Milky Way, and more than one and a half times bigger. This infrared photo was taken by the Spitzer Space Telescope. (NASA/JPL/STScI)

Rotating Disk

Jan Oort Discovers the Differential Rotation of the Milky Way

Jan Hendrik Oort is one of the greatest astronomers of the twentieth century. His groundbreaking contributions help unravel the mysteries of the Milky Way, he is one of the pioneers of radio astronomy, and also one of the initiators of the European Southern Observatory. He explains the origin of comets and studies superclusters of galaxies, but, for his whole life, Jan Oort regrets what he later calls his own oversight: the fact that he himself should have seen the solution to the mystery that he describes in his 1926 thesis.

Oort was born in 1900 and studies astronomy in Groningen under Jacobus Kapteyn, his great role model. As early as 1904, Kapteyn discovers that the motions of stars in the sky are not evenly distributed, but he does not have a conclusive explanation for his two 'star streams'. On the basis of his star counts, Kapteyn concludes that the Sun is more or less at the center of a flat galaxy with a diameter of around 40,000 lightyears.

Oort's thesis addresses Kapteyn's star streams. But the promising young astronomer does not become acquainted with the publications of the Swede Bertil Lindblad, which finally set him on the trail of a solution, until after he has been awarded his Ph.D.

Like Harlow Shapley, Lindblad believes that the Milky Way is much larger than the 'Kapteyn system' and that the Sun is a long way from its center. In 1925, he proposes a model in which the Milky Way consists of a number of different systems, each flattened to a different degree and with its own rotational velocity. He claims that the stars in the flat disk of the Milky Way rotate the fastest, while the system of globular star clusters turns more slowly.

In 1927, Oort publishes an article in which he elaborates on Lindblad's theories. They offer an excellent explanation for the star streams, as long as you assume that the rotational velocity of stars in the Milky Way also depends on their distance from the center. Oort concludes that the stars in the vicinity of the Sun have an orbital velocity of around 270 kilometers per second, and that they are at about 20,000 lightyears from the center.

A year later, the young astronomer from Leiden writes a letter to Lindblad, enclosing a copy of the article, and Lindblad immediately sees the great importance of Oort's discovery. Finally, it gives astronomers a real grasp of the structure and dynamics of the Milky Way, a tough task when you can only look at it from the inside. Oort is also one of the first to correctly conclude that the disk of the Milky Way must contain a lot of absorptive interstellar material, which explains why Kapteyn underestimated its size.

Most spiral galaxies rotate in just one direction. M64, however, has an oppositely rotating core, possibly the result of a collision with another galaxy. The thick, dark belt of dust was probably also created in the aftermath of the collision. (NASA/ESA/Hubble Heritage Team)

M74 is an impressive spiral galaxy at a distance of around 30 million lightyears, in the constellation of Pisces. It is in many ways similar to our own Milky Way and rotates in a clockwise direction. It rotates so slowly, however, that it does not observably change its orientation in the course of a human lifetime. (NASA/ESA/Hubble Heritage Team)

1927

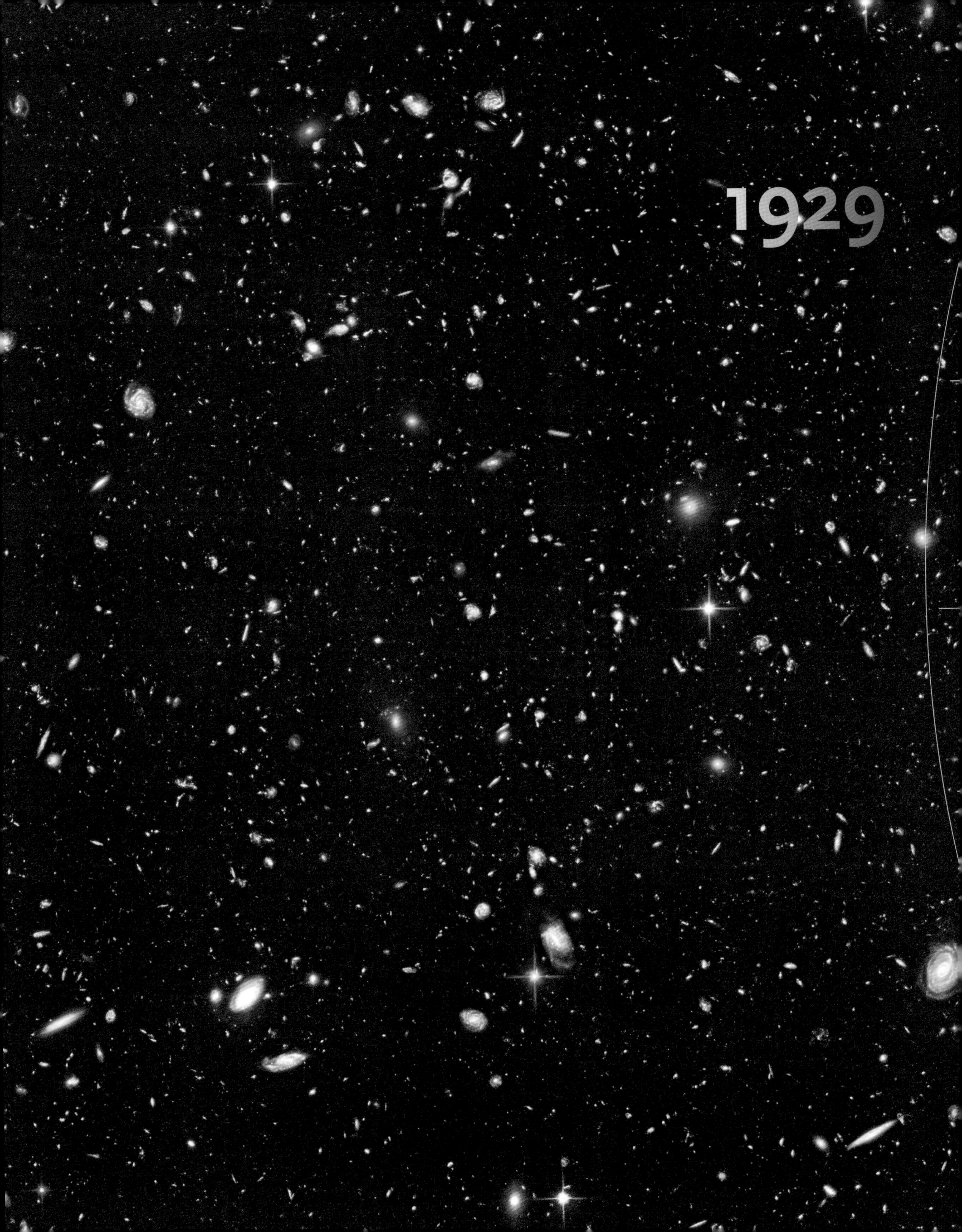
1929

Inflated Space

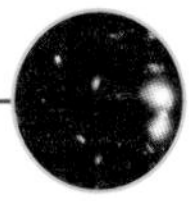

Edwin Hubble Discovers the Expansion of the Universe

In 1912, Vesto Slipher of Lowell Observatory in Flagstaff, Arizona, is the first to measure the radial velocity of a spiral nebula. Slipher photographs the spectrum of the Andromeda Nebula and sees that the dark spectral lines have all shifted to shorter wavelengths. If this blue shift is interpreted as a Doppler effect, the Andromeda Nebula is moving towards us at a speed of some 300 kilometers per second; scientists have never before measured such high speeds.

Slipher photographs other nebular spectra, but the Andromeda Nebula proves to be an exception – most nebulae show a red shift and are therefore moving away from us, and all of them are moving at extremely high speeds. James Keeler and William Campbell of the Lick Observatory also measure red shifts in nebular spectra. Yet even after Edwin Hubble shows in 1924 that spiral nebulae are located far beyond the Milky Way, astronomers still have no inkling of why they are moving away at such high speeds.

At the end of the 1920s, Hubble joins forces with his assistant Milton Humason, two years his junior. Humason never completed his secondary school and joined the staff at Mount Wilson Observatory after having worked as a mule-driver, janitor and night assistant. He is a meticulous observer, recording other nebular spectra with the 2.5-meter Hooker Telescope.

The distances of many of these galaxies have since been calculated, because Cepheids have been discovered in them, and the distances can be deduced from the observed pulsation periods of these variable stars. On the basis of available observations of 46 galaxies, Hubble makes a sensational discovery in 1929: distant galaxies are on average moving away at very high speeds, while those closer to us have a smaller red shift.

This observed connection between distance and red shift supports the theory of the Russian cosmologist and mathematician Alexander Friedman. In 1992, Friedman proposes a solution for the field equations of Albert Einstein's general theory of relativity, which suggest that space is expanding. Hubble's discovery is the first observational evidence for this cosmic expansion, and forms the basis of the Big Bang theory.

Hubble leaves the cosmological interpretation of his discovery to others. It is the Belgian cosmologist and Jesuit priest Georges Lemaître who, around, but the the same time, first presents the theory that the universe originated billions of years ago in an extremely hot and compact state: the Big Bang.

Hubble's name remains forever linked to the discovery of the expanding universe. In the 1980s, NASA decides to call the first large space telescope after him, and the Hubble Space Telescope is launched in 1990.

➋ Two galaxies in the constellation of Eridanus become distorted under the influence of each other's gravitational pull. In the early youth of the universe, encounters between galaxies were far more common than now, because all objects in the cosmos were closer together than they are now. (Gemini Observatory/Travis Rector)

➌ This image of the Hubble Ultra Deep Field combines the observations of two cameras on board the Hubble Space Telescope. The countless small specks of light are all individual galaxies many billions of lightyears away, showing us the universe as it was in the remote past. (NASA/ESA)

Lilliputian Planet

Clyde Tombaugh Discovers the Dwarf Planet Pluto

At the beginning of the twentieth century, no one bats an eyelid anymore when unknown objects are discovered in the solar system. The number of asteroids is growing steadily, partly due to the development of photography, and new planetary moons are found on a reasonably regular basis. Perhaps, then, there is still a full-fledged planet hiding somewhere out in the outer regions of the solar system?

Percival Lowell thinks so because even when you take the gravitational disturbance of Neptune into account, Uranus still seems to display tiny orbital deviations. According to Lowell, these are caused by an as yet unknown 'Planet X'. Like Le Verrier before him, he tries to prove his theory through calculation. At Lowell Observatory in Arizona, he takes photographs of the 'suspect' areas in the night sky, and by comparing images taken on different days, he tries to track down a celestial body moving through the sky far beyond the orbit of Neptune.

In 1911, to facilitate the search, Lowell buys a 'blink comparator', a machine that allows you to look at two images of the night sky alternately, so that a moving speck of light will seem to jump to and fro and therefore be easier to see. But new observations of Uranus continue to produce different predictions for the position of the unknown planet, and Lowell's search comes up with nothing. He dies in 1916 without ever knowing whether Planet X exists or not.

At the end of the 1920s, Lowell's successor, Vesto Slipher, resumes the search for Planet X. Slipher has a special telescope built with a huge field of view, and hires farmer's son and amateur astronomer, Clyde Tombaugh, to expose the photographic plates and operate the blink comparator. Tombaugh realizes how uncertain the positional predictions are and decides to search the entire solar system systematically. In the afternoon of Tuesday February 18, 1930, he gets lucky. On images made on January 23 and 29, he discovers a small, faint star that slowly changes position. A few weeks later it is clear that the object is orbiting the Sun beyond the orbit of Neptune.

The discovery of Planet X is announced on March 13, Lowell's birthday. At the suggestion of 11-year-old Venetia Burney from Oxford, the new planet is named Pluto, after the Roman god of the underworld.

It is clear from the outset that Pluto is not a normal planet. Its orbit is highly inclined compared to those of the other planets, and it is very eccentric: once during each revolution Pluto even moves closer to the Sun than Neptune. More remarkable still, even through the world's largest telescopes, Pluto is nothing more than a rather unimpressive speck of light. That suggests that it is very small, and therefore not massive enough to noticeably disturb the orbital motion of other planets.

Pluto does not lose its planetary status until more than 75 years later when in August 2006 the International Astronomical Union decides that Pluto should henceforth be called a dwarf planet.

In 2005, two new, small moons were discovered around Pluto, which were given the names Nix and Hydra. This Hubble photo shows the two moons as faint specks of light. The brighter objects are Pluto and its large moon, Charon, which was discovered earlier in 1978. (NASA/ESA/STScI)

Illustration of the dwarf planet Pluto and its relatively large moon Charon. The Sun, some 6 billion kilometers way, is little more than an extremely bright star. The two bodies consist mainly of ice, and orbit each other in a little over 6 days. (ESO)

1930

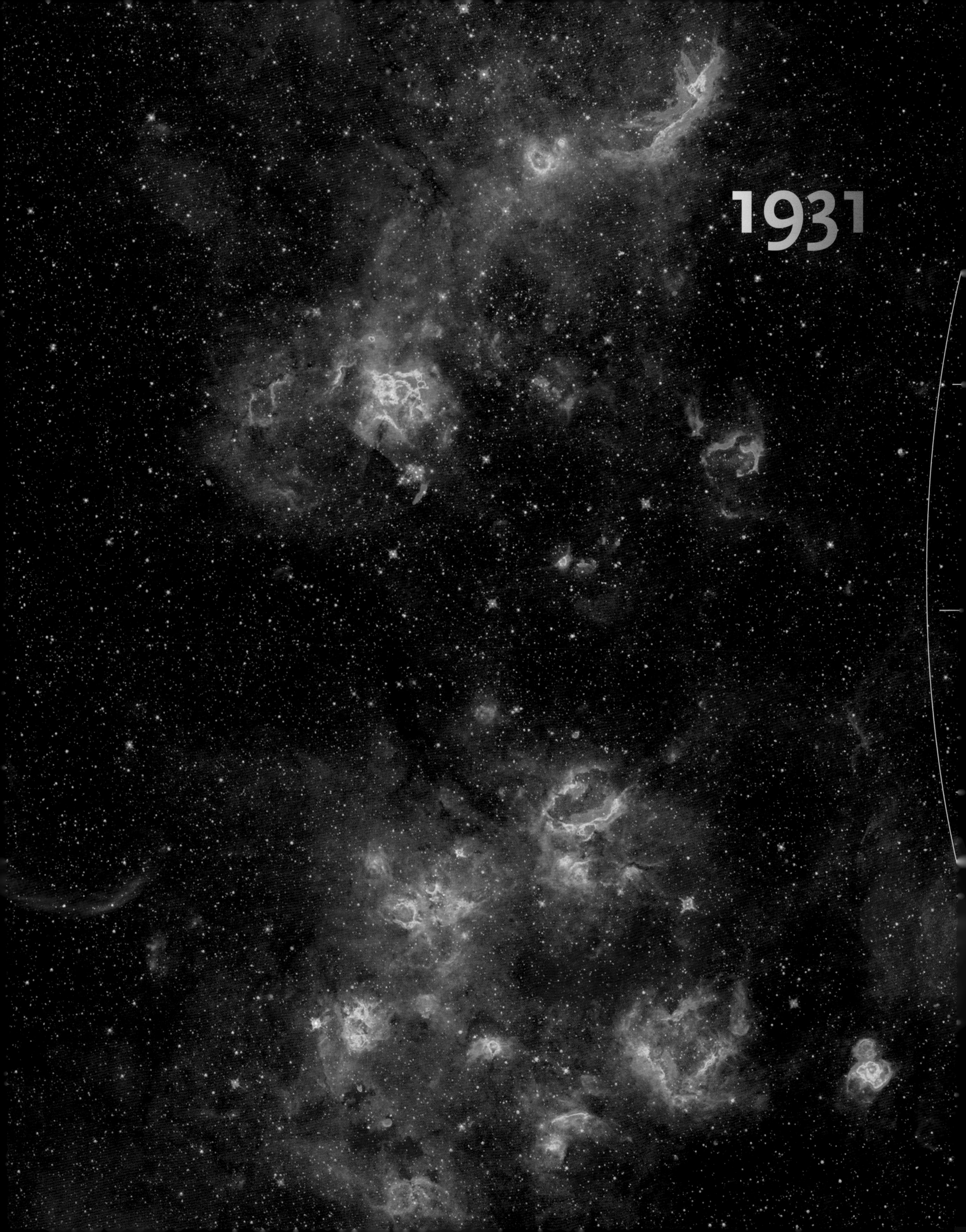
1931

Galactic Podcast

Karl Jansky Discovers Cosmic Radio Waves

Since time immemorial astronomy has been a 'visual' science. Astronomers use telescopes to study visible light from planets, stars, nebulae and galaxies. Although it is known that other forms of electromagnetic radiation exist – William Herschel discovers infrared radiation as far back as 1800 – the fact that such invisible forms of 'light' originate from the universe is a new insight, and the American Karl Jansky throws the door to modern multispectral astronomy wide open.

Jansky is the descendant of Czech immigrants. He studies physics and, at the age of 22, is given a job as a radio technician at Bell Telephone Laboratories in Holmdel, New Jersey. His task is to study the origins of interference on maritime telephone links. Specifically for this purpose, Jansky builds a gigantic antenna installation 30 meters in diameter and 7 meters high, and fitted with four wheels from a Model T Ford, so that it can be turned.

'Jansky's merry-go-round', as the contraption is called, enables the direction from which radio interference is received to be determined very precisely. In the summer of 1931, it is clear that much interference (with a wavelength of around 15 meters) is caused by lightning discharges, also at greater distances from the antenna. In addition, Jansky discovers a hissing signal that increases in strength and then decreases again every day.

At first it seems that the radio hiss is coming from the Sun, but after some weeks it becomes clear that the cycle of the interference is shifting slightly. The broad peaks in the signal do not follow each other at periods of 24 hours, but at intervals of 23 hours and 56 minutes, exactly the rotational period of the Earth with respect to the stars. The source of the interference thus seems to lie somewhere in the universe.

In the spring of 1932, Jansky is convinced that the signal comes from the center of our Milky Way, in the constellation of Sagittarius. He publishes his findings in December of that year, and on May 5, 1933, the *New York Times* devotes a front-page article to the discovery.

It takes a long time for the discovery of cosmic radio waves to penetrate the world of astronomy. The first simple radio telescopes are not built until the early 1940s, and radio astronomy does not really take off until the 1950s. Jansky does not live to see that: he dies of a heart condition in 1950 at the age of 44. His name lives on in the unit used by radio astronomers to designate the strength of radio sources: 1 jansky equals a flux of 10^{-26} watts per square meter per hertz.

Cosmic radio waves do not contain much energy. All the radiation captured by terrestrial radio telescopes contains less energy than a christmas tree light burning for one minute.

Karl Jansky was the first to discover radio waves from the center of the Milky Way. On a blackboard showing the constellations of Ursa Major, Ursa Minor, Cassiopeia and Cygnus, he points out the direction from which the most powerful radio waves originate. (NRAO/AUI)

Panorama of the Milky Way in the constellations of Scutum and Aquila, composed on the basis of radio observations (*red*) and infrared measurements (*green*). The conspicuous red arcs are the remnants of supernova explosions, which produce a relatively large quantity of radio waves. (NRAO/AUI)

Invisible Stuff

Jan Oort and Fritz Zwicky Discover Dark Matter

It is not the first time that astronomers have deduced the presence of unknown celestial bodies from gravitational effects. The planet Neptune was found on the basis of the disturbances it causes in the orbital movement of Uranus, and the existence of the white-dwarf companions of Sirius and Procyon is betrayed by the wobbling motions of the two bright stars. In 1932, Jan Oort goes a step further: from gravitational measurements he concludes that there must be twice as much material in the neighborhood of the Sun than has been discovered using telescopes.

At the end of the 1920s, Oort unravels the mystery of the rotation and dynamics of the Milky Way, largely a matter of measuring the motions of stars in the flat central plane of the galaxy. But what about motions perpendicular to that plane, upwards and downwards? Oort realizes that these movements will shed light on the average matter density of the Milky Way.

The principle is simple – if there is a very large amount of matter in the Milky Way plane, a star will need to move at a very high vertical velocity to escape to any great distance above or below the plane. If the density – and therefore the gravitational field – is much smaller, you should be able to put a considerable distance between yourself and the Milky Way at a lower speed.

Oort gathers information on the positions and velocities of a large number of stars in the vicinity of the Sun, and calculates the average density at approximately one solar mass of matter per 375 cubic lightyear. However, if you calculate the total mass of all the known stars in the neighborhood of the Sun, you cannot account for even half of that density. Hence there appears to be more matter in the Milky Way plane than you would say at first glance.

Oort is one of the first to use the term 'dark matter', in a 1932 publication. A year later, Swiss astronomer Fritz Zwicky, who is associated with the California Institute of Technology, discovers that there must also be enormous quantities of dark matter elsewhere in the universe. Zwicky measures the velocities of individual galaxies in clusters, and they prove to be exceptionally high. That means the clusters must also contain large quantities of invisible matter; the combined gravity of the galaxies themselves is far too weak to explain the high speeds.

Oort's dark matter consists partly of faint dwarf stars and clouds of interstellar gas and dust. Zwicky's dark matter can be partly explained by the hot, rarefied gas in the space between galaxies, but this by no means explains all the measurements. In the 1970s and 1980s, astronomers discover that the universe also contains enormous quantities of dark matter which is not made up of normal atomic nuclei. At present, the true nature of this matter remains a mystery.

The blue areas on this photograph of a distant cluster of galaxies show where most of the dark matter is located, on the basis of gravity measurements. It suggests that two galaxy clusters collided here at some time in the past. (NASA/ESA/CXC)

The two superclusters Abell 901 and 902 contain countless galaxies, but also a lot of dark matter (purple). The distribution of the matter is derived from measurements of tiny distortions of the light coming from distant background galaxies. (NASA/ESA/STAGES Collaboration)

1932

1933

Terminal Explosions

Fritz Zwicky and Walter Baade Discover the True Nature of Supernova Explosions

New stars have been appearing in the sky from time to time since the dawn of history, and sometimes they are so bright that for a few weeks they can also be seen during the day. Tycho Brahe wintesses a *Stella Nova* in 1572 and Johannes Kepler in 1604, and they are also observed in 1006 and 1054. In 1892, Edward Emerson Barnard discovers that novae are not new stars but explosions of existing stars. Yet little is known about the distance and true nature of these nova explosions.

At the end of the 1920s, German astronomer Walter Baade makes a distinction between *novae* and what he calls *Hauptnovae*. When Edwin Hubble calculates the distance to the Andromeda Nebula in 1924, Baade realizes that the nova observed in the nebula in 1885 must have been extremely bright.

In 1931, Baade emigrates to the United States, where he works closely with the Swiss astronomer Fritz Zwicky at Mount Wilson Observatory. The two European astronomers are convinced that there is a fundamental difference between novae and supernovae – the American name for Baade's *Hauptnovae*.

After James Chadwick discovers the neutron – one of the particles making up an atomic nucleus – in 1932, Zwicky proposes the seemingly far-fetched theory that a supernova explosion occurs when a regular star implodes to become a 'neutron star' – a small, extremely compact ball of densely packed neutrons, in effect a colossal atomic nucleus.

Zwicky and Baade present their model at a scientific conference in 1933, and publish it a year later. Historical research by Baade suggests that the 'new stars' that appeared in the sky in 1006, 1054, 1572 and 1604 were also supernovae – the terminal explosions of stars, during which most of the star's mantle is blown into space at enormous velocity, and the core implodes to become a hypothetical neutron star. The first actual neutron star is not found until 1967.

From 1936 on, Zwicky uses the Schmidt Telescope at Palomar Observatory to hunt for supernovae in other galaxies. Eventually he discovers a record 120 of them. In a galaxy like our own Milky Way a supernova explosion occurs on average once every 50 years, but if you keep a close eye on a few hundred galaxies, you will find one every couple of months.

Incidentally, despite their enormously powerful luminosity, most supernovae in the Milky Way are impossible to see because their light is absorbed by clouds of dust. The only supernova visible to the naked eye in the twentieth century flares up in February 1987 in the Large Magellanic Cloud, a companion to the Milky Way at a distance of about 160,000 lightyears.

➐ In 1994, a supernova flared up in the outer regions of the dusty galaxy NGC 4526. The supernova – the explosion of a massive star that has reached the end of its life – produced almost as much energy as all the other stars in the galaxy combined. (NASA/ESA/STScI)

➌ Cassiopeia A is the expanding remnant of a star destroyed during a catastrophic explosion at the end of the seventeenth century. This photograph is composed of visible light, infrared and X-ray images. Cassiopeia A is around 10,000 lightyears away. (NASA/JPL/STScI/CXC/SAO)

Near Miss

Karl Reinmuth Discovers the Earth-Grazer Hermes

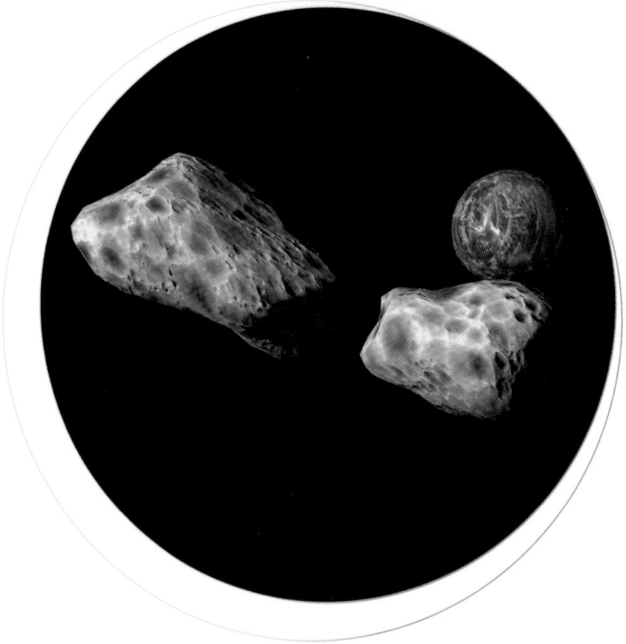

After the discovery of Ceres in 1801 by Giuseppe Piazzi, the number of asteroids increases rapidly in the course of the nineteenth century, and by 1900 the count has risen to 463. With the introduction of photography, the search speeds up even more. German asteroid hunter Karl Reinmuth, who discovers his first asteroid in 1914 at the age of 22, eventually finds nearly 400.

Most asteroids go around the Sun between the orbits of Mars and Jupiter, but some of Reinmuth's objects cross the Earth's orbit. The first of these 'Earth-grazers' is found on April 24, 1932. Because it passes closer to the Sun than other asteroids, Reinmuth names it Apollo, after the Greek god of the Sun. It is, however, not possible to determine Apollo's orbit accurately and it is not seen again until 1973.

On October 30, 1937, Reinmuth strikes lucky again. On photographs made two days earlier, he discovers a small asteroid passing the Earth at high speed and at quite a short distance: 741,000 km – about twice as far away as the Moon. Never before has a celestial object been observed that approaches so close to the Earth. A few days later the small lump of cosmic rock is so far away that it can no longer be seen.

The object is given the provisional name 1937 UB. Although asteroids are officially not given a proper name until their orbits have been calculated accurately, Reinmuth calls this Earth-grazer Hermes, after the swift messenger god of the Greeks.

Hermes retains the record for being the object to approach closest to the Earth until 1989, when the asteroid Asclepius flies past at a distance of 700,000 km. Only later is it discovered that Hermes broke its own record in 1942. The small asteroid is rediscovered by Brian Skiff of Lowell Observatory on October 15, 2003. It is given the official asteroid number 69230. Accurate orbital calculations show that in 1942 Hermes flew past the Earth unnoticed at a distance of only 600,000 kilometers.

In 2003, astronomers observe Hermes with the 300-meter radio telescope at Arecibo in Puerto Rico. The radar measurements show that the earth-grazer consists of two separate lumps of rock each about 300 meters in diameter, circling each other at a distance of about a kilometer. If Hermes should ever collide with the Earth at some time in the future, it could mean the destruction of an entire continent.

Karl Reinmuth names many of 'his' asteroids after his colleagues and members of their family. He even honors the German asteroid expert Gustav Stracke, who does not want a celestial object named after him: the first letters of the names of asteroids 1227–1234 spell out the name GSTRACKE. Asteroid 1111 is named Reinmuthia, for Reinmuth himself.

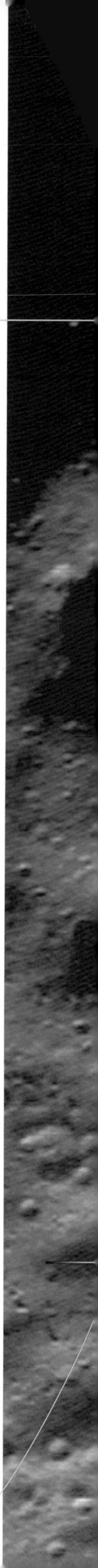

Illustration of the binary asteroid Hermes when it approached the Earth in 1937. Many asteroids, including Hermes, consist of two objects that orbit each other at a very close distance. (Keith Cowing)

Close-up of the surface of the asteroid Eros, made in September 2000 by the American space probe NEAR-Shoemaker. Eros is an elongated lump of rock (34 × 11 × 11 kilometers) in an elongated orbit around the Sun, which may collide with the Earth in the very distant future. (NASA/JPL/JHU-APL)

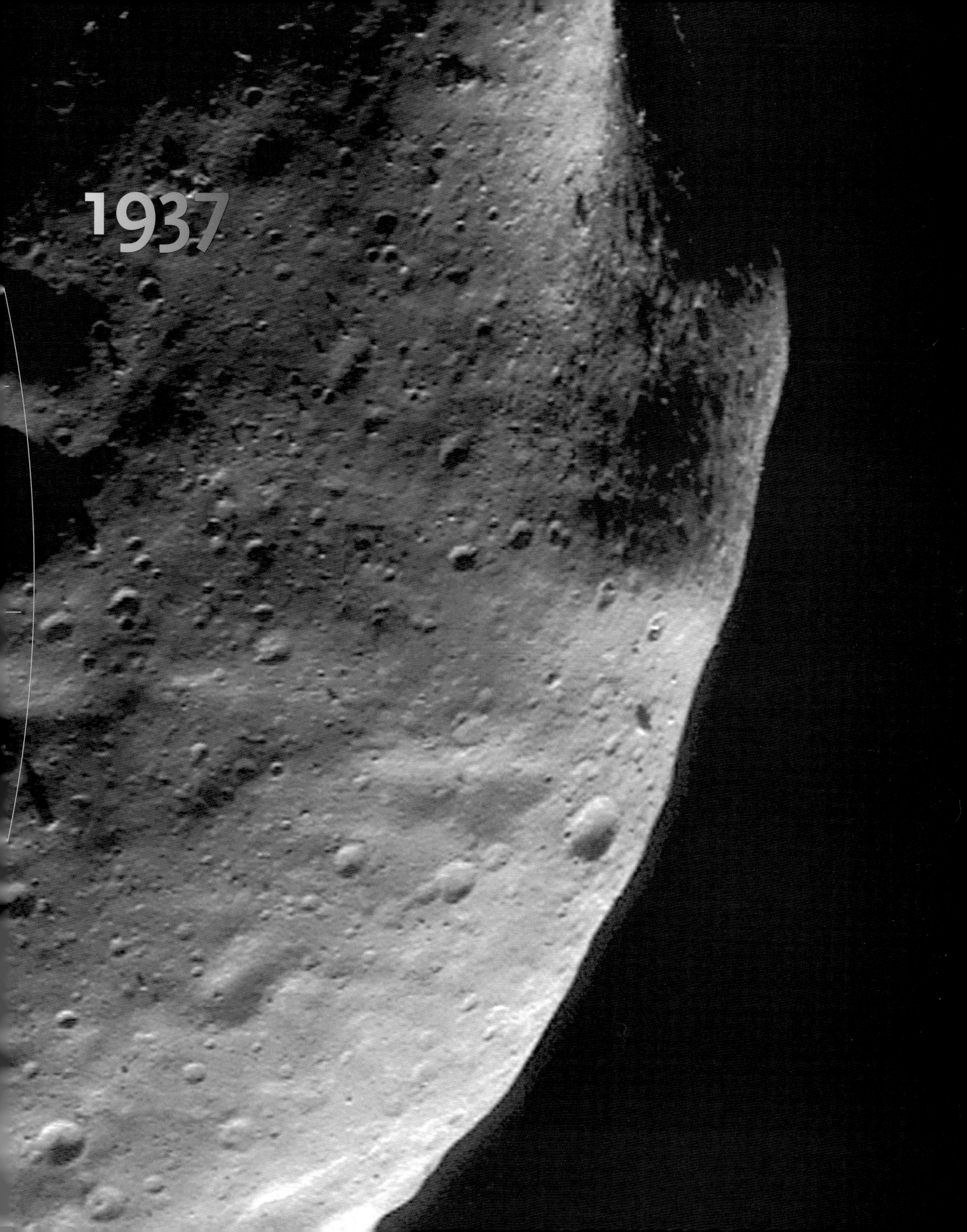
1937

1938

Sunny Train Ride

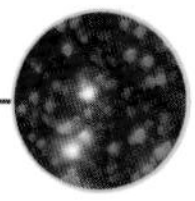

Hans Bethe Discovers the Energy Source of the Stars

In the nineteenth century, physicists unravel the fundamental laws of thermodynamics. Terms and concepts that have been in daily use for centuries – like energy, labor and capacity – suddenly acquire a clearly defined, scientific meaning. At the same time it becomes clear that, in nature, there is no such thing as a free lunch: every form of energy must come from somewhere. So how does that work with the Sun and the stars?

There are various ideas about this in circulation at the end of the nineteenth century. According to one theory, the Sun radiates energy because it is slowly contracting. Another sees the Sun as a gigantic coal-fueled power plant, and yet another claims that the Sun acquires its energy from meteorite impacts.

It is the British astrophysicist Arthur Eddington who is the first, in 1920, to present a feasible argument that the Sun acquires its energy from nuclear reactions in its interior. Stars consist largely of hydrogen; in their interiors pressure and temperature are so high that the hydrogen is spontaneously converted into helium, releasing energy in the process. Eddington's theory becomes increasingly popular in the 1920s and 1930s, but no one knows exactly what kind of nuclear reactions are driving the stars.

In April 1938, Russian-American physicist George Gamow convenes a conference on the source of stellar energy in Washington. One of those present is German physicist Hans Bethe. Bethe fled Nazi-Germany in 1933 at the age of 26, and after two years in England emigrated to the United States. He decides to solve the puzzle of the nuclear reactions in the Sun himself during the train ride back from Washington to his home institute, Cornell University in Ithaca, New York. He finds the answer just before the conductor invites the passengers to the restaurant car for dinner.

Bethe realizes that there are two ways in which hydrogen can be converted into helium. In the first, the proton–proton cycle, two hydrogen nuclei (protons) fuse to become a single deuterium nucleus. After an intermediate stage, during which lithium is produced, the two deuterium nuclei fuse to create helium. In the second reaction, the CNO cycle, the conversion takes place via the catalyst elements carbon (C), nitrogen (N) and oxygen (O). Both kinds of reactions occur in the Sun, while in massive stars, the CNO cycle dominates.

Bethe is not the only one to solve the nuclear fusion mystery. In Germany the CNO cycle is discovered independently of Bethe in 1938 by physicist Carl von Weizsäcker, the older brother of the former German president Richard von Weizsäcker. It is not until the late 1950s that Geoffrey and Margaret Burbidge, Willy Fowler, and Fred Hoyle explain the many other nuclear fusion reactions that occur in the interior of stars, leading to the formation of practically all elements in the periodic table.

During a gigantic explosion on the Sun, more than a billion tons of matter is expelled into space at speeds of millions of kilometers per hour. Nuclear fusion reactions in the interior of the Sun are the underlying cause of all high-energy phenomena on the surface. (NASA/ESA)

The globular star cluster Omega Centauri, seen here on an infrared photo made by the Spitzer Space Telescope, contains several millions of stars like the Sun – each one a cosmic nuclear plant generating energy by converting hydrogen into helium. (NASA/JPL/NOAO/AURA/NSF)

Celestial Radio Beacon

Grote Reber Discovers the Brightest Radio Source in the Sky, Cygnus A

Pioneers have the place to themselves. For a decade, American radio engineer Grote Reber is the only radio astronomer in the world. With a self-built dish antenna, he is the first to map the sky at radio wavelengths.

Reber was born in Wheaton, a suburb of Chicago, at the end of 1911. He studies radio engineering, and is fascinated at a young age by Karl Jansky's discovery in 1931 of radiation coming from the center of the Milky Way. He writes to the Bell Telephone Laboratories, where Jansky works, asking for a job, but they have nothing for him. He remains resolute that he wants to study cosmic radio waves himself, and preferably with a large parabolic antenna that focuses the faint signals from the universe on one point, where they can be picked up by a sensitive receiver.

Reber asks a local construction company how much it would cost to build such an antenna, and they quote him $7,000. Because the young engineer doesn't have that kind of money, he decides to build the radio telescope himself in his backyard. It is a gigantic contraption, nine meters in diameter, that can be directed at various altitudes above the southern horizon. It takes Reber four months to build the telescope in his spare time, and a total of $1,300 – at that time, about the price of a car.

In 1937, Reber conducts the first observations with the telescope, at a wavelength of 9 centimeter, but without success. He tries again, at a wavelength of 33 centimeter, but again he detects no cosmic radio waves. Finally, Reber builds a receiver for radiation at a wavelength of 1.87 meters and, in 1938, records for the first time the 'radio static' from the center of the Milky Way that Jansky had discovered earlier.

Reber's parabolic antenna can, however, be pointed much more precisely than Jansky's 'merry-go-round'. Night after night Reber can be found in his backyard, mapping the radio waves from the skies. In theory he can measure the radiation during the day, too, but then there is a lot of interference from terrestrial sources. Furthermore, he has a day job at an electronics business in Chicago.

In 1939, Reber discovers a very powerful source of radio waves in the constellation of Cygnus, which he calls Cygnus A. Later it proves to be a distant galaxy that is emitting enormous amounts of radio waves, possibly because there is a supermassive black hole at its core.

Reber's radio maps of the sky mark the beginning of radio astronomy, but he remains the only practitioner of this new discipline until 1947. In the 1960s, a replica of his backyard telescope is installed at Green Bank, West Virginia, which now houses the largest fully steerable radio telescope in the world, with an elliptical dish measuring 100 x 110 meters.

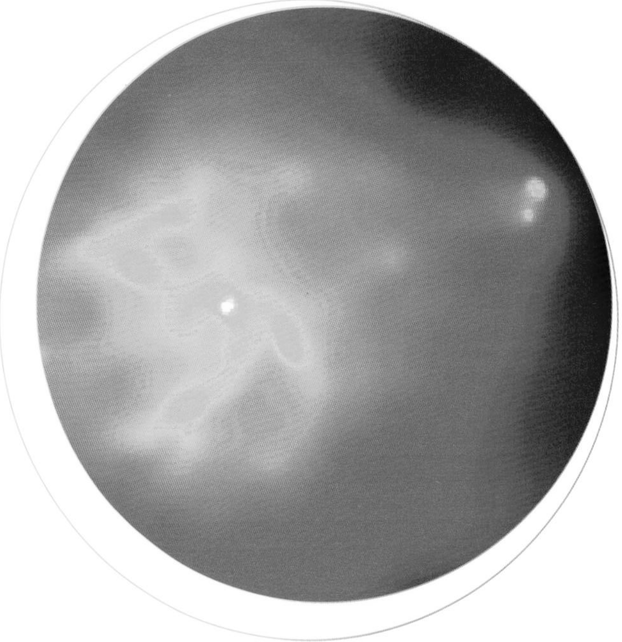

⬉ X-ray image of Cygnus A, made by the Chandra X-ray Observatory. The core of the radio galaxy expels two jets of high-energy particles into space, creating a cavity in the hot gas in which Cygnus is embedded. (NASA/UMD)

➔ The two jets of electrically charged particles expelled from the core of Cygnus A are slowed down by thin intergalactic material, creating two large lobes of hot gas at the outer ends. This radio 'photo' shows the brightness of the radio waves in different colors: red is bright, blue is relatively faint. (NRAO/AUI)

1939

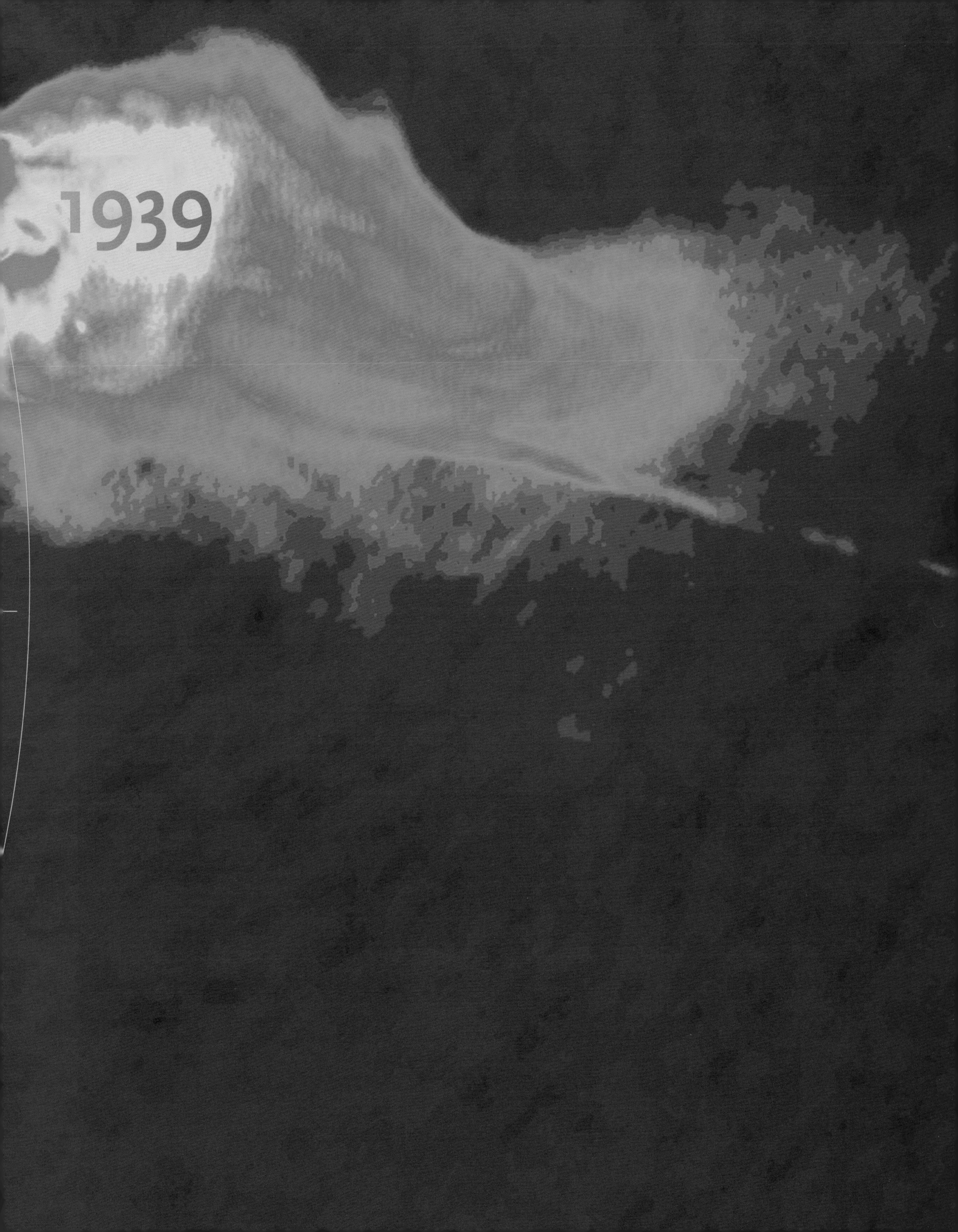

1944

Discoveries in Wartime

Walter Baade Discovers that There Are Two Star Populations

German astronomer Walter Baade emigrates to the United States in 1931, where he becomes a staff astronomer at Mount Wilson Observatory. With the 2.5-meter Hooker Telescope, he makes extremely detailed photographs of nebulae and galaxies, partly thanks to the special photographic emulsions developed by Kodak for astronomical use. Baade's images of the night sky are the best in the world.

During the Second World War, many American scientists are put to work for defense programs, like the Manhattan project which developed the first atomic bomb. Baade is, however, a German citizen and the American government officially sees him as a potential enemy of the state. He is not permitted to leave the area around Mount Wilson, and is one of the few astronomers to remain at the observatory, where for a long time he has the Hooker Telescope all to himself.

The photographs Baade makes during the war are of unprecedented quality. That is largely due to the blackout regulations, which mean that he has no interference at all from Los Angeles 'light pollution'. He is the first to photograph individual stars in the densely populated center of the Andromeda Galaxy.

❷ The sparkling star cluster in this Hubble photo is part of a large star forming region in the Small Magellanic Cloud, at a distance of 210,000 lightyears from the Earth. At the bottom of the image is a much older group of stars. (NASA/ESA/A. Nota)

❸ Large clouds of gas and dust in a star forming region in the Large Magellanic Cloud can produce both low-mass, Sun-like stars that can reach great ages and massive giant stars that explode as supernovae after a few tens of millions of years. (NASA/ESA/Hubble Heritage Team)

Baade uses two different emulsions for his photographs; one is especially sensitive to red light and the other to blue light. In 1994, by comparing the different images of the Andromeda Galaxy, he makes a remarkable discovery. The galaxy's spiral arms contain primarily blue stars, while the center has mainly red stars. Because many stars swell up to become red giants at the end of their lives, Baade concludes that the stars at the core of the galaxy are old, while those in the arms are much younger.

In the early 1950s, Baade makes another important discovery that both stellar populations turn out to contain Cepheids – stars that vary in brightness in a predictable manner. In 1912, Henrietta Leavitt discovered that there is a link between their pulsation period and their average luminosity, but the period-luminosity law for young Cepheids in population I appears to be completely different to that for the older Cepheids in population II.

Edwin Hubble was not aware of the existence of the two stellar populations when he determined the distance of the Andromeda Galaxy at approximately one million lightyears in 1924. Baade recalculates the distance and concludes that it is twice as far away, and therefore twice as large as had always been assumed. The same applies to other galaxies. Baade therefore doubles the size of the universe in one fell swoop. The age of the universe is also adjusted upwards, so that it no longer conflicts with the geological age of the Earth.

Veiled Moon

Gerard Kuiper Discovers the Atmosphere Around the Saturnian Moon Titan

In 1655, Dutch physicist and astronomer Christiaan Huygens sees a small speck of light moving around the distant planet Saturn. This is the first discovery of a moon around another planet since Galileo discovered four moons around Jupiter in 1610. The Saturnian moon is not given its name, Titan, until 1847. It is a very appropriate name: Titan is an extremely large moon, with a diameter larger than that of the planet Mercury.

In 1903, Spanish astronomer Josep Comas Solà sees Titan as a tiny sphere, which seems to be a little fainter at the rim than in the middle. Solà believes that this may indicate that Titan has an atmosphere, like the Earth and Venus.

It is the Dutch-American planetary scientist Gerard Kuiper who first shows that Titan does indeed have an atmosphere. Kuiper has been working in the United States since 1933, and is associated with University of Chicago's Yerkes Observatory. Like many of his colleagues, he is sent to work during the Second World War in a laboratory for radio and radar research, but in the winter of 1943/1944, he gets a few months leave.

During that winter he studies the large moons of the outer planets with the 2.1-meter Struve Telescope at the McDonald Observatory in Texas, which at that time is the second largest telescope in the world. Kuiper's spectroscopic observations, with one of the best instruments of his time, leave no room for doubt: there is methane gas on Titan. The moon clearly does have an atmosphere – something that has never before been discovered on a planetary moon.

Kuiper makes many other remarkable discoveries in the outer regions of the solar system. In 1948, he discovers the small Uranian moon Miranda, followed a year later by the Neptunian moon Nereid. Around 1950, Kuiper predicts the existence of a broad, flat belt of comets, beyond the orbit of Neptune. The first Kuiper Belt object is found in 1992, and we now know for certain that Pluto is part of this belt of frozen celestial bodies.

In 1980, Titan is studied close up for the first time by the American space probe Voyager 1. Voyager discovers that the atmosphere is much denser than was previously thought, and consists primarily of nitrogen – the same gas that makes up most of the atmosphere of the Earth. The surface cannot be seen, because of thick layers of haze. Voyager 1 also discovers that the size of Titan has always been overestimated because of its thick atmosphere. It proves to have a diameter of 5,150 kilometer – a little less than the Jovian moon Ganymede.

The nitrogen-rich atmosphere of Titan is studied on the spot for the first time in January 2005 by the European probe, Huygens. Kuiper does not live to see it; the 'titan' of planetary research dies of a heart attack at the end of 1973.

➊ Image of the Saturnian moon Titan made in January 2008 by the American planetary explorer Cassini. On photographs taken in visible light, nothing can be seen of Titan's surface; it is concealed by layers of haze in the atmosphere. (NASA/JPL/SSI)

➋ Infrared cameras make it possible to penetrate the clouds and layers of smog in Titan's atmosphere and photograph formations and color variations on the surface. This mosaic image was made by the Cassini space probe. The large, dark area is known as Xanadu. (NASA/JPL/SSI/Mattias Malmer)

1944

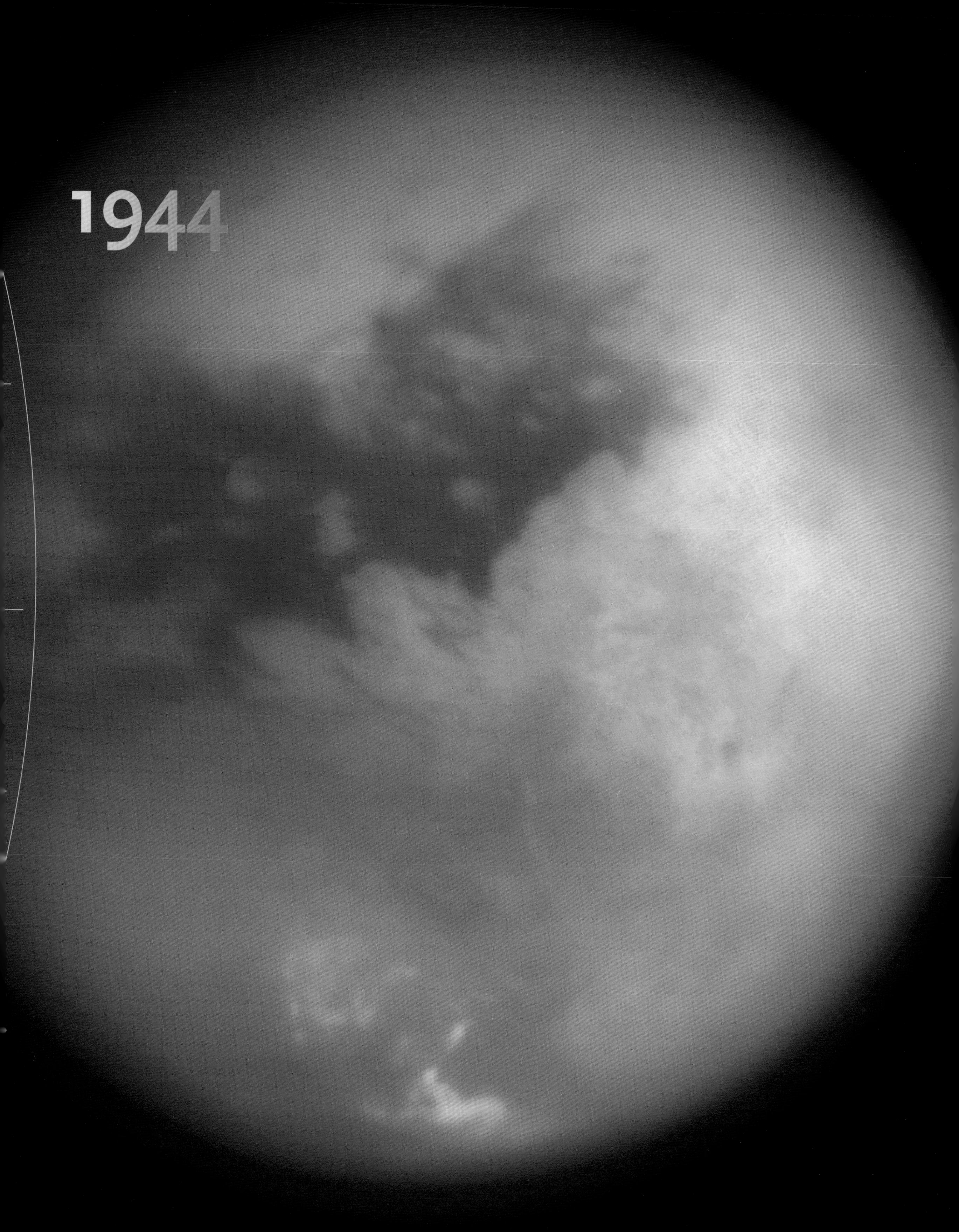

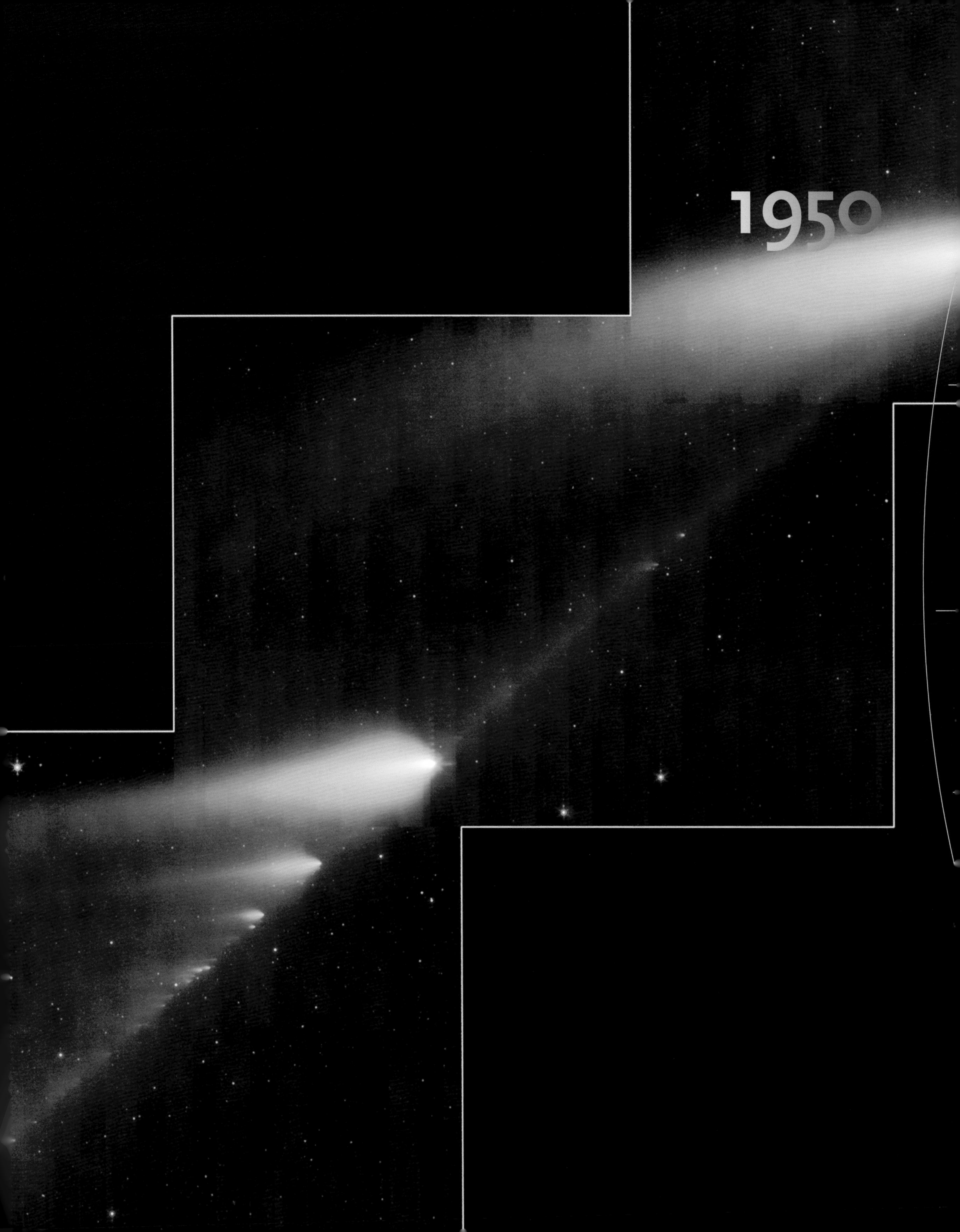
1950

Icy Cloud

Jan Oort Discovers the Origin of Long-Period Comets

Comets have captured the imagination since times of old. Edmund Halley shows in 1705, for example, that the impressive 'stars with tails' move in highly elongated orbits around the Sun and are therefore, like the Earth and the other planets, part of the solar system. But little is known about the nature and origin of comets until far into the twentieth century.

According to the English astronomer Raymond Lyttleton, comets are tenuous clouds of dust and rubble. Russian Sergej Vsekhsvyatskij believes that they are condensations of volcanic gas expelled by active volcanoes on planets and planetary moons. Astronomers have as yet no explanation for the difference between short-period and long-period comets. Short-period comets, with an orbital period of up to about 200 years, move in the same direction around the Sun as the planets and move more or less in the same plane. Long-period comets have orbital periods of millions of years, and their extremely elongated orbits display all possible orientations.

Leiden professor Jan Oort becomes interested in the phenomenon through the Ph.D. research of his student, Adriaan van Woerkom, who studies the orbits and possible origins of long-period comets. Van Woerkom subsequently shows that Vsekhsvyatskij's theory cannot be correct, and he also shows that long-period comets do not originate in interstellar space.

Oort starts to make calculations since the orbits of 19 long-period comets have been determined very accurately. Although they pass through the inner regions of the solar system at different distances from the Sun, the furthest point in the orbits of most of them is at roughly the same distance: around 3 trillion kilometers, or 20,000 times the distance between the Earth and the Sun. This can only be explained satisfactorily if there is a gigantic reservoir of comets at that distance, from which a comet now and then falls into the inner regions of the solar system. That happens, for example, when its orbit is disturbed by a passing star.

The Oort Cloud, as the comet reservoir is named, must contain many trillions of frozen cometary nuclei and extend halfway to the nearest star. No one has ever observed the Oort Cloud directly, but there are few who doubt its existence: there is simply no better explanation for the origin of long-period comets. Besides, computer simulations by the Canadian theorists Martin Duncan, Scott Tremaine and Thomas Quinn at the end of the 1980s show that the Oort Cloud is a natural by-product of the formation of the solar system.

In 1932, the existence of the comet cloud was also predicted by the Estonian astronomer Ernst Öpik. Although Oort's analysis was more thorough, the cloud is sometimes known as the Öpik-Oort Cloud.

➋ Comet Hyakutake, which flew past the Earth at a short distance in the spring of 1996 and was very clearly visible from the northern hemisphere, has an orbital period of more than a hundred thousand years. Long-period comets like Hyakutake come from the Oort Cloud. (NASA/Retiono Virginian)

➌ After spending millions of years in the cold, dark Oort Cloud, cometary nuclei can experience orbital disturbances and fall towards the inner regions of the solar system, where they evaporate and disintegrate in a relatively short time, as happened to comet Schwassmann-Wachmann 3. (NASA/JPL)

Hydrogen Hiss

Harold Ewen Discovers the 21-centimeter Line of Neutral Hydrogen

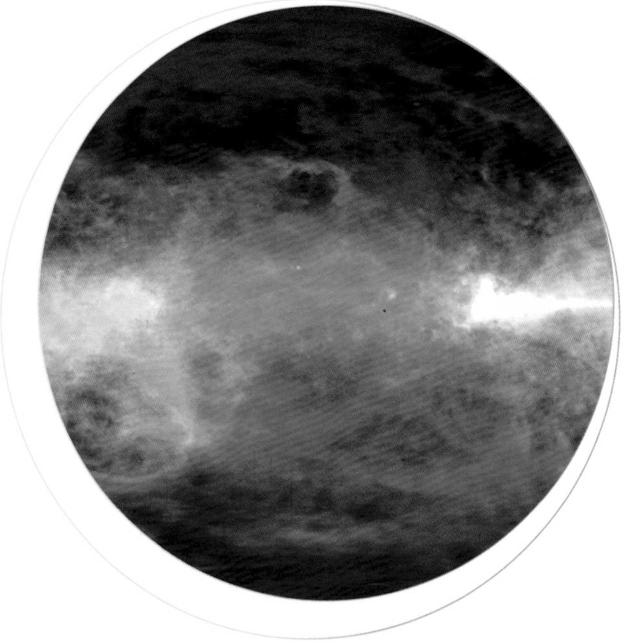

Harold Ewen thinks that he is competing with the Russians when, around 1950, he tries to find the line radiation of hydrogen. According to the Russian Iosif Shklovsky, the radiation should be relatively easy to detect, but it is not until he has discovered the 21-centimeter line that Ewen finds out that Dutch astronomers nearly beat him to it; if he had known that they were also searching for the line in the Netherlands, he would never have started looking.

The fact that neutral hydrogen emits radiation is deduced theoretically during the Second World War by astronomer Henk van de Hulst in Leiden, at the request of his mentor Jan Oort, who is one of the first to recognize the importance of the discoveries of Karl Jansky and Grote Reber and realizes that radio astronomy opens up a completely new window to the universe.

Twenty-five year old Van de Hulst – the son of teacher and author of children's book W.G. van de Hulst – predicts that a hydrogen atom emits a small quantity of radiation at a wavelength of 21 centimeters when its electron starts spinning in the opposite direction. He presents his conclusions at a meeting of the Dutch Astronomers' Club on April 15, 1944.

With the help of radio engineer Lex Muller, Oort starts looking for the faint line radiation of hydrogen. He uses a converted German Würzburg radar antenna with a diameter of 7.5 meters at Radio Kootwijk in the Veluwe national park. However, in the spring of 1950, the program is delayed by a fire in the receiver equipment.

At Harvard University, American Harold ('Doc') Ewen is also searching for the 21-centimeters line. As part of his Ph.D. research, he builds a horn antenna and a radio receiver but discovers in 1950 that the design needs to be radically modified. His supervisor, Edward Purcell, pays the $300 for the modifications out of his own pocket. Half a year later, on March 25, 1951, Ewen strikes lucky. The radiation of neutral hydrogen gas, predicted seven years previously by Van de Hulst, has finally been found.

Immediately after his discovery, Ewen hears that Van de Hulst is visiting Harvard College Observatory. Ewen informs him of his discovery personally, and also has a long telephone conversation with Oort during which he describes his method in detail. On May 11, the 21-centimeter radiation is also detected in Kootwijk. A few weeks later, the feat is repeated by Australian astronomers, and the three teams publish their results in the scientific weekly *Nature* on September 1.

Hydrogen is the most common element in the universe, and that makes the 21-centimeter line one of the most important aids for radio astronomers. By mapping the radiation of neutral hydrogen, they can study the structure and dynamics of the Milky Way and other galaxies.

Radio map of the sky, based on observations with the 25-meter radio telescope in Dwingeloo. The distribution of cool hydrogen gas is recorded at radio wavelengths. The colors show the different radial velocities of the gas clouds. (Dap Hartmann)

The Triangulum Galaxy, more than 2 million lightyears away from the Milky Way, is one of the nearest large galaxies. This image of the cool hydrogen gas in the outer regions of the galaxy (shown here in purple) is based on observations made with radio telescopes in the United States and the Netherlands. (NRAO/AURA/NSF)

1951

1957

Elementary Stardust

Fred Hoyle and his Colleagues Discover the Origin of the Elements

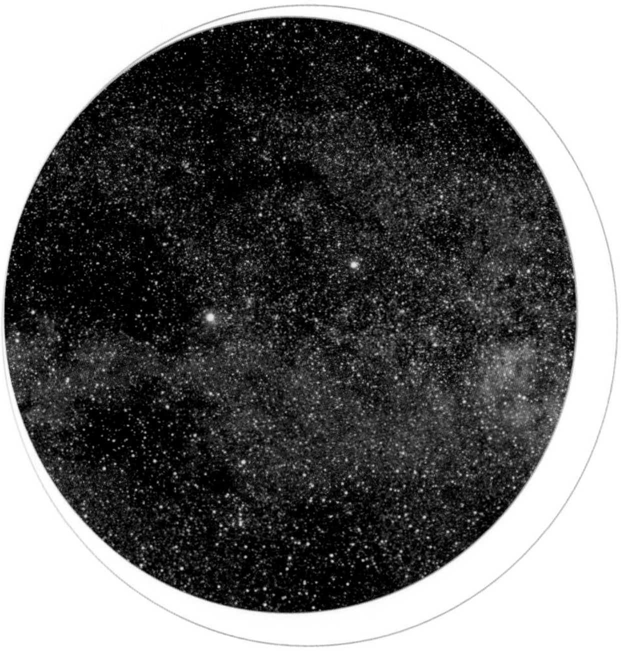

In 1983, Fred Hoyle misses out on his share in the Nobel Prize for Physics because of his controversial ideas on cosmological issues as perhaps do Geoffrey and Margaret Burbidge. The prestigious prize goes to Willy Fowler who, in 1957, discovers the origin of the chemical elements in the universe, together with Hoyle and the Burbidges.

Perhaps Dutch astrophysicist Bruun van Albada also deserves a share in the prize. In 1946, at almost the same time as Hoyle, he publishes an article in which he sets out his theory that most of the 'heavy' elements in the universe (those heavier than helium) are created in the interior of stars. It has been known for some time that stars generate energy by fusing hydrogen into helium, thanks to the work of Hans Bethe and Carl von Weizsäcker, but astronomers are still in the dark about the origin of the heavier elements.

Russian-American physicist George Gamow, one of the leading proponents of the Big Bang theory, proposes that all these elements were created during nuclear fusion reactions in the early youth of the universe. According to Hoyle and Van Albada, however, nucleosynthesis (the production of new atomic nuclei) can only occur in the hot interior of stars. Speaking on a radio program for the BBC in March 1949, Hoyle pokes fun at Gamow's theory of the primal explosion, during which he is the first to use the term 'Big Bang'.

Hoyle elaborates his ideas in an article in 1954, but it goes relatively unnoticed by nuclear physicists, however, because it is published in the first issue of a new astrophysical magazine. In 1957, together with Margaret and Geoffrey Burbidge and the renowned nuclear physicist Willy Fowler of the California Institute of Technology, Hoyle finally publishes a monumental article in the *Review of Modern Physics*. The article is invariably referred to as B^2FH, after the initials of the four authors.

Hoyle and his colleagues claim that the wide variety of nuclear reactions by which nearly all elements in the periodic table are created occur in the hot interior of stars, where pressure and temperature are extremely high. Gradually the implications of this become clear: the atoms in our bodies that are heavier than helium, such as carbon, calcium and iron, were all formed in stars. In other words, we are all made of stardust.

Why the Nobel Committee only honors Fowler in 1983 is not clear. The official reading is that he is (incorrectly) seen as the leader of the team. The more widely held opinion, however, is that Fred Hoyle's unconventional ideas played an important role, and he continues to oppose the Big Bang theory until his death in 2001. Together with Geoffrey and Margaret Burbidge (and with astrophysicists Chandra Wickramasinghe and Halton Arp) he also questions the cosmological interpretation of the redshift of galaxies and the expansion of the universe.

Alpha and Beta Centauri are among the closest stars in the Milky Way. Nearly all the other stars in this wide-angle photo are much further away. In the interiors of these stars, carbon, oxygen, nitrogen and other elements are created by nuclear fusion reactions of lighter elements. (ESO/Claus Madsen)

Illustration of a protoplanetary disk around a young protostar. In a gas and dust disk like this, over a period of several million years, planets can form. All elements in the disk that are heavier than helium were created at some time in the interiors of other stars. (NASA/JPL)

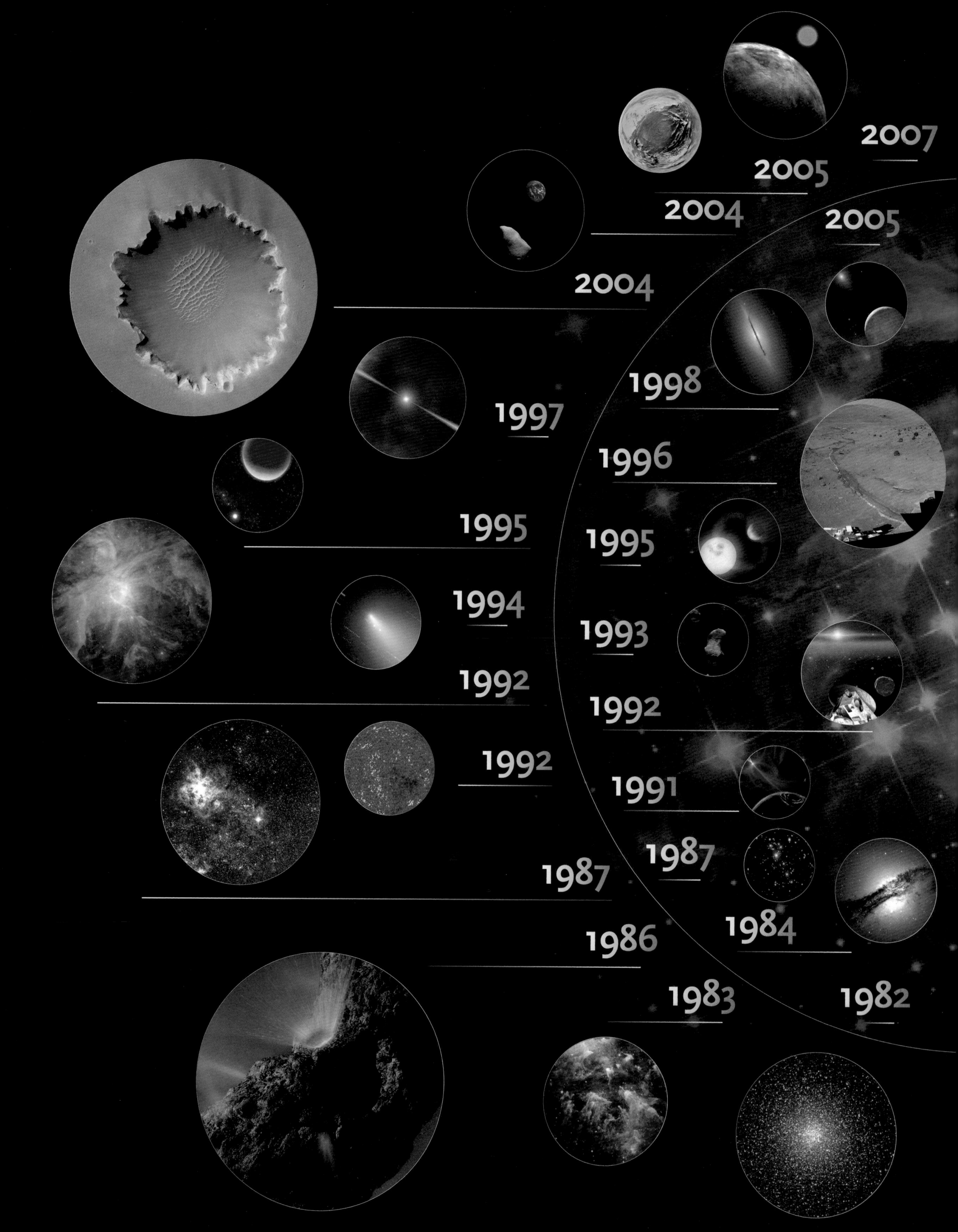
2007
2005
2004
2005
2004
1998
1997
1996
1995
1995
1994
1993
1992
1992
1992
1991
1987
1987
1984
1986
1983
1982

1958 - 2008

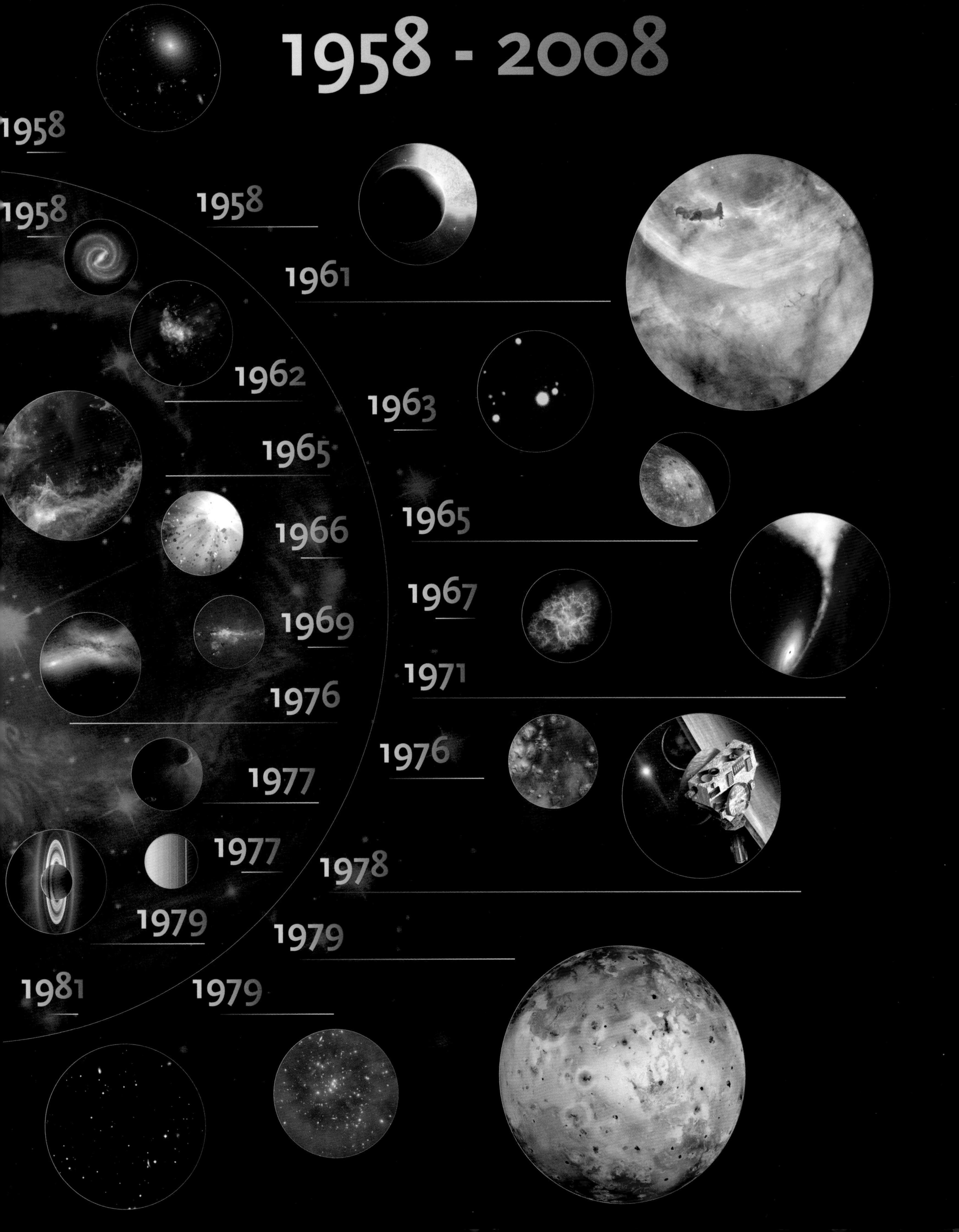
1958
1958
1958
1961
1962
1963
1965
1965
1966
1967
1969
1971
1976
1976
1977
1977
1978
1979
1979
1981
1979

Around the year 2000, there are more astronomers active worldwide than ever lived before 1950, and new methods and instruments have never been developed at a rapid pace. No wonder that by far the most revolutionary discoveries in astronomy date from recent decades. That is not a distorted view of history but a hard fact; as a scientific discipline, astronomy is currently undergoing faster growth and development than ever before.

Revolutionary Discoveries

At the same time it is becoming clearer than ever that progress in science is only possible with progress in methods and instrumentation. Our current knowledge of the universe would never have been possible without the three technological pillars on which modern astronomy rests.

The advent of space travel has enabled us to place telescopes beyond the disturbing influence of the terrestrial atmosphere and to visit other objects in the solar system. The unraveling of the electromagnetic spectrum means that we can not only capture and study visible light and radio waves from the universe, but also infrared radiation, ultraviolet light, X-rays and even high-energy gamma ray radiation. Computer technology enables astronomers to analyze enormous quantities of observational data and create

in a Mysterious Universe

theoretical models, and makes possible the construction of gigantic telescopes and the development of new observation methods.

Yet in 2010 there is no place in astronomy for human overconfidence and arrogance. The more we discover, the more new puzzles present themselves. By far the largest part of the universe remains one big, unsolved mystery.

Cosmic Gathering

Elizabeth Scott and Jerzy Neyman Discover Superclusters

Galaxies are not evenly distributed across the sky, and many more are visible in some directions than in others. If you do not know the exact distance to each galaxy, that says little about their actual distribution throughout space. Furthermore, it is difficult to quantify the 'lumpiness' of the universe, and to determine the significance of the clustering. In 1958, with their discovery of superclusters, Elizabeth Scott and Jerzy Neyman take the first revolutionary steps towards statistical cosmology.

Neyman is a Russian mathematician who flees his native country after the 1917 revolution because he is of Polish descent and considered an enemy of the state. Via Warsaw and London he arrives in the United States at the end of the 1930s, where he sets up the statistics faculty at the University of California in Berkeley.

Elizabeth Scott, born in the year of the Russian revolution, studies astronomy and, at a young age, publishes articles on cometary orbits and binary star systems. After completing her studies, she increasingly focuses on the mathematical fundations of astronomy, partly because at the time women are still not allowed to use the big telescopes at Mount Wilson Observatory. In 1951, two years after being awarded her Ph.D., she is appointed associate professor at Berkeley.

Scott and Neyman become interested in the research being conducted by Donald Shane and Carl Wirtanen at the Lick Observatory. In 1954, Shane and Wirtanen publish a much-discussed article about the unequal distribution of galaxies in the sky. They divide the sky into small 'cells' and count the number of galaxies in each cell. Their maps prove conclusively the existence of clusters, but even larger structures seem to be visible.

Scott and Neyman apply their statistical expertise to Shane and Wirtanen's data. They devise methods to quantify the degree of clustering and show convincingly that there are indeed gigantic superclusters of galaxies in the universe. In 1958, they present their results at a meeting of the Royal Statistical Society in London.

That same year, George Abell publishes a catalog of 4,073 'normal' clusters of galaxies which have several hundred members and are tens of millions of lightyears across. Superclusters, on the other hand, can contain tens of thousands of individual galaxies, are often elongated and can extend to hundreds of millions of lightyears. In 2003, the discovery of the Sloan Great Wall, nearly 1.4 billion lightyears long, is announced.

Later in her career, Elizabeth Scott conducts statistical research into the status of women in the academic world. In the early 1980s, she shows unequivocally that there is still a long way to go before emancipation is achieved in the world of science.

Stars in our own Milky Way are filtered out in this mosaic image of the southern sky. The colored dots show the distribution of more than three million galaxies at various distances. Such a two-dimensional image already shows clearly that galaxies are not evenly distributed throughout the universe. (APM Survey)

Hubble image of a cluster of galaxies at a distance of 450 million lightyears in the constellation of Centaurus. The individual galaxies in a cluster are held together by their mutual gravity. Clusters are in turn grouped in gigantic superclusters. (NASA/ESA/Hubble Heritage Team)

1958

1958

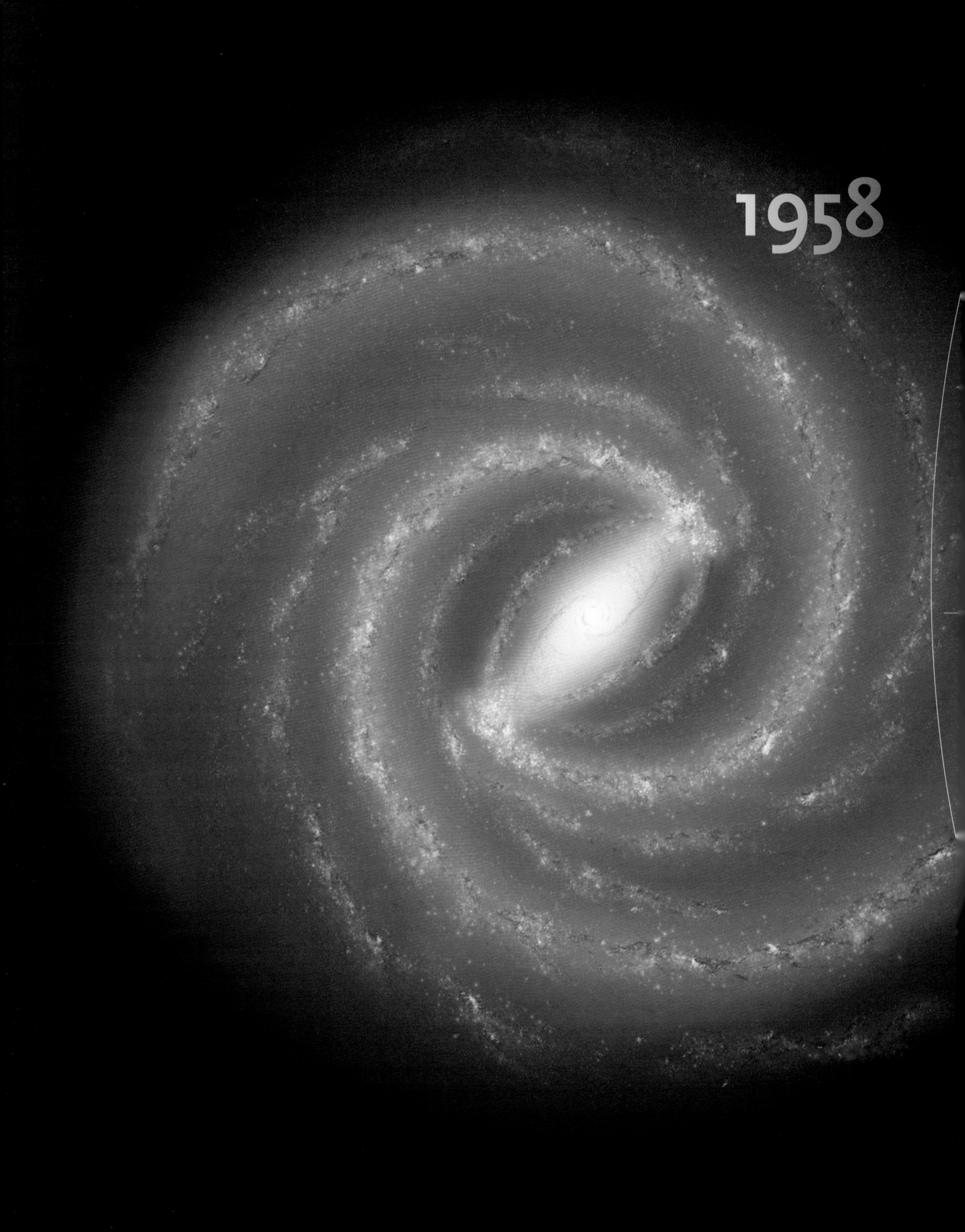

Majestic Spiral

Jan Oort Discovers the Spiral Structure of the Milky Way

When Jan Oort becomes a professor of astronomy in Leiden in 1935, he says in his inaugural speech that 'we are still unable to read the language of spiral galaxies'. Less than 25 years later, the Rosetta Stone has been found and astronomers have started to decipher the galactic script.

Not all galaxies have a spiral structure; there are also elliptical galaxies with an elongated or somewhat flattened spherical form. We have known for a long time, however, that our own Milky Way is very flat, and many astronomers assume that it is a stately spiral galaxy, like the Andromeda Galaxy. Because the Sun and the Earth are located in the flat, central plane, the spiral structure is very difficult to see.

In 1951, the American astronomers Don Osterbrock, Bill Morgan and Stewart Sharpless show that the brightest stars in the interior regions of the Milky Way are not distributed at random, but are arranged along spiral-shaped arcs – something the Amsterdam astrophysicist Anton Pannekoek noted earlier. Yet the radio maps published a few years later by Oort and his colleagues are much more convincing – they show the spiral structure of the Milky Way in detail for the first time.

In the summer of 1952, after the discovery of the 21-centimeter line radiation of neutral hydrogen gas, the Dutch astronomers start a systematic radio survey of the Milky Way plane mostly done by hand using the 7.5-meter radar antenna at Radio Kootwijk. A second more detailed survey is conducted between November 1953 and August 1955; interim results are published in the *Bulletin of the Astronomical Institutes of the Netherlands*.

In the meantime, Oort has received a government grant to build a much larger radio telescope with a diameter of 25 meters. The Dwingeloo Telescope is officially brought into operation in the spring of 1956, and, for about a year is the largest in the world. The telescope is also used to map the distribution of hydrogen gas in the Milky Way plane.

In 1958, Jan Oort, Frank Kerr and Gart Westerhout publish the results in a large review article in the *Monthly Notices of the Royal Astronomical Society*. The radio observations show that the Milky Way is just such an impressive spiral galaxy as the Andromeda Galaxy. Furthermore, Oort and his colleagues succeed in measuring rotation speeds in the inner regions of the Milky Way and discover turbulent motions in its center.

Measurements of the distribution of stars in the center of the Milky Way conducted with infrared telescopes in the late 1980s show that the Milky Way is not a normal spiral galaxy, but a barred spiral galaxy. At its center there is an elongated structure of primarily old stars; the two main spiral arms of the Milky Way extend from the ends of this 'bar'.

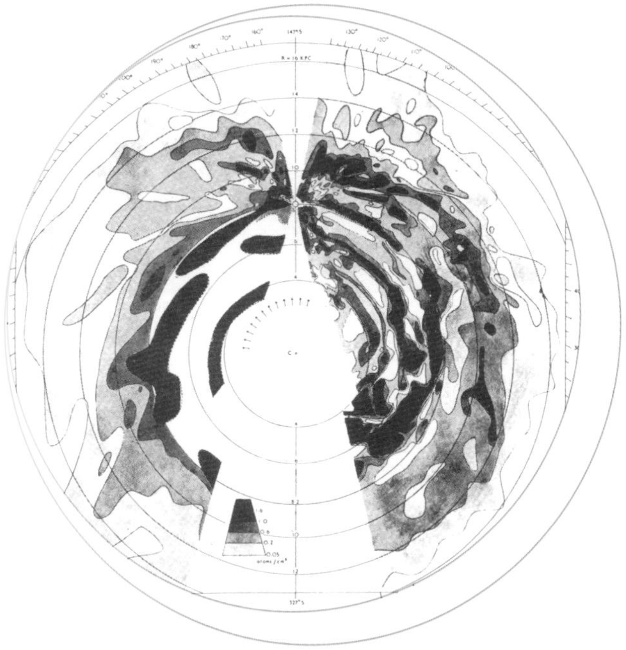

➋ Map showing the distribution of hydrogen gas in the Milky Way, based on observations with the radio telescopes in Kootwijk and Dwingeloo. This was the first map clearly showing the spiral structure of the Milky Way. (Leiden Observatory)

➌ Illustration of the Milky Way, based on measurements by the Spitzer Space Telescope and others. The Milky Way has a somewhat elongated central section (the 'bar'), from the outer ends of which great spiral arms protrude. The pink spots are star forming regions. (NASA/JPL)

141 — 142

Interplanetary Weather

Eugene Parker Discovers the Solar Wind

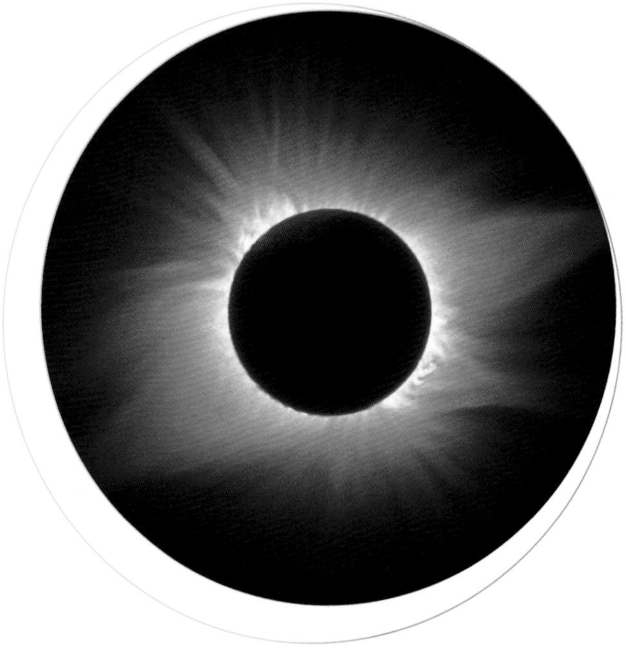

In 1619, Johannes Kepler is the first to explain why the tails of comets always face away from the Sun. According to Kepler, comet tails consist of particles that are carried along by the light from the Sun. Nearly two and a half centuries later, Richard Carrington suggests that it is not so much a matter of the sunlight exerting an influence on other celestial bodies, but of less rapidly moving 'solar particles'. This also explains why the effects of the solar flare that Carrington observes in 1859 are only noticeable a few days later.

Laboratory experiments by the Norwegian physicist Kristian Birkeland show that auroras are indeed caused by electrically charged particles. Birkeland suggests in 1916 that the Sun emits 'beams of electrical particles' and that the space between the planets is therefore not empty. It is not until 1958, however, that American solar physicist Eugene Parker provides a theoretical basis for what he calls the 'solar wind'. Parker bases his work on that of English mathematician Sydney Chapman, who in the 1950s is the first to conduct calculations on the corona of the Sun.

The corona is a deep layer of thin gas surrounding the Sun, which is only visible during a total solar eclipse. Measurements on the structure of this 'solar atmosphere' from the 1930s have already shown that the gas in the corona must have an unimaginably high temperature of about one million degrees. Chapman deduces that such hot gas conducts heat very efficiently. That means that even at a great distance from the Sun it has such a high temperature that it is hardly kept in check by the Sun's strong gravitational pull.

Parker elegantly fits all the pieces of the puzzle together; because of the high temperature of the corona, gas continually flows out into space. These electrically charged particles, predominantly protons and electrons, are carried along the lines of the Sun's magnetic field which, according to Parker, are spiral-shaped because of the rotation of the Sun. The solar wind is responsible for the gas tails of comets and for the auroras in the Earth's atmosphere.

Parker's solar wind article is initially rejected by two reviewers in 1958, but thanks to the intervention of Subrahmanyan Chandrasekhar, editor-in-chief of *The Astrophysical Journal*, it is eventually published. A few months later the Russian space probe Luna 1 proves the existence of the solar wind – a continuous flow of electrically charged particles traveling at a speed of several hundred kilometers a second. In 2003, Parker is awarded the prestigious Kyoto prize for his discovery.

Each second, the Sun loses more than a million tons of matter through the solar wind. That is equal to the mass of the Earth every 150 million years. But compared to the enormous mass of the Sun, that is nothing more than a drop in the ocean.

➊ The Sun continually expels electrically charged particles into space, losing more than a million tons of matter per second in the process. The tenuous outer atmosphere of the Sun, the corona, is only visible during a total solar eclipse, when the Sun's bright surface is concealed by the Moon.

➋ Illustration of a massive exoplanet in close orbit around its parent star. Such 'hot Jupiters' have been found around many Sun-like stars, sometimes at distances of less than a few million kilometers. They are exposed to extremely high temperatures and a powerful stellar wind. (ESO)

1958

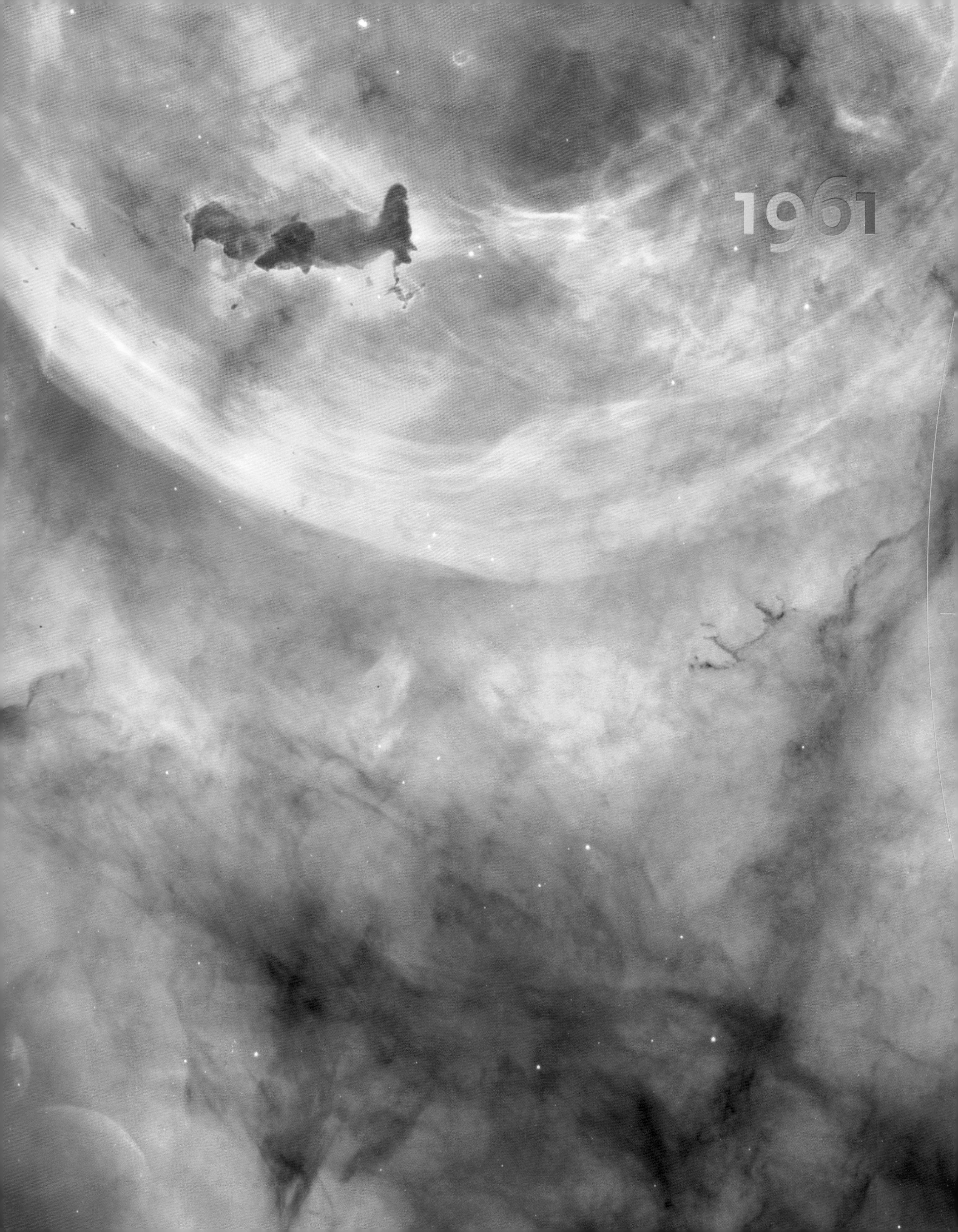
1961

Cosmic Evolution

Martin Ryle Discovers That the Universe Is Evolving

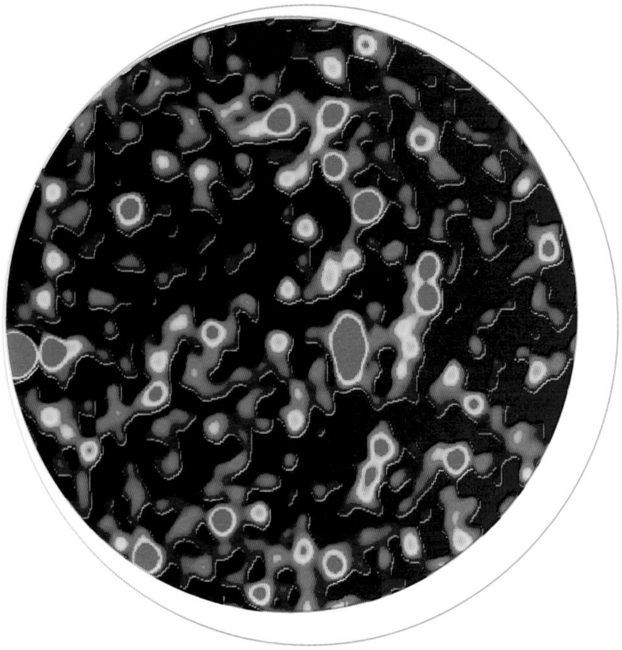

Early in 1961, Fred Hoyle gets a call from the Mullard Company in London, the main sponsor of the Mullard Radio Astronomy Observatory at Cambridge University, asking if he would like to attend a sensational presentation by radio astronomer Martin Ryle the following week. When Hoyle arrives at the headquarters of the electronics company, he is ceremoniously led to a stage, where he finds himself in the full glare of the spotlights and faced with a row of news-hungry journalists. The curtain then rises on Ryle, who explains how his latest radio observations from the universe completely wipe the floor with Hoyle's cosmological theories. Hoyle is then allowed to respond to the press. The same day, the headlines in the London newspapers announce that Hoyle's ideas are ripe for the trash-can.

Not all astronomers treat their colleagues in such an uncivilized manner. But Martin Ryle is known for being hotheaded, and he has been at loggerheads with Hoyle for many years. In 1948, together with Hermann Bondi and Thomas Gold, Hoyle publishes an alternative to the Big Bang theory. According to this Steady State theory (also known as the theory of continuous creation) the universe may be expanding, but it does not change in appearance noticeably over time; because new atoms are continually being created, the average density remains stable.

Ryle has no time for the Steady State theory and does everything he can to discredit Hoyle's ideas. The results that he presents in London in 1961, and which are published later that same year in the *Monthly Notices of the Royal Astronomical Society*, make it clear that the universe is definitely evolving, and precisely in the way suggested by the Big Bang theory.

Using the radio telescopes at Mullard Observatory, Ryle and his colleagues chart hundreds of more or less point-like radio sources. Many are identified with distant galaxies, but there is something strange going on. The number of faint radio sources – in all probability galaxies at great distances from the Earth – is larger than you would expect. Because at great distances in the universe you are also looking far back in time, it can only mean one thing: long ago there were more 'radio galaxies' than there are now. In other words, the universe is evolving.

Ryle has already come to the same conclusion at the end of the 1950s, when he publishes his first catalog of radio sources. In 1961, however, he is 100% certain, and embarrasses Fred Hoyle in public. Hoyle remains unconvinced and continues to doubt the Big Bang theory until his death.

Ryle not only shows that the universe is evolving and in the 1940s develops the aperture synthesis method, by which different radio telescopes can be linked to achieve an extremely high spatial resolution. In 1974, together with his colleague Antony Hewish, he is presented with the first Nobel Prize for Physics awarded for astronomical research.

The Westerbork Synthesis Radio Telescope in Drenthe, the Netherlands, has been used to map the radio waves emitted by extremely distant galaxies in the Hubble Deep Field. There were more radio galaxies in the youth of the universe than there are now – proof of cosmic evolution. (ASTRON)

The irregularly shaped dark spots on this Hubble photograph are embryonic stars in the Carina Nebula. The origins of the Sun, the Earth and life are part of a great cosmic cycle and cannot be seen in isolation from the evolution of the universe. (NASA/ESA/Hubble Heritage Team)

Penetrating Look

Riccardo Giacconi Discovers the First Galactic X-Ray Source, Scorpius X-1

Astronomers have only just become accustomed to the many new possibilities offered by radio astronomy when a new area of the science presents itself. In the summer of 1962, the first source of cosmic X-rays is discovered, heralding the birth of high-energy astrophysics. The discovery earns Riccardo Giacconi the Nobel Prize for Physics in 2002.

Giacconi was born in Milan, where he later studies physics and conducts research into cosmic rays. In 1956, at the age of 25, he emigrates to the United States. Shortly afterwards, the Soviet Union launches its first artificial satellite, Sputnik 1, and astronomers become interested in the scientific opportunities offered by space exploration. Giacconi's fellow Italian Bruno Rossi tells him at a party that American researchers are especially interested in cosmic X-rays.

Giacconi sets up a research group at the American Science & Engineering company, and within two and a half years the group has grown from three to seventy members. A period follows with many secret military projects and – in between – a number of scientific experiments, and on June 18, 1962, a Geiger counter is sent into space for the first time, on board the US Air Force's Aerobee sounding rocket.

The rocket flight, from White Sands Missile Range in New Mexico, lasts less than six minutes, with the rocket reaching a maximum height of about 130 kilometers. Yet in those few minutes the Geiger counter does register X-rays coming from the direction of the constellation Scorpius. It is already known that the Sun emits X-rays, but Scorpius X-1 is the first X-ray source outside the solar system.

It is not until several years later that a small variable star is discovered at the position of Scorpius X-1. It appears to be an X-ray binary star system, with a small, compact neutron star orbiting a low-mass 'normal' star. Gas from the companion flows to the neutron star and becomes so hot in the process that it emits X-rays. Scorpius X-1, which is at a distance of about 10,000 lightyears, turns out to emit a hundred thousand times more energy than the Sun.

Giacconi continues to play a pioneer's role in the new discipline. He is closely involved in the development of the new X-ray satellite Uhuru, the later satellite Einstein, and the current flagship of American X-ray astronomy, the Chandra X-ray Observatory. In addition, he is the first director of the Space Telescope Science Institute and is director-general of the European Southern Observatory for six years.

X-ray observations now have an established place in astronomy, offering astronomers a penetrating look at extreme objects like pulsars and black holes. Scorpius X-1 remains the strongest source of X-rays in the heavens.

Illustration of an X-ray binary system, in which matter from a Sun-like star flows to a compact companion, like a neutron star. The gas accumulates in a spinning 'accretion disk' and becomes so hot in the process that it emits X-rays. (ESA)

In the core of the Milky Way, there is a supermassive black hole that emits powerful X-rays. On this X-ray photo of the center of the Milky Way, made by the Chandra X-ray Observatory, many X-ray binary systems – small, point-like X-ray sources – are visible. (NASA/CXC/MIT)

1962

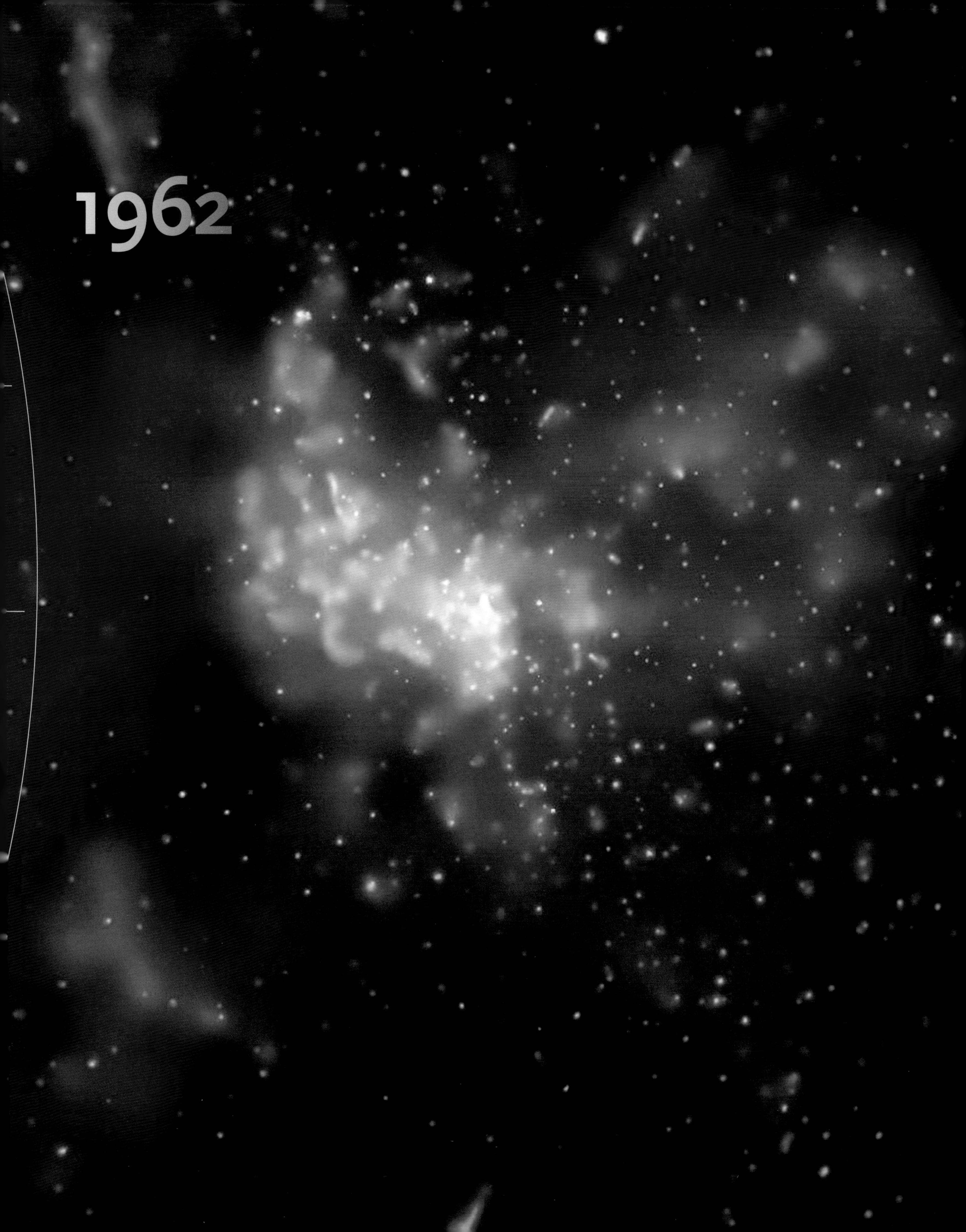

1963

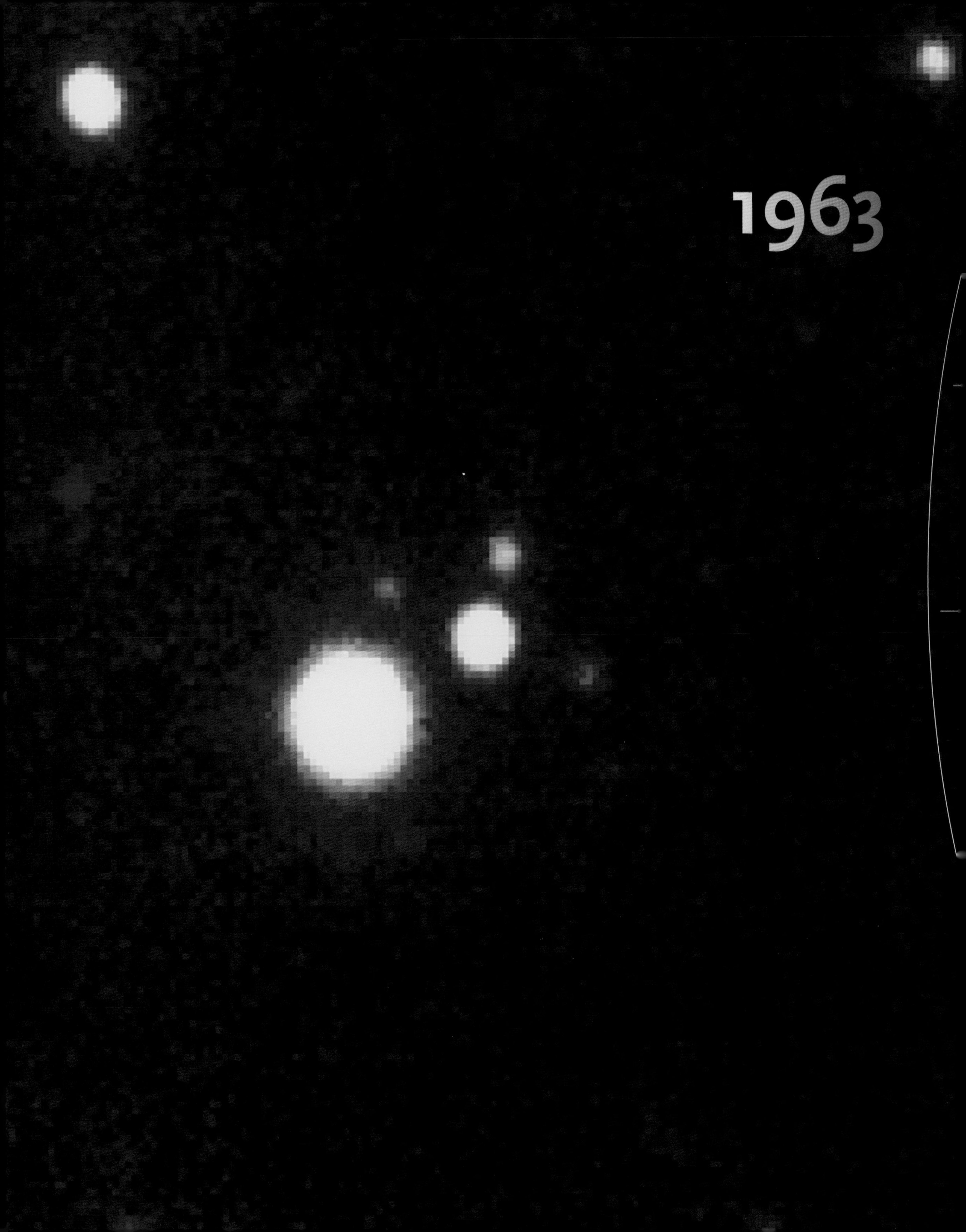

Far-Off Shores

Maarten Schmidt Discovers Quasars

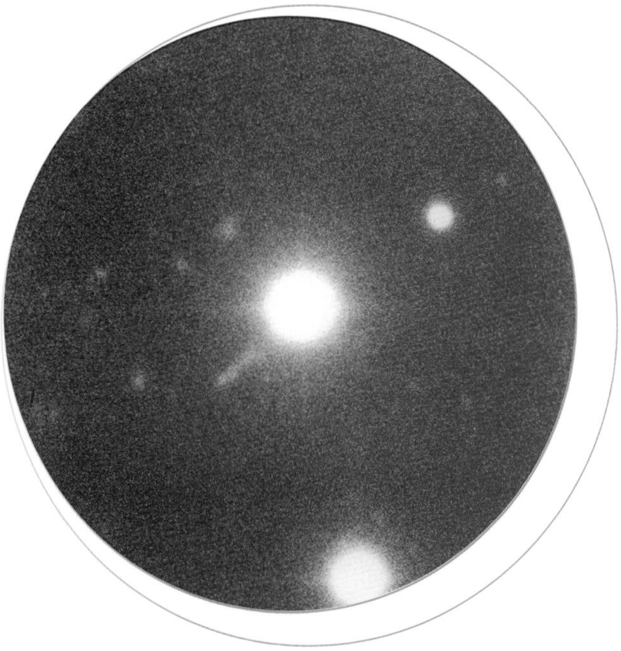

At the end of the 1950s, at Palomar Observatory in California, Rudolph Minkowski hunts for the optical counterparts of recently discovered 'radio stars' – perfectly pointlike sources of radio waves in the cosmos. Often, at the location of a radio star, Minkowski discovers a distant galaxy, but in some cases there is nothing of note to see through an optical telescope.

When Minkowski retires in 1960, his work is taken over by Maarten Schmidt, a young astronomer from the Netherlands who studied under Jan Oort in Leiden and moved to the United States in 1959. Schmidt becomes intrigued by the unidentified radio stars. He feels that if their positions in the sky could be determined more accurately, it might be possible to find an optical counterpart.

That proves possible in 1962 with 3C273 (the 273rd object in the third Cambridge catalog of radio sources). The radio star is eclipsed by the Moon, and by measuring the time of the occultation, Australian astronomers manage to determine its position accurately. That leads to the discovery that the radio source coincides with an inconspicuous speck of light - apparently a small star in our own Milky Way. But why would the star be such a powerful source of radio waves?

◐ This photo of the quasar 3C273, made with the 5-meter Hale Telescope at Palomar Observatory, shows the remarkable jet expelled into space by the quasar. 3C273 was the first quasi-stellar radio source whose distance was determined. (Caltech/Maarten Schmidt)

◑ Large terrestrial telescopes have discovered a trio of quasars in the constellation of Virgo, at a distance of around 10.5 billion lightyears. The three quasars – each of which is a galaxy with an active black hole at its center – will probably fuse to form a single super quasar at some time in the future. (ESO)

At the end of 1962, using the 5-meter Hale Telescope on Palomar Mountain, Schmidt photographs the star's spectrum. It doesn't resemble the spectrum of a normal star at all, although a number of spectral lines can be distinguished. It is not until February 5, 1963, that Schmidt realizes that they are the spectral lines of hydrogen but with an enormous redshift; the remarkable radio star is apparently a distant galaxy, whose light waves have been strongly stretched by the expansion of the universe.

Other point-shaped radio sources at enormous distances are soon identified, and astronomers call them quasi-stellar radio sources, or quasars for short. The fact that they are relatively easy to observe despite their great distances, means that they must generate enormous amounts of energy. In 1974, Roger Blandford and Martin Rees publish their theory that this energy comes from the immediate vicinity of a supermassive black hole at the core of a distant galaxy. The gas that disappears into one of these black holes is heated to a very high temperature, causing it to emit large amounts of radio waves, visible light and X-rays.

Quasars are visible at very great distances because of their enormous luminosity. Schmidt and his colleagues continue to hunt quasars at record distances from the Earth until the 1990s. In 2000, in data from the Sloan Digital Sky Survey a quasar is discovered at a distance of 12.8 billion lightyears. The light from this distant object displays a redshift of no less than 550%.

Ancient Light

Arno Penzias and Robert Wilson Discover the Afterglow of the Big Bang

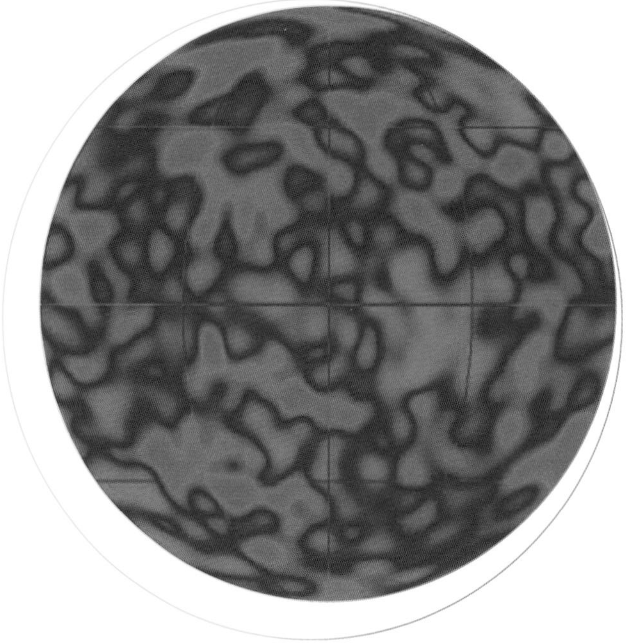

In 1964, American radio engineers Arno Penzias and Robert Wilson are hunting pigeons with an air rifle. The birds have filled the massive horn antenna at Bell Laboratories in Holmdel with poop, and Penzias and Wilson suspect that this is the cause of the mysterious interference that they are picking up. The interior of the antenna has been given a thorough cleaning, but now they have to get rid of the pigeons.

Penzias and Wilson have no idea at all that the microwave interference does not come from bird poop but from the Big Bang. Without knowing what they are looking for, they discover the cosmic background radiation – the cooled off and severely weakened remnant of the high-energy radiation with which the universe was filled a few hundred thousand years after its birth. In 1978, they are awarded the Nobel Prize for Physics for their discovery.

The existence of this background radiation was predicted at the end of the 1940s by George Gamow, Ralph Alpher and Robert Herman. According to Alpher and Herman it is an extremely faint signal, consistent with a temperature slightly above absolute zero, and at this time it seems impossible that the radiation will ever be detected.

In 1964, however, Russian cosmologists Yakov Zel'dovich and Igor Novikov come to the conclusion that it must be possible to detect the background radiation with very sensitive radiometers. That same year, Americans Robert Dicke, David Wilkinson and Peter Roll build a microwave receiver at Princeton University, with which they also set out to find the 'echo of the Big Bang'.

While the search for the cosmic background radiation continues at Princeton without success, the faint signal is detected by accident fifty kilometers to the east, in Holmdel. The publications by Penzias and Wilson and by Dicke's team appear in 1965 in the same issue of *The Astrophysical Journal*. Some astronomers have doubts about the interpretation of the observations at first, but the discovery is very soon accepted by almost everyone as persuasive 'proof' of the Big Bang theory.

In the 1970s, cosmologists predict that the cosmic background radiation must display very slight temperature variations of a ten-thousandth of a degree. They are caused by density fluctuations in the new-born universe, out of which, over the course of billions of years, the current large-scale structure of the cosmos has evolved. In 1990, the American Cosmic Background Explorer (COBE) satellite maps these temperature variations. The COBE map is known as the 'baby photo of the universe'.

In the early years of the twenty-first century, precision observations of the cosmic background radiation, conducted with instruments in space or on high mountain tops on Earth, lead to the current standard cosmological model.

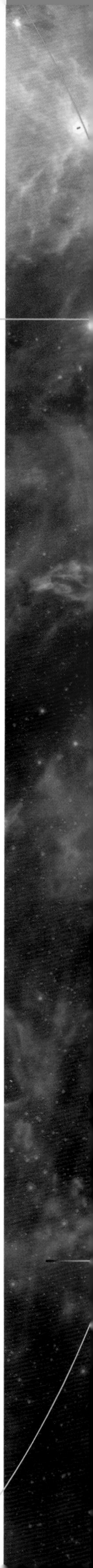

Small temperature variations in the cosmic background radiation were first measured in detail by the Cosmic Background Explorer. They indicate the existence of density fluctuations in the newborn universe, from which in the course of time clusters and superclusters formed. (NASA)

Infrared photo of the star forming region W5, made by the Spitzer Space Telescope. Without the original density fluctuations in the young universe (visible as temperature variations in the cosmic background radiation) stars would never have been born, and there would never have been life in the universe. (NASA/JPL/CfA)

1965

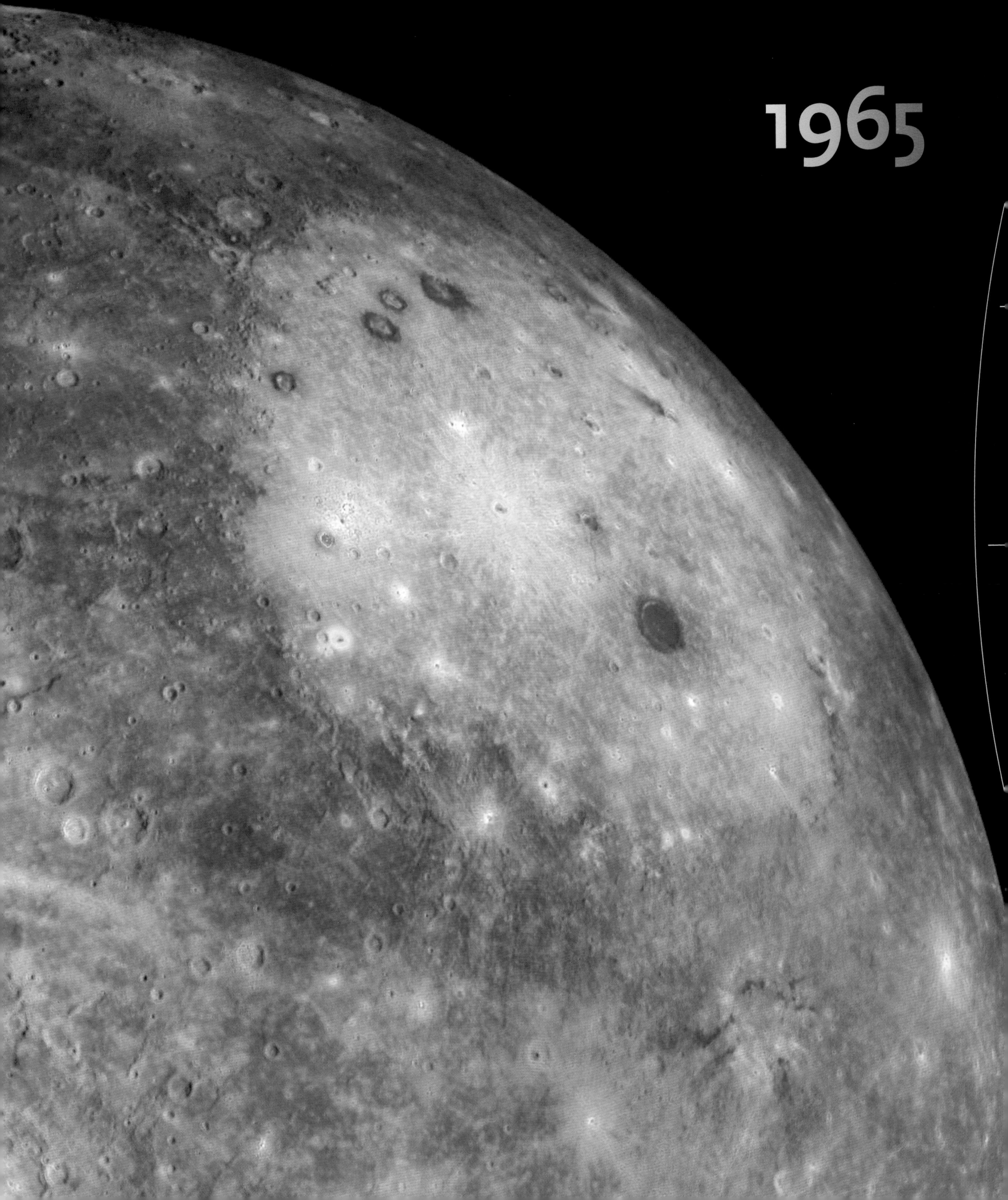

1965

Radar Revelations

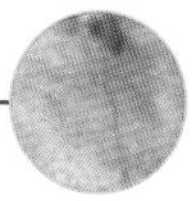

Gordon Pettengill Discovers the Rotation Period of Mercury

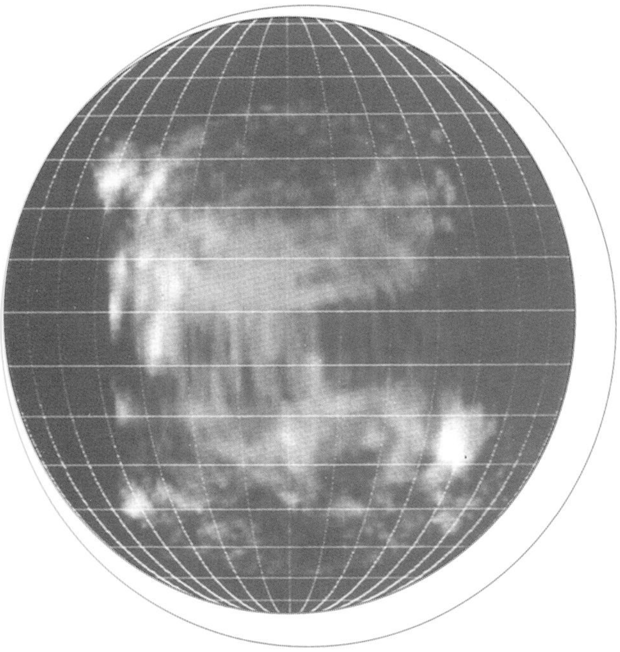

In 1659, Christiaan Huygens is the first to determine the rotation period of another planet when he sees a dark spot on the surface of Mars that disappears and reappears as regular as clockwork. Mars seems to be spinning on its axis every 24.5 hours, and later astronomers measure the rotation periods of the giant planets Jupiter and Saturn in a similar way.

For the small planet Mercury, however, this proves more difficult. Mercury is always more or less close to the Sun in the sky and is less than 5,000 kilometers across. Some astronomers claim to be able to see surface details, but the observations are not conclusive. It is generally assumed that Mercury spins on its axis every 88 days – the same time it takes the planet to complete one orbit around the Sun. That means that Mercury always shows the same face to the Sun, like the Moon to the Earth.

Gordon Pettengill of the Massachusetts Institute of Technology discovers in 1965 that the small planet actually spins on its axis much more rapidly, not by mapping features on its surface but by studying radar echoes from Mercury. With the 300-meter radio dish at Arecibo in Puerto Rico, he sends a powerful radar signal in the direction of the planet. A few minutes later he receives an extremely faint echo back, and doppler shifts in the radar signal provide information on the rotation speed of the surface.

◐ One of the first radar images of Venus, obtained with the aid of American radio telescopes. The light areas (called Alpha Regio and Beta Regio) are mountainous highlands on the planet. (NASA)

◑ The Caloris Basin on the small planet Mercury, photographed by the American planetary explorer Messenger in early 2008. The colors indicate differences in mineralogical composition. The basin was created billions of years ago by a cosmic impact. (NASA/JHU-APL/ASU/CIW)

At the end of the 1950s, Pettengill is one of the first to use radar for astronomical purposes. Radar echoes have already been received from the Moon, but never from another celestial body. That is not so surprising: because of the much greater distance, a radar echo from the planet Venus – the planet that can come closest to the Earth – is ten million times fainter than an echo from the Moon.

In the spring of 1961, Pettengill leads one of the four teams (in the United States, Great Britain and Russia) that succeed, practically simultaneously, in detecting radar echoes from Venus. The measurements confirm earlier suggestions that Venus has a very slow rotation period of 243 days and spins on its axis in the 'wrong' direction. They also make it possible to determine the distance to Venus very accurately. Without that information, the first American interplanetary space probe, Mariner 2, would have passed Venus without acquiring any useful data. After the Arecibo dish has come into operation in 1964, Pettengill and his colleague, Rolf Dyce, aim their radar sights at Mercury. The planet turns out to spin on its axis not in 88 days but in 59 – two-thirds of its orbital period. That means that there is a kind of 'synchronized rotation', but in a completely different way than has always been assumed. Later, Pettengill also conducts radar observations of the asteroid Eros, the nucleus of comet Encke and the large moons of the giant planets.

Optical Illusion

Martin Rees Discovers the Explanation for Superluminal Velocities

Martin Rees is still 23 when he writes his much-quoted article on 'The Appearance of Relativistically Expanding Radio Sources' for *Nature*. That is quite a mouth full to say, but what it boils down to is that Rees predicts superluminal velocities – velocities faster than the speed of light.

The speed of light is first determined by Danish astronomer Ole Rømer in 1676, on the basis of observations of Jupiter's moons. When Jupiter is closer to the Sun, its light takes less time to reach us than when it is much further away. Because of that difference, occultations, transits and eclipses of Jupiter's moons do not always occur at the calculated time, and the measured variations can be used to deduce the speed of light.

Rømer's method is not very accurate, and additionally, distances in the solar system are not yet known very precisely. Later, the speed of light is determined at 300,000 (299,792.458 to be exact) kilometers per second. According to Einstein's special theory of relativity, that is also the highest speed possible in nature. Nothing can move faster than the speed of light.

Yet in 1996, Martin Rees predicts that velocities can be observed in the cosmos that far exceed the speed of light. These superluminal velocities occur when matter moves more or less in the direction of the Earth at a 'relativistic speed' (nearly as fast as the speed of light). That results in a relativistic optical illusion, whereby an observer on Earth sees the matter move sideways at an apparent velocity which can be several times faster than the speed of light.

Four years after Rees' prediction, in 1970, a superluminal velocity is measured for the first time in a distant source of radio waves. The observations are made using the Very Long Baseline Interferometry (VLBI) method, which involves linking radio telescopes at very great distances from each other to obtain extremely sharp images.

By 2010, countless superluminal radio sources have been identified. In nearly all cases, they are high-energy jets of electrically charged particles expelled into space in two directions from the cores of radio galaxies and quasars, where they emit radio waves. When one of these jets is more or less aimed at the Earth, superluminal speeds are observed.

Rees' explanation for this cosmic illusion is actually very simple. A gas cloud that emits radio waves and comes towards us at a small angle at almost the speed of light travels, say, 98 lightyears in 100 years. At the end of that period the travelling time of the light is therefore 98 years shorter. It therefore seems that its sideways motion in the sky has only taken 2 years instead of a hundred, and this creates the illusion of an extremely high speed.

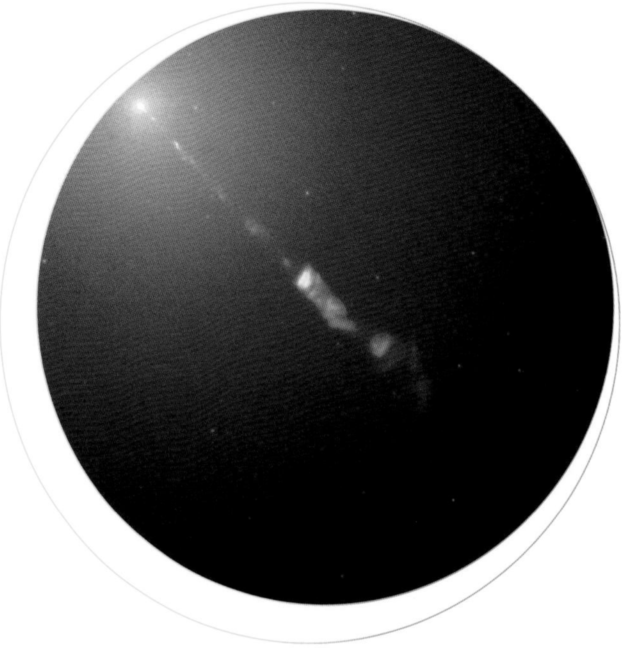

↑ The Hubble Space Telescope took this photograph of the high-energy jet of the galaxy M87, at a distance of 50 million lightyears in the constellation of Virgo. Superluminal velocities have also been registered in this jet of electrically charged particles, which is directed towards the Earth at a small angle. (NASA/ESA/Hubble Heritage Team)

→ Superluminal velocities are primarily observed in quasars. This illustration shows how, in the high-energy particle wind of a quasar, dust particles can be formed with diverse mineral compositions, including microscopic precious stones, like sapphires and rubies. (NASA/JPL)

1966

1967

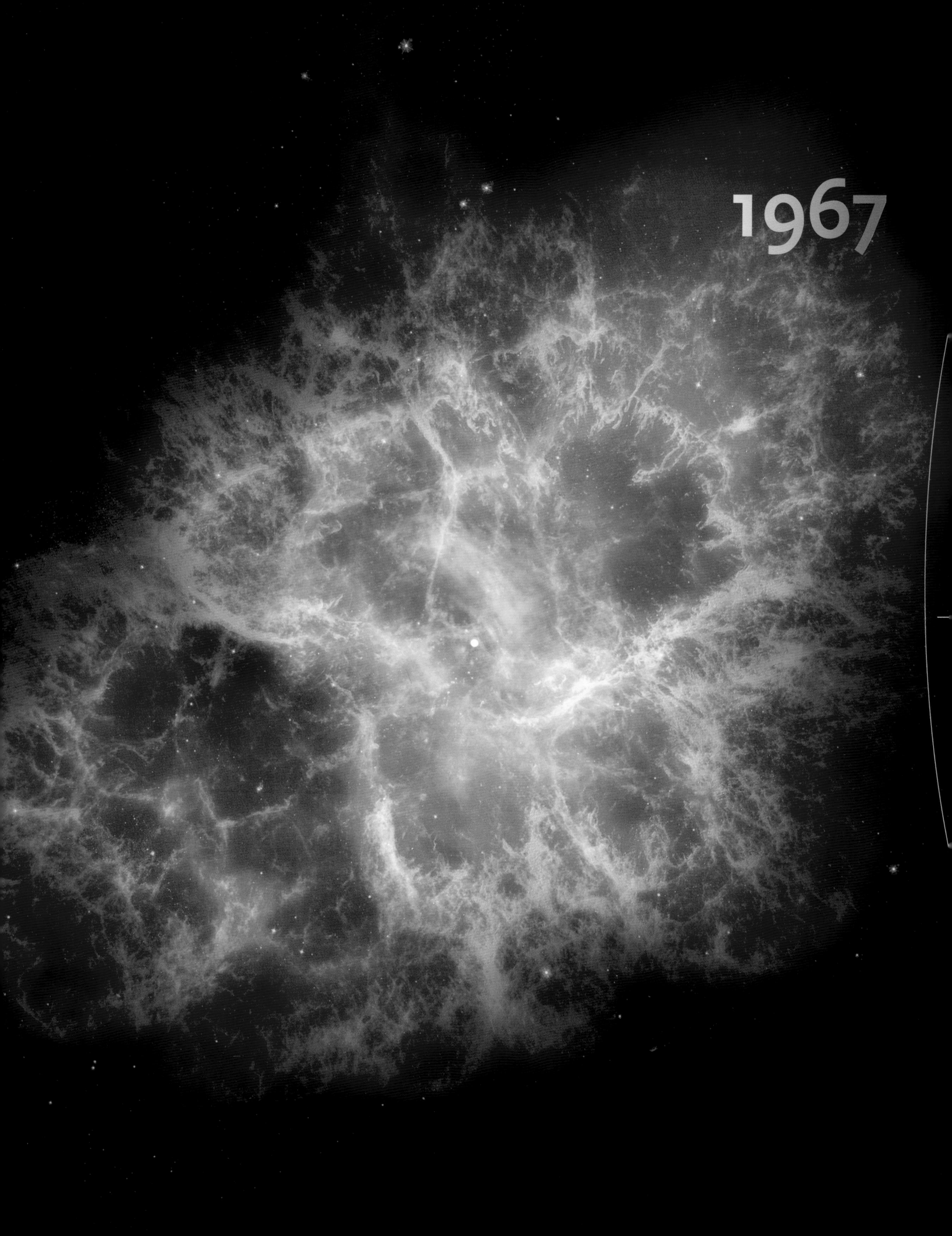

Extraterrestrial Beeps

Jocelyn Bell Discovers Pulsars

When Jocelyn Bell discovers regular radio beeps from the cosmos in 1967, her Ph.D. supervisor, Antony Hewish, gives serious thought to the possibility that they are artificial signals from an extraterrestrial civilization. That does not last long, however, because before the year is out, more of the beeping signals are discovered, and it quickly becomes clear that they are spinning neutron stars.

Bell grows up in Belfast where her father is an architect and is involved in the construction of the Armagh Planetarium in Northern Ireland. She becomes fascinated by the universe, studies physics at the University of Glasgow, and conducts her Ph.D. research under radio astronomer Hewish at Cambridge. In the same year, she is in put in charge of a remarkable radio telescope specially built to hunt quasars.

The 'telescope' covers an area of almost 2 hectares and consists of more than a thousand poles, between which a total of 20,000 kilometers of antenna wire is hung. The intention is to trace quasars and other point-like radio sources. The signals are then registered on a roll of paper with a pen recorder. Bell analyzes the results – 30 meters of paper a day – by hand and by eye.

At the end of August, she sees an anomalous signal – the pen marks on the paper do not look like the pattern you would expect from a quasar, or from interference caused by terrestrial radio sources. Even more strange each time the signal seems to come from the same direction in the sky.

In November, Bell installs another pen recorder with a higher paper speed. It is now clear to see that the signal consists of short pulses, at regular intervals of 1.33 seconds. At first, Hewish is convinced that it must be a terrestrial signal because no known celestial body displays such rapid variations. In December, however, Bell discovers another three similar radio sources, with slightly different pulsation periods. One of them corresponds to the already known radio source, Cassiopeia A. These remarkable pulsating stars are given the name pulsars.

At the start of 1968, the results are submitted for publication in *Nature*. Just before the article appears, Hewish presents the discovery at a symposium in Cambridge. British astronomer Fred Hoyle suggests that the mysterious pulsars are the remnants of supernova explosions: very rapidly rotating neutron stars, extremely small in size but with exceptionally high density. (The existence of neutron stars was already predicted in 1933 by Walter Baade and Fritz Zwicky; it took 44 years for them to be discovered.)

Antony Hewish is presented with the Nobel Prize for Physics in 1974 for this discovery of pulsars. Many astronomers still believe, however, that the prize should actually have gone to his student, Jocelyn Bell.

◐ Some pulsars are surrounded by disks or rings of dust particles, the heat radiation of which can be detected with an infrared telescope like the Spitzer Space Telescope. The dust comes from the exploded star and will, in the course of time, fall back towards the pulsar again. (NASA/JPL)

◐ The Crab Nebula, in the constellation of Taurus, was created nearly a thousand years ago by the supernova explosion of a massive star. The core of the star imploded to become a compact, rapidly spinning neutron star, emitting pulses of light and radio waves – a pulsar. (NASA/JPL/ESA/CXC)

Military Spin-Off

Ray Klebesadel Discovers Gamma-Ray Bursts

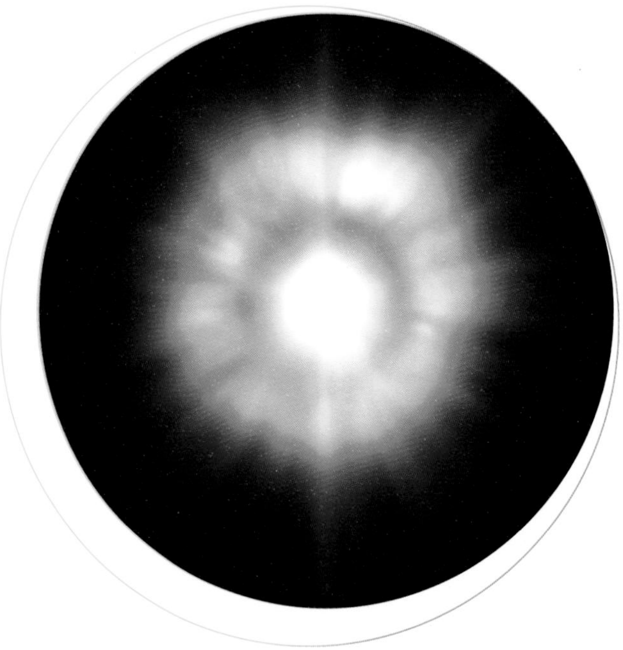

Gamma-ray bursts are the most powerful explosions in the universe. They emit as much energy in a fraction of a second as the Sun in ten billion years, and they are observable over distances of billions of lightyears. Every day, one or two occur somewhere in the cosmos, and you never know in advance exactly where and when, and they cannot be seen from the Earth: cosmic gamma rays don't penetrate the Earth's atmosphere and can only be observed from satellites.

Gamma-ray bursts are discovered by secret military satellites. Vela 4A and 4B are launched in April 1967, and like earlier Vela satellites, they are designed to search for traces of forbidden nuclear tests by the Soviet Union. Enormous amounts of high-energy gamma radiation are released during nuclear explosions, and the detectors on board the Vela satellites record the resulting gamma rays.

In March 1969, Ray Klebesadel and Roy Olson of Los Alamos National Laboratory are working their way through enormous piles of observational data from Vela 4A and 4B. They discover that on July 2, 1967, both satellites registered a remarkable double burst of gamma rays. It looks completely different to the gamma signal you would expect from a nuclear test nor can the burst have come from a supernova explosion or a solar flare.

Klebesadel and Olson decide to keep their unexpected discovery to themselves until the Vela 5 and Vela 6 satellites have also been launched. These are much more sensitive and can also determine from what direction the gamma-ray explosions are coming. Eventually, Klebesadel, Olson and their colleague Ian Strong discover sixteen bursts, which are distributed randomly across the sky.

Klebesadel, Olson and Strong finally announce their discovery on June 1, 1973. The true nature of the high-energy explosions, however, remains an unsolved mystery for more than twenty years. Over the course of time, gamma detectors on board balloons and satellites record hundreds of gamma-ray bursts. In particular, NASA's Compton Gamma Ray Observatory, a gigantic gamma-ray satellite launched in April 1991, conducts a lot of research into the strange explosions.

Nothing is known, however, about how far away the gamma-ray bursts are. Their celestial positions cannot be determined accurately enough, and the short-lived bursts cannot therefore be identified with any known celestial bodies. They could be explosions on stars in our own Milky Way, or extremely high-energy explosions in remote galaxies. There is something to be said for both options.

On April 22, 1995, a public debate on the nature and distance of gamma-ray bursts is organized in Washington, in the same room where Harlow Shapley and Heber Curtis crossed swords on the distance to spiral nebulae 73 years previously. The mystery of the gamma-ray bursts is not solved until 1997 when Dutch astronomers Paul Groot and Titus Galama show that they are indeed the most powerful explosions in the cosmos.

Illustration of a gamma-ray burst. Gamma-ray bursts are the most powerful explosions in the universe since the Big Bang. In less than a second, they produce as much energy as the Sun in ten billion years. They occur when massive, rapidly rotating stars implode into black holes. (ESA/ECF)

In the core of the Milky Way, seen here on an image produced by the Spitzer Space Telescope, there are a number of extremely bright and massive stars – many times more massive and millions of times brighter than our Sun. At the end of their short lives, these stars will probably generate gamma-ray bursts. (NASA/JPL)

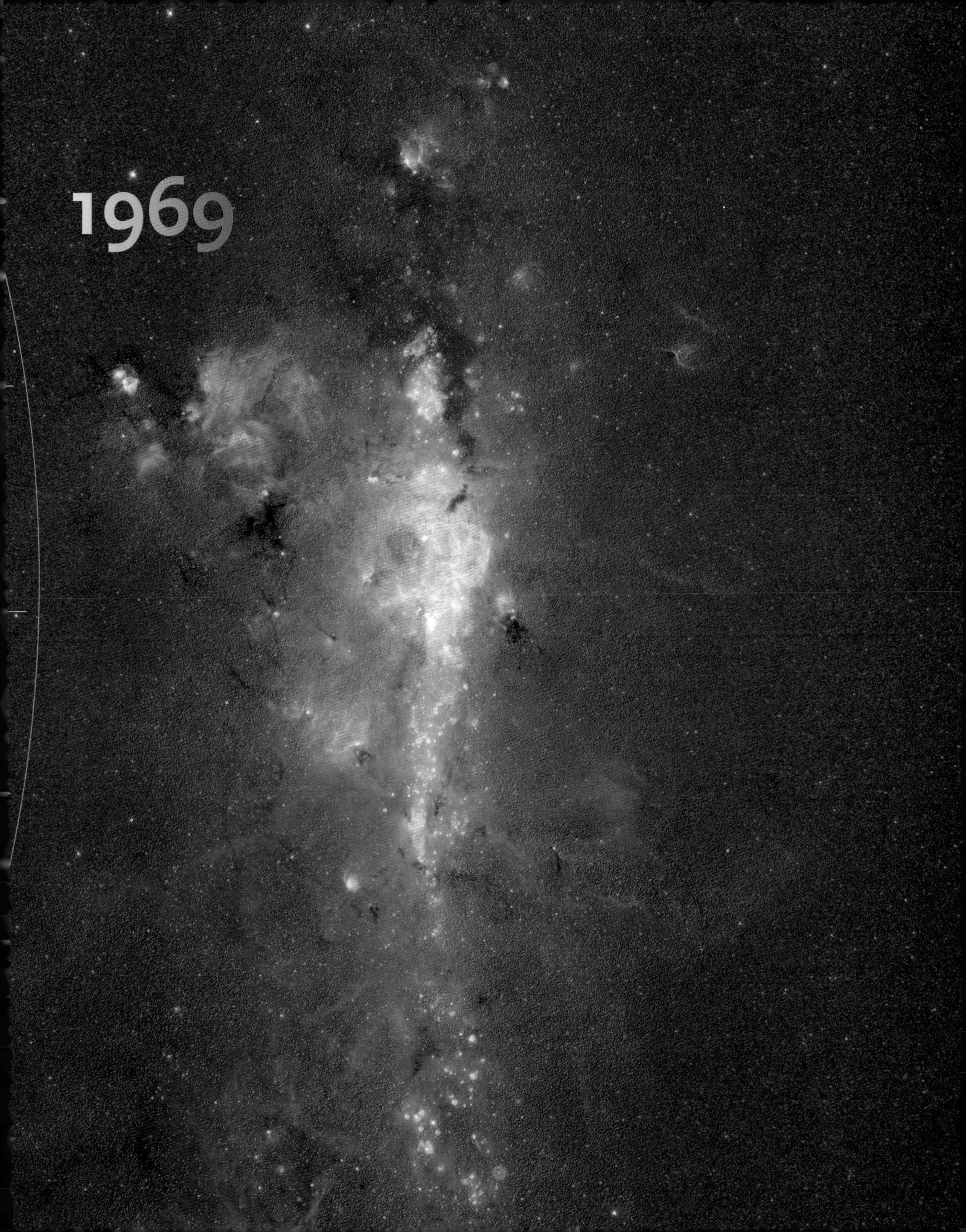
1969

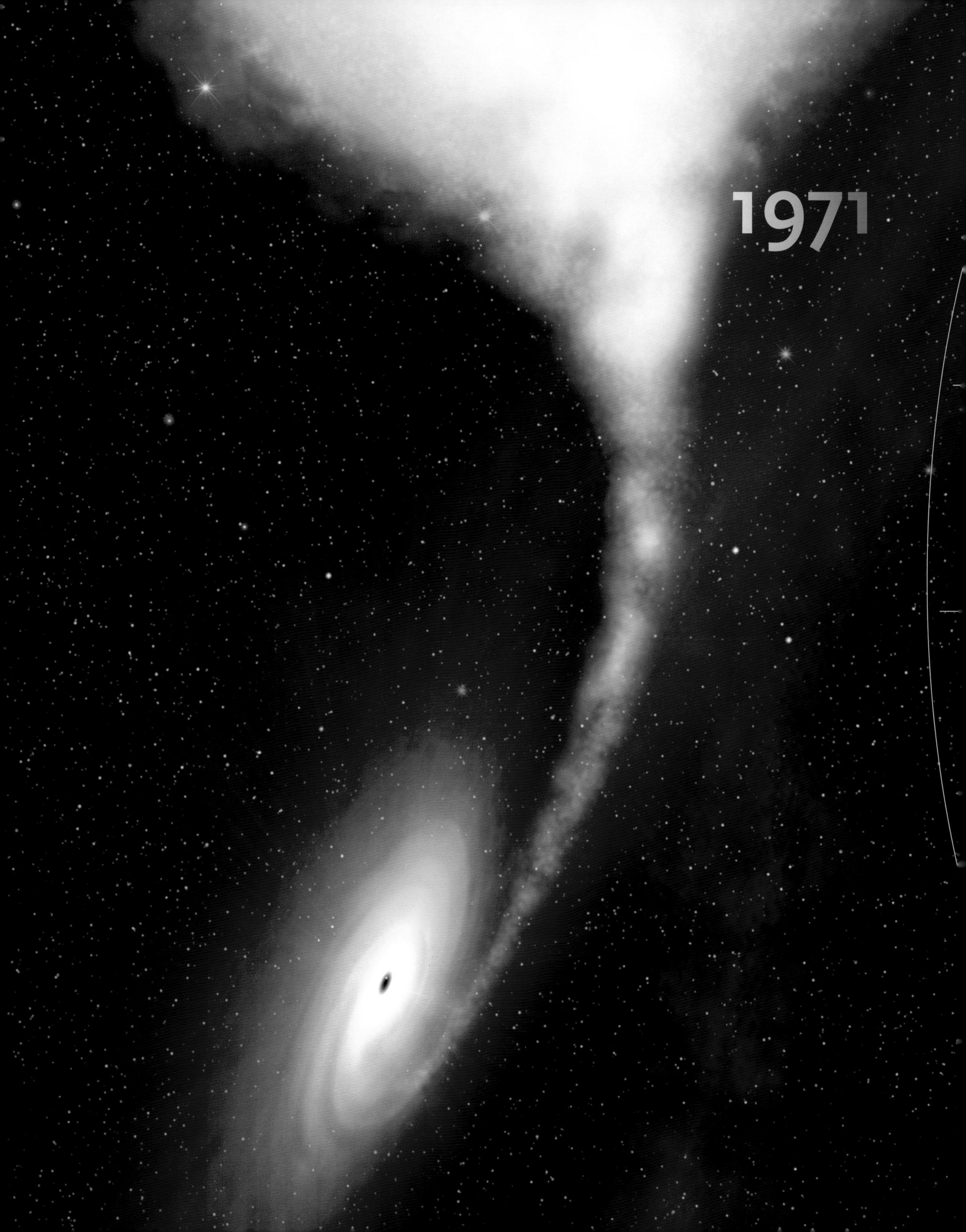

1971

Bright Shell, Black Core

Louise Webster and Paul Murdin Discover that Cygnus X-1 Is a Black Hole

With a Geiger counter on board a rocket, Riccardo Giacconi discovers the first cosmic X-ray source, Scorpius X-1, in 1962. Many more follow soon after. In 1964, once again using an Aerobee sounding rocket launched from White Sands Missile Range in New Mexico, a team led by Herbert Friedman discovers a bright source of X-rays in the constellation of Cygnus, which is given the name Cygnus X-1.

With the launch of the Uhuru satellite in December 1970, Giacconi and Friedman take a significant step towards identifying Cygnus X-1. In 1971, the satellite detects rapid variations in the X-ray brightness of the high-energy source, indicating that it must be relatively small. Later that same year, astronomers discover a source of radio waves in the same part of the sky, which changes in brightness in exactly the same way. The position of the radio source is measured accurately and proves to be identical to that of a variable star at around 6,000 lightyears distance.

The star, known as HDE 226868, is a blue supergiant about thirty times more massive than the Sun, twenty times larger, and a few hundred thousand times brighter. Although the supergiant has a surface temperature of more than 30,000 degrees Celsius, it cannot be the source of the observed X-rays. Is there perhaps an 'X-ray' star in orbit around it?

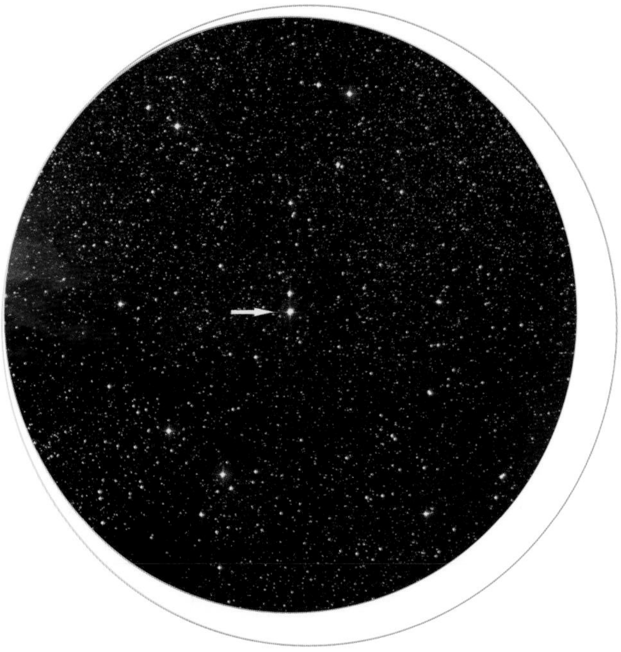

↗ The arrow on this photograph of the constellation of Cygnus points to the inconspicuous star HDE 226868. This massive star is accompanied by an invisible stellar black hole. Gas flowing from the star towards the hole produces powerful X-rays. (ESA/ECF)

→ Illustration of Cygnus X-1. Gas from the hot, blue giant star HDE 226868 flows towards the black hole, where it accumulates in a rotating accretion disk before disappearing into the hole. The disk is so hot that it emits high-energy X-rays. (NASA/ESA/M. Kornmesser)

In the fall of 1971, Louise Webster and Paul Murdin of the Royal Greenwich Observatory in London photograph the spectrum of the star for several weeks. The spectral lines turn out to shift a little back and forth every 5.6 days meaning means that the star alternately moves towards us and then away again. Doppler measurements show that HDE 226868 is accompanied by another object that must be more than eight times more massive than the Sun.

The companion cannot be a normal star; a star weighing eight solar masses would simply have to be visible. Nor can it be a compact neutron star because neutron stars cannot be more massive than three solar masses. Webster and Murdin conclude that the invisible companion must be a stellar black hole. The gas from the blue supergiant flows towards the black hole, accumulates in a swirling disk, and emits an enormous amount of X-rays.

The two British astronomers publish their observations in *Nature* in early January 1972. Canadian astronomer Thomas Bolton comes to the same conclusion, independently of Webster and Murdin, and his article appears a month later.

Not everyone is immediately convinced of the existence of black holes. In 1975, the renowned British physicist Stephen Hawking makes a bet with his American colleague Kip Thorne; Hawking does not believe that Cygnus X-1 is a black hole. But the evidence keeps on piling up and, in 1990, Hawking admits defeat and Thorne wins an annual subscription to the monthly *Penthouse*.

Asymmetric Expansion

Vera Rubin and Kent Ford Discover the Proper Motion of the Milky Way

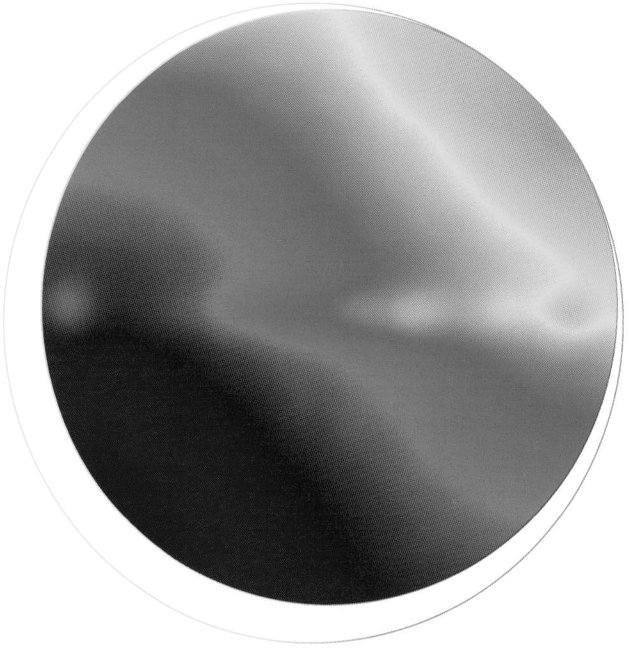

As a young girl, Vera Rubin is already fascinated by the night sky. At night she looks out her bedroom window and sees how the stars move around the Pole Star. At 14, she builds her first telescope and resolves to become an astronomer. She studies at Cornell University, with famous physicists like Richard Feynman and Hans Bethe, and does Ph.D. research under the supervision of cosmologist George Gamow, the father of the Big Bang theory.

In December 1950, Vera gives her first presentation at a meeting of the American Astronomical Society (AAS). She is 22 and has just given birth to her first son. She addresses a hall full of prominent astronomers on her ideas about the rotation of the universe. Her statistical study of cosmic redshifts indicates that observed galactic motions cannot exclusively be attributed to the expansion of the universe, but also suggest rotation on a grand scale.

The distinguished audience stop just short of laughing Rubin out of the hall. It is a controversial argument, with little scientific foundation, and no one takes her seriously. After being awarded her Ph.D. in 1954, Rubin takes a break from research. She starts teaching and devotes a lot of energy to raising her family.

A quarter of a century after her AAS presentation, however, Rubin returns with a similar claim. Together with her colleague Kent Ford of the Carnegie Institution in Washington, where Rubin has been working since 1965, she measures the recessional velocities of dozens of spiral galaxies distributed throughout the cosmos. The distances of these galaxies are deduced from their observed brightness, on the assumption that they all have the same actual luminosity. By then comparing distance and speed, Rubin and Ford come to the conclusion that the expansion of the universe does not appear to occur equally fast in all directions. It is as though the Local Group of galaxies, to which the Milky Way belongs, has a proper motion with respect to the expansion of the universe.

According to Rubin and Ford, that proper motion is about 450 kilometers per second. They publish their results in 1976, and again meet with much disbelief and criticism, especially regarding the distance calculations and their statistical significance. No one wants to believe that there are such large-scale motions in the universe.

Three years after publication of the 'Rubin-Ford effect', as the asymmetry of the expansion of the universe is called, George Smoot of the Lawrence Berkeley National Laboratory discovers that the cosmic background radiation displays a similar asymmetry: it is warmer in one direction than in the other, and that, too, is the consequence of the motion of the Milky Way. The measurements of the cosmic background radiation make it possible to determine the proper motion of the Local Group much more accurately: it is moving at more than 600 kilometers per second relative to the expansion of the universe.

↑ The cosmic background radiation – the 'echo' of the Big Bang – is warmer in one direction than in the other. That minute temperature difference, shown here in different colors, is caused by the motion of the Milky Way through the universe. (NASA)

→ Because of their spatial motion, galaxies occasionally collide. NGC 520, 100 million lightyears away, is an example of such a cosmic collision, estimated to have occurred some 300 million years ago. (NASA/ESA/Hubble Heritage Team)

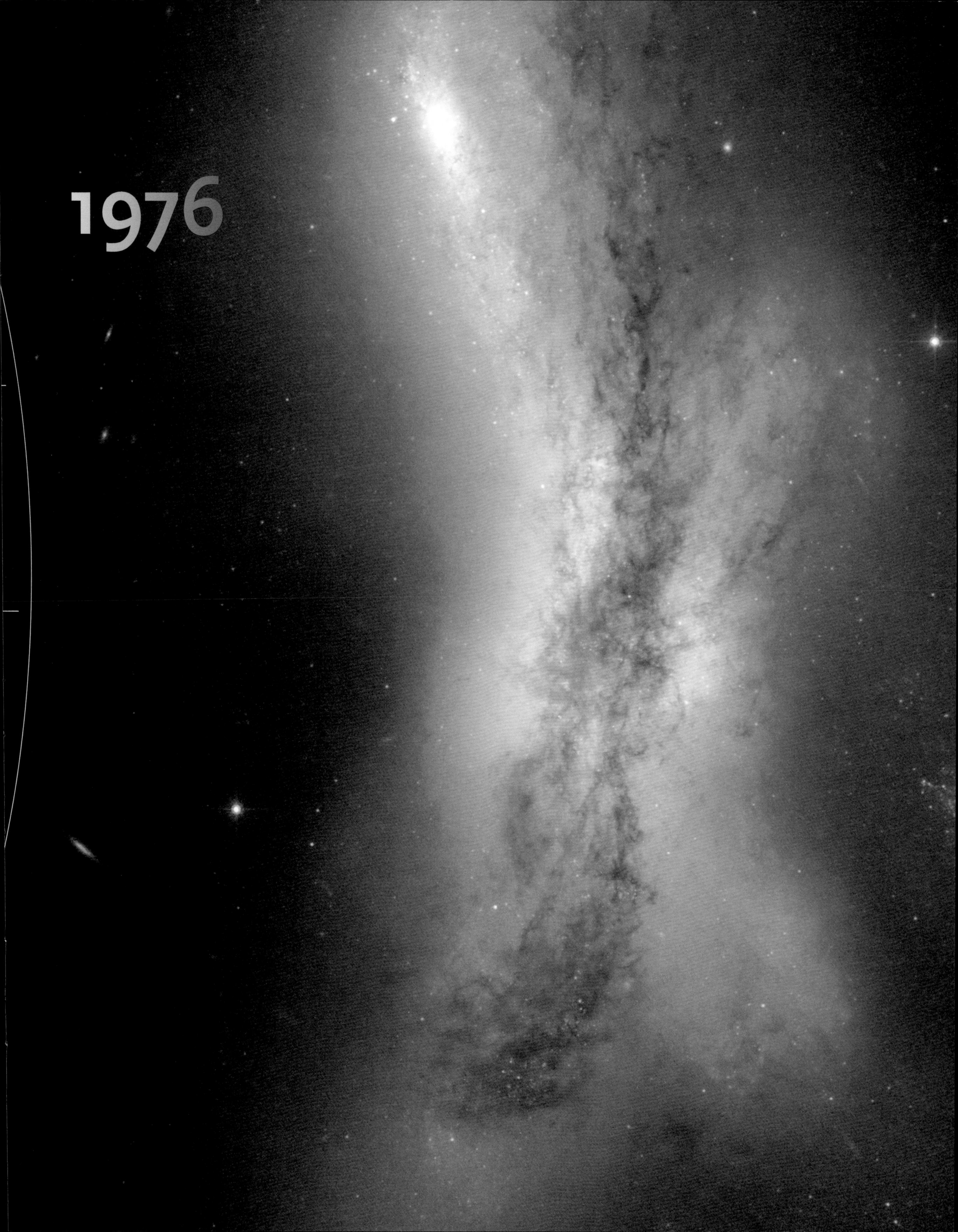
1976

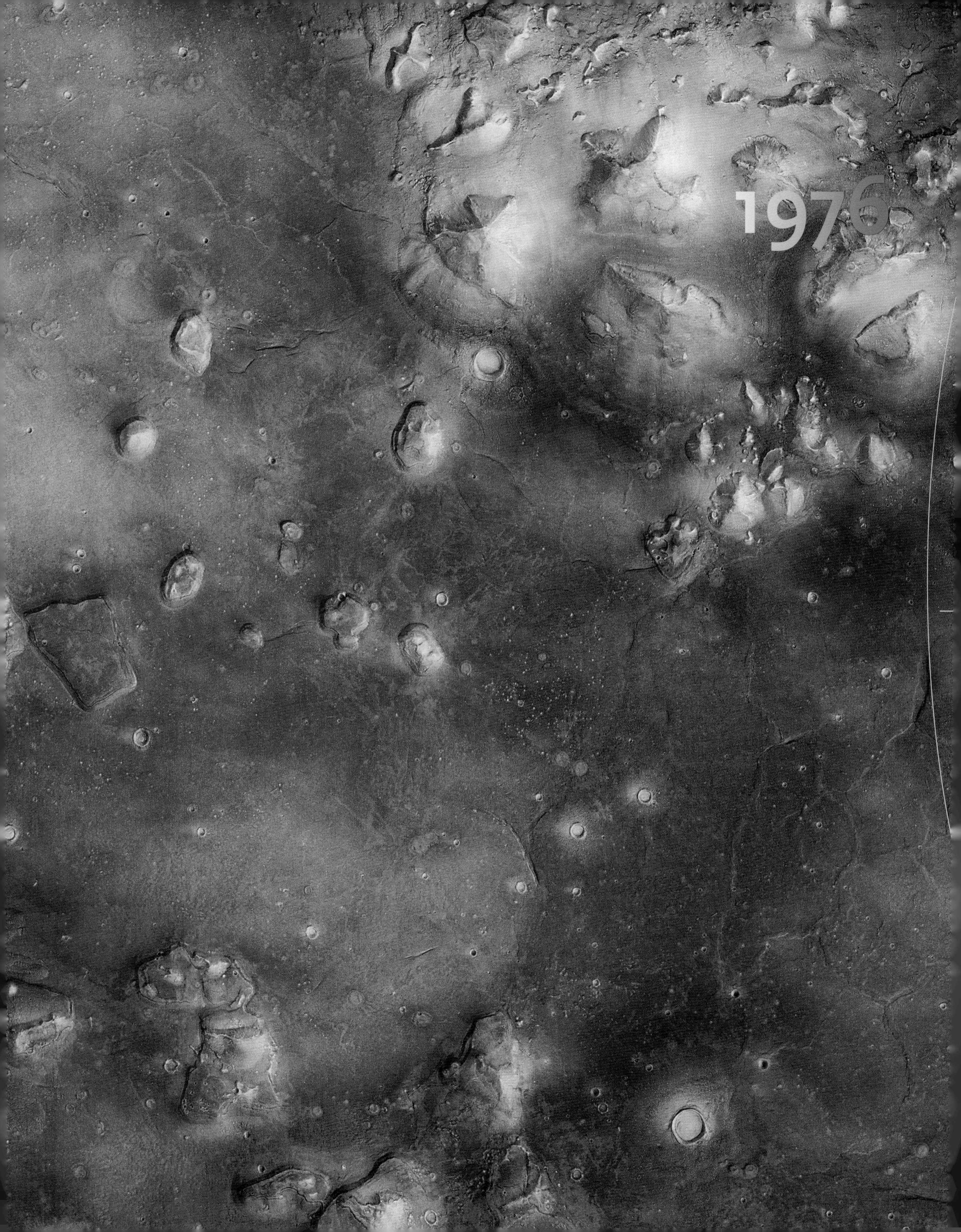
1976

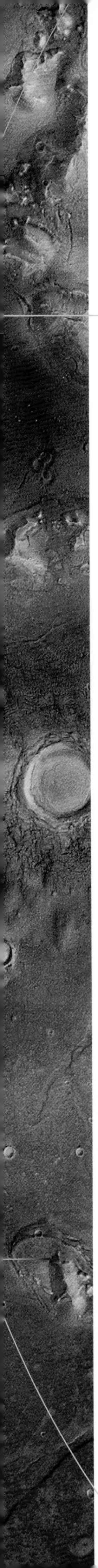

Striking Similarity

Viking 1 Discovers the Face on Mars

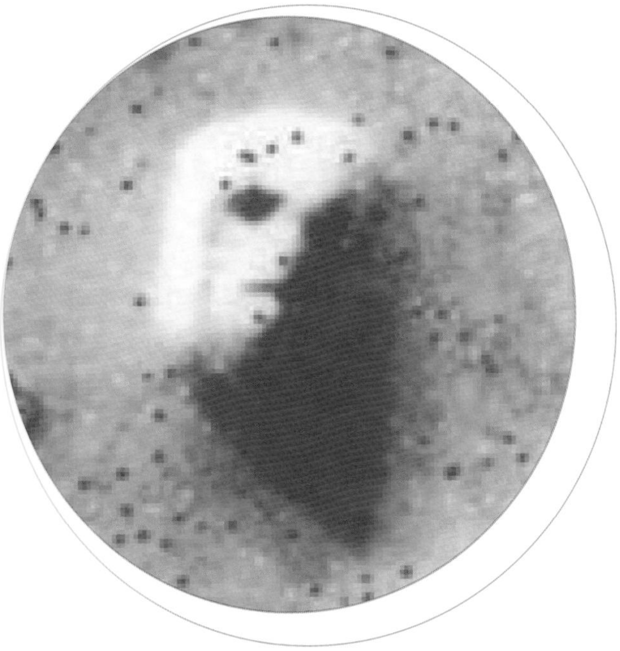

The American Viking project, conducted in the 1970s, is one of the most ambitious planetary research programs of the twentieth century. Two large space probes are put into orbit around Mars, and both send landers to the surface, which descend with the help of parachutes and retro-rockets. The Viking landers search for micro-organisms (without success), while the motherships make detailed maps of the planet.

Viking 1 is launched on August 20, 1975, and enters orbit around Mars on June 19, 1976. The landing in Chryse Planitia takes place on July 20. The Jet Propulsion Laboratory in California – NASA's flight control center for unmanned planetary research – issues daily press releases with spectacular new photos of the red planet, both from the surface and from space.

On July 30, NASA issues a press release with an extraordinary photograph taken six days earlier by the Viking 1 Orbiter from an altitude of just under 2,000 kilometers. Remarkable flat-top mesas can be seen in the Cydonia region, one of which bears a striking resemblance to a human face. The hill is about one and half kilometers across.

A year later, computer scientists Vincent DiPietro and Gregory Molenaar of the Goddard Space Flight Center claim that the 'Face on Mars' is probably not a natural formation, but a gigantic sculpture created by a lost Martian civilization. DiPietro and Molenaar interpret other geological structures in Cydonia as pyramids and ruined cities. They claim that their conclusions are born out by computer-enhanced images of the original Viking 1 photographs.

Journalist Richard Hoagland gives a great deal of publicity to DiPietro and Molenaar's ideas as hundreds of articles, dozens of books and countless websites discuss the Face on Mars. NASA explains time and time again that it is simply an optical illusion caused by a trick of the light, but is accused of trying to cover up the discovery of extraterrestrial life. A new conspiracy theory is created.

In the early twenty-first century, the Face on Mars is finally photographed in greater detail, including by the Mars Orbiter Camera on board the Mars Global Surveyor probe, the High Resolution Stereo Color camera on board the European Mars Express (which even makes a 3D image), and the High Resolution Imaging Science Experiment, part of NASA's Mars Reconnaissance Orbiter. The new photographs are nearly 20 times sharper than the Viking image. They show an intriguing, but undoubtedly natural, flat-top mountain, but Hoagland and his followers are, however, not convinced; pseudo-science clearly captures the imagination better than geology.

➋ Enlargement of the original Viking image of the Face on Mars. The many black pixels are caused by interference in the radio connection; one of them happens to be exactly where you would expect a nostril. (NASA/JPL)

➌ Color photograph of the Cydonia region, made by the European planetary explorer Mars Express. The 'Face on Mars' can be seen left of center (upside down). Many other strange rock formations can be seen in the same area. (ESA/DLR/G. Neukum)

Hybrid Intruder

Charles Kowal Discovers the First Centaur, Chiron

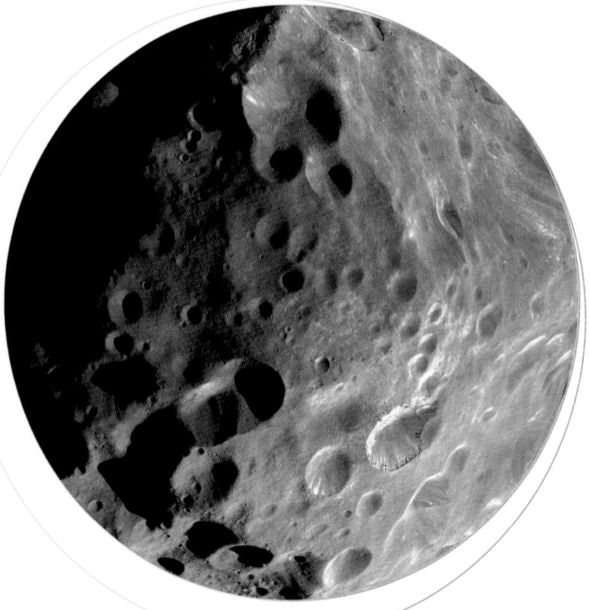

Charlie Kowal is 16 when he leaves his parent's house in the state of New York in 1957, and goes to study astronomy in sunny California. He is crazy about the stars, and has set his sights on a job at Mount Wilson Observatory; his wish comes true four years later. Practically every night, Kowal can be found at the eyepiece of one of the telescopes on Mount Wilson or on Palomar Mountain. He measures the brightnesses of stars, and at the request of Fritz Zwicky, searches for supernova explosions in distant galaxies.

On the photos that he makes with the 1.2-meter Schmidt Telescope at Palomar Observatory, Kowal also discovers asteroids and comets, becoming interested in the solar system. In the mid-1970s, he finds two small moons around the giant planet Jupiter, and in the summer of 1977, on his own initiative, he starts searching for an unknown object beyond the orbit of Pluto. What Clyde Tombaugh succeeded in doing in 1930 – discovering a new planet – must be possible again with larger telescopes and better photographic emulsions.

Kowal's hunt for Planet X ultimately lasts eight years, and he never finds an object beyond Pluto. Yet right at the beginning of his search, in October 1977, he discovers a remarkable object in the constellation of Aries. 'Object Kowal', as it is called for some time, follows an elongated orbit that extends from just inside the orbit of Saturn to close to that of Uranus and has an estimated diameter of 140 kilometers.

Kowal realizes that there are probably more hybrid objects like this – large, asteroid-like bodies in comet-like orbits in the outer regions of the solar system. He calls them centaurs, after the half-horse, half-human beings from Greek mythology, and object Kowal is given the name Chiron.

Orbital calculations show that Chiron must have passed close to Saturn in 1664. In the future, its orbit will undoubtedly be disturbed again and it will either end up in the inner solar system or will be flung into space. In 1988, Chiron suddenly becomes a lot brighter; the remarkable object displays a comet-like explosion and is shrouded in a cloud of gas and dust. It is given the asteroid number 2060, and the cometary designation 95P.

The second centaur is not discovered until January 1992, by David Rabinowitz using the University of Arizona's Spacewatch Telescope. This object is given the name Pholus, after another centaur. Pholus, too, moves in an elongated orbit, which is also highly inclined to the plane of the ecliptic. It is larger than Chiron and has a striking red color, possibly caused by the effect of sunlight on its icy surface.

Dozens of centaurs have since been found. They come from the Kuiper Belt, beyond the orbit of Neptune, and their stay in the outer regions of the solar system is temporary. The existence of the belt is ultimately demonstrated in August 1992, when the first object beyond the orbit of Pluto is discovered.

The small Saturnian moon Phoebe is probably a captured centaur – an irregularly shaped lump of ice and rock, originating from the outer regions of the solar system. Phoebe is estimated to be one and a half times larger than Chiron, the first centaur to be discovered. (NASA/JPL/SSI)

Chiron must have passed close to Saturn in 1664, and will do so again in the distant future. This artist's impression shows the icy object starting to evaporate a little under the influence of the weak sunlight, like a comet. (William K. Hartmann)

1977

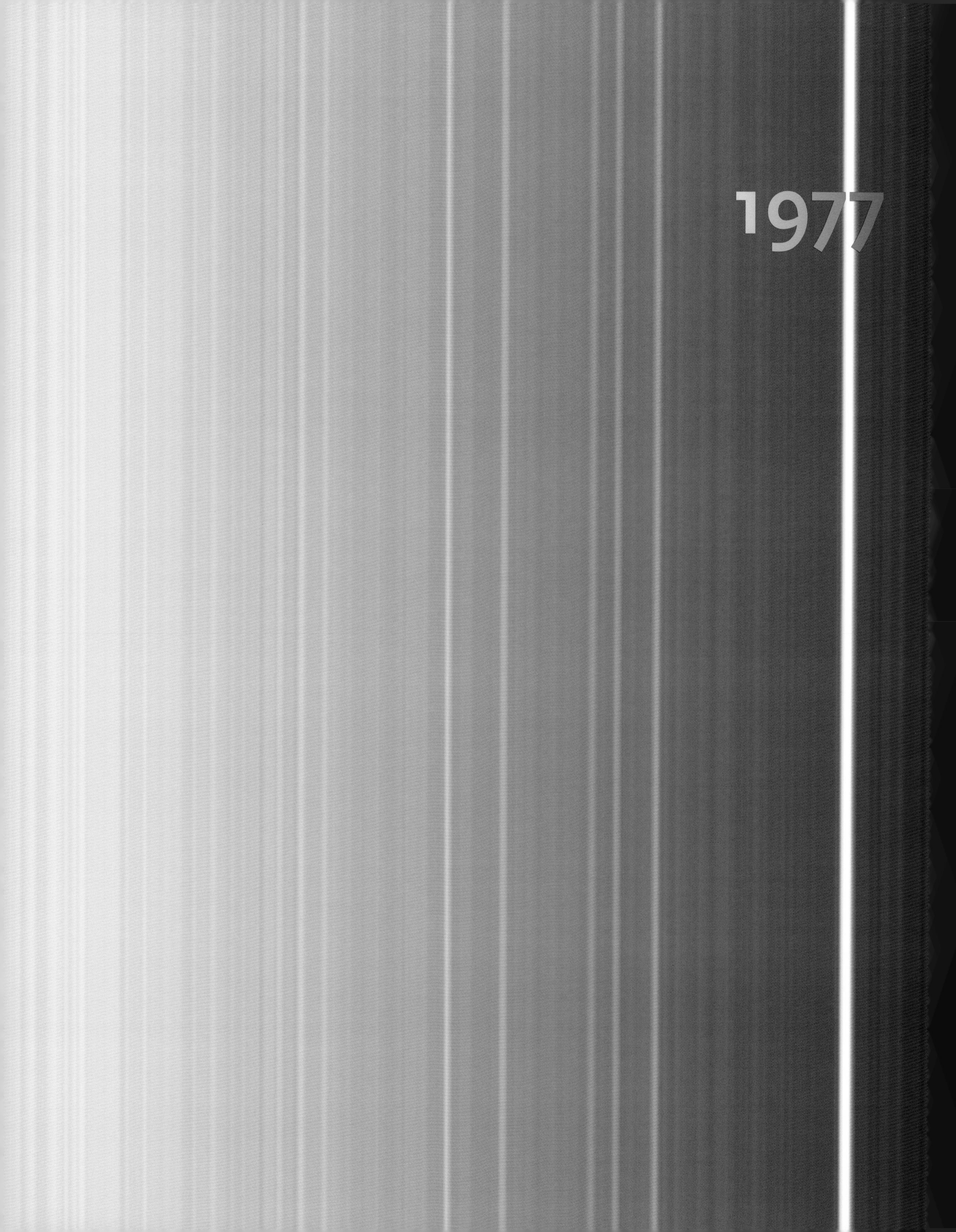

1977

Skinny Hoops

Jim Elliot and his Colleagues Discover the Rings of Uranus

Eight years after William Herschel discovers the planet Uranus, he thinks he can see a ring around the planet. On February 22, 1789, he makes a sketch of it, which is not published until 1797. Herschel's observations are never confirmed by anyone else, and astronomers therefore assume that he imagined the ring.

Nearly 200 years later in 1977, Uranus' ring system is discovered by coincidence. On March 10, three astronomers from Cornell University – Jim Elliot, Ed Dunham and Doug Mink – observe the occultation of a star by Uranus. During the occultation, the star disappears behind Uranus, and the starlight shines briefly through the planet's atmosphere which enables astronomers to learn a lot about the composition and density of the atmosphere.

Because the occultation is only visible from an area on Earth where there are no major observatories, Elliot, Dunham and Mink take a flight on the Kuiper Airborne Observatory, a NASA cargo plane converted into a flying observatory. It has a 90-centimeter telescope on board, named after the Dutch-American planetary researcher Gerard Kuiper.

Just before the star (SAO 158687) is occulted by Uranus, they see it 'flip' off and on again five times. After the occultation, the same thing happens again – Uranus appears to be surrounded by five thin rings which momentarily blocked the starlight, making Uranus the second planet to have rings after Saturn. The astronomers publish their discovery in *Nature* on May 26. A detailed analysis of the measurement data later reveals another four narrow rings, and the rings are designated with numbers and Greek letters.

On January 26, 1986, the American planetary explorer Voyager 2 flies by Uranus at a short distance and close-ups are made of the planet and its moons for the first time. Voyager 2 photographs the ring system, finds another two as yet unknown rings and also discovers a number of small moons. Two of them, Cordelia and Ophelia, flank the Epsilon ring. The gravitational pull of these two 'shepherd moons' ensures that the material in the ring is not dispersed into space.

The rings of Uranus consist of relatively large, very dark lumps of rock, although some parts of the system also contain finer dust particles. The rings were probably created when a small moon was pulverized during a cosmic collision and are estimated to be a few hundred million years old.

Between 2003 and 2005, the Hubble Space Telescope discovers another two very thin rings of dust, at a great distance from the planet, bringing the total number of Uranian rings to thirteen. The outer of the two rings is strikingly blue, while the inner one is very red. It is, however, impossible that any of these rings were the one that William Herschel saw more than 200 years ago.

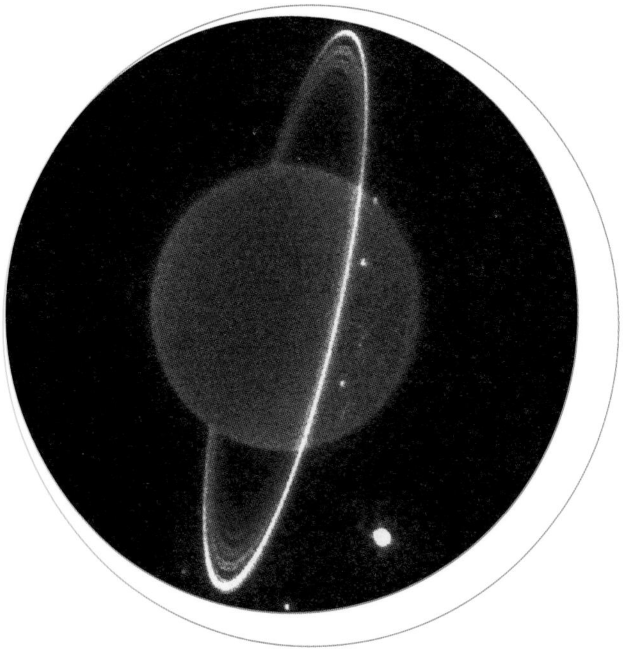

The ring system of Uranus is captured by a sensitive infrared camera on the Keck II Telescope on Mauna Kea, Hawaii, which detects the slight heat radiation of the dust particles in the rings. (SSI/University of California, Berkeley)

Color image of the rings of Uranus, composed from photographs taken by the American planetary explorer Voyager 2 in January 1986. The subtle color differences between the individual rings are enormously enhanced here; they tell us something about the composition of the ring particles. (NASA/JPL)

Double Planet

Jim Christy Discovers the Plutonian Moon Charon

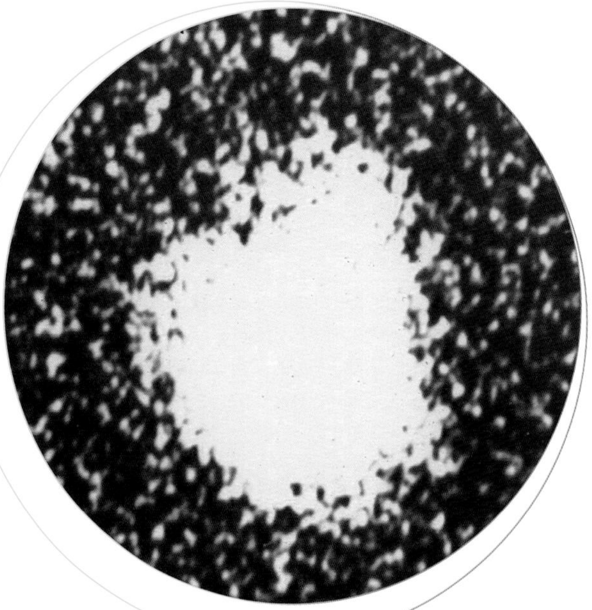

Jim Christy could have discovered Pluto's moon back in 1965. Christy was working at the U.S. Naval Observatory in Flagstaff, Arizona, where he was an expert in analyzing photographic plates. His colleague, Otto Franz, of the neighboring Lowell Observatory had made some photos of Pluto, but the small speck of light didn't seem to be completely round. Franz asked Christy to take a look, but he didn't notice anything out of the ordinary except that the plates were a little over-exposed.

Thirteen years later, Christy is working in Washington at the headquarters of the Naval Observatory. In the summer of 1978, his boss Bob Harrington asks him to analyze images of Pluto made earlier that year in Flagstaff. On June 22, Christy discovers that the grainy pictures of Pluto show strange protrusions, in different directions at different times, while the stars on the photos are all perfectly round dots. The following day he gets out the old photographic plates made by Franz and they show the lump, too. Christy concludes that there is a moon in orbit around Pluto.

The discovery of Pluto's moon is announced on July 7. Its orbital period allows Pluto's mass to be calculated. This is the first conclusive evidence that it is a small, low-mass dwarf planet, and can never have been responsible for the orbital perturbations of Uranus and Neptune, as Percival Lowell suspected in the early twentieth century. Christy calls the moon Charon, after the ferryman of the Greek underworld, but also after his second wife, Char, who he married in 1975.

It is immediately clear that Charon is a very special moon. It is less than 20,000 kilometers from Pluto, and has an orbital period of 6 days, 9 hours and 18 minutes – exactly the same as Pluto's rotation period. That means that the two objects always show each other the same face. With a diameter of 1,170 kilometer, Charon is half as big as Pluto; in fact, together they make up a binary system.

Between 1985 and 1990, Charon's orbit around Pluto is oriented in such a way that the two bodies pass in front of each other in turn. The variations in brightness observed during these occultations enable astronomers to draw up rough maps. Later, detailed photographs are made of Pluto and Charon with the Hubble Space Telescope, and on March 8, 1996, Pluto experts Alan Stern and Marc Buie present the first 'world maps' of the small planet and its moon.

Charon is no longer Pluto's only moon. In 2005, Max Mutchler of the Space Telescope Science Institute discovers two small moons in orbit at a greater distance from Pluto, and they are given the names Nix and Hydra.

On January 19, 2006, the American space probe New Horizons is launched, and on July 14, 2015, after a journey of nine and a half years, New Horizons will fly past Pluto and Charon at a distance of 10,000 kilometers and make the first detailed maps of the dwarf planet and its largest moon.

↑ Photograph on which Charon, Pluto's large moon, was discovered. The grainy, diffuse image of Pluto on this photographic plate is not exactly round; the protrusion rotates around it in a little more than six days. Astronomers did not succeed in seeing Pluto and Charon separately until much later. (USNO)

→ The American space probe New Horizons will arrive at Pluto and Charon in the spring of 2015. During the flyby of the dwarf planet, both objects will be studied in detail. New Horizons will then continue on its journey to study two ice dwarfs in the Kuiper Belt. (JHU-APL/SwRI)

1978

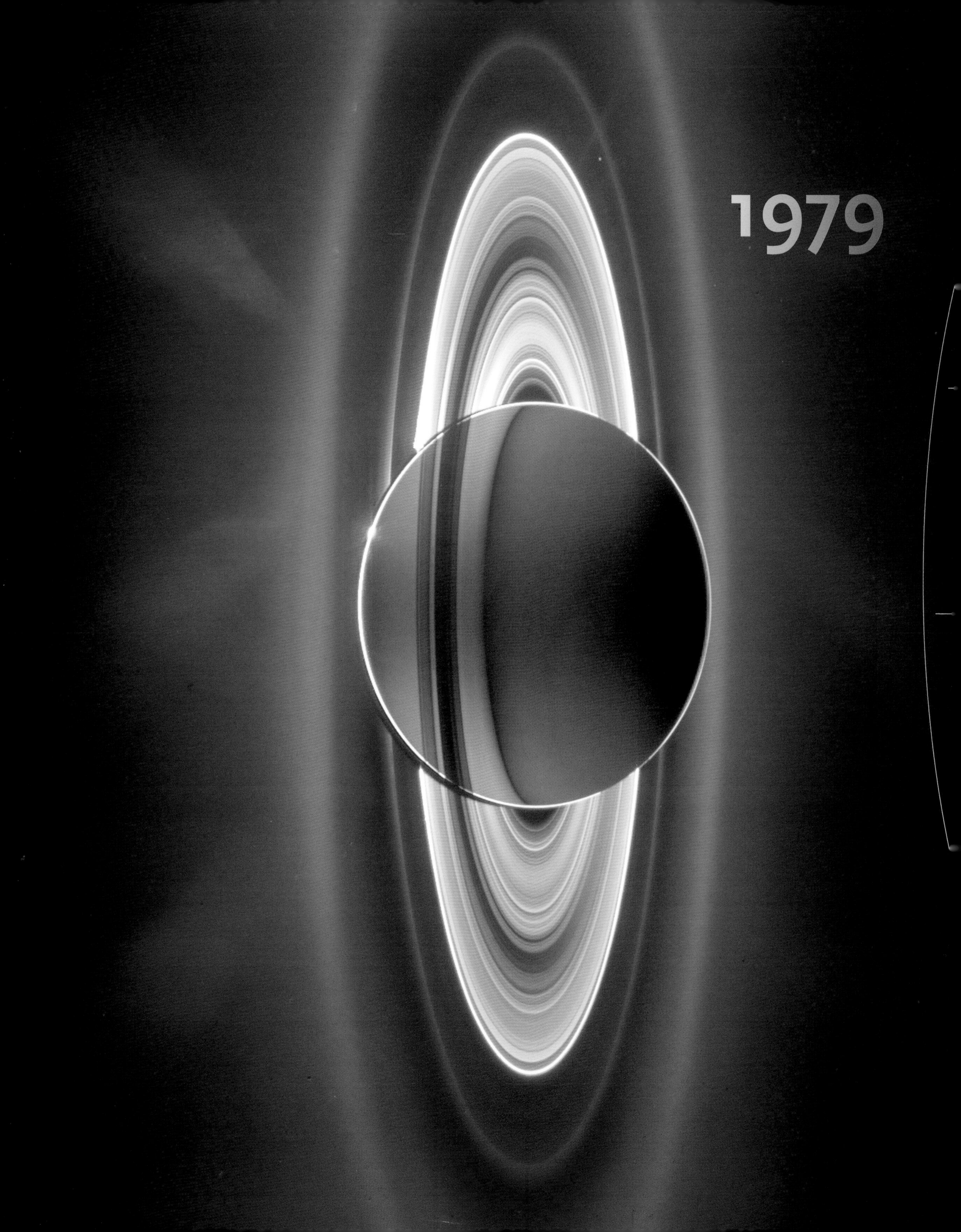
1979

Dusty Belt

Voyager 1 Discovers the Ring Around Jupiter

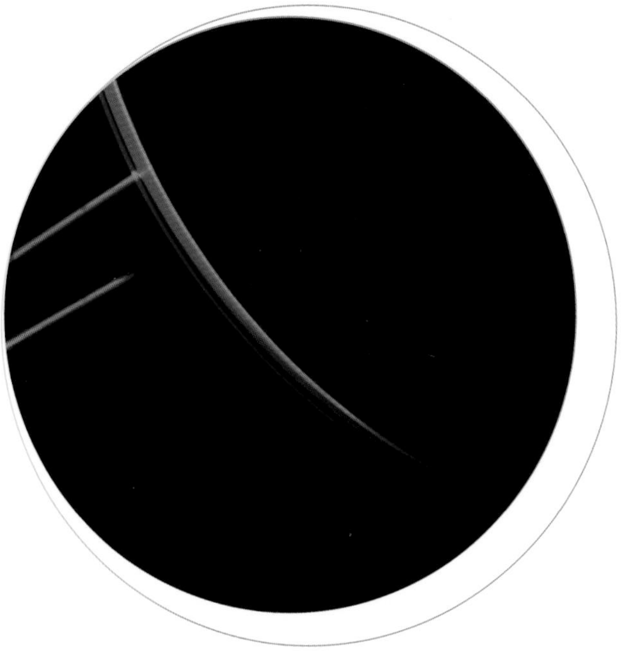

Pioneer 11 already discovers that there is something strange about the giant planet Jupiter, and on December 4, 1974, the unmanned planetary explorer flies past Jupiter at a short distance, a year after its twin brother Pioneer 10. At just under 130,000 kilometer from the planet, Pioneer 11's measuring instruments register a dip in the quantity of electrically charged particles that have accumulated in Jupiter's radiation belt. It is not until several years later that it becomes clear that this is caused by a thin, narrow ring around the planet.

In 1974, Saturn is still the only planet around which rings have been discovered. After the discovery of the thin, dark rings around Uranus in 1977, planetary researchers bear in mind that Jupiter may also have a ring system, though it will of course be much more difficult to observe than that of Saturn. The space probe Voyager 1, which flies past Jupiter in early March 1979, is programmed to make a long-exposure photograph of the planet at the moment it crosses the equatorial plane. The photo, taken from a distance of 1.2 million kilometers with an exposure time of more than 11 minutes, clearly shows a thin ring. The discovery is announced on March 7, 1979, and less than three months later the Voyager team publishes the results in *Science*.

Backlight image of Jupiter's ring system, made by Voyager 2 from a distance of 1.45 million kilometers. The sunlight illuminates the giant planet's atmosphere and the microscopically small particles in the narrow ring. The photograph was made while Voyager 2 was in Jupiter's shadow. (NASA/JPL)

Saturn's impressive ring system also contains many microscopically small particles. They are visible as hazy belts in this backlight image, made by the Cassini space probe. Saturn's main rings also contain larger ring particles and fragments of rock. (NASA/JPL/SSI)

It does not take long for astronomers to realize that Jupiter's ring was also detected indirectly by Pioneer 11 in 1974. Voyager 2's observation program is modified and, when the probe flies past Jupiter on July 9, 1979, it takes detailed photographs of the ring system. It turns out to consist of a flat, quite narrow main ring with a strikingly sharp outer edge, a much wider but less dense 'halo ring,' and another tenuous outer ring that extends past the orbit of the small moon Amalthea.

The Galileo space probe, which arrives in orbit around Jupiter at the end of 1995 and observes its clouds, moons and rings, discovers that the tenuous outer ring consists of two parts. The microscopic dust particles in these two wide rings come from the moons Amalthea and Thebe. Particles in the main ring spiral inwards and, within a few centuries, disappear into the giant planet's atmosphere. Apparently, new material is consistently produced, probably by collisions between particles measuring from a few centimeters to a few hundred meters at the most.

A few days after the discovery of Jupiter's ring has been announced, astronomers succeed in detecting the minuscule heat radiation from its particles with a sensitive infrared telescope on Earth. Later, the rings are studied routinely using large terrestrial telescopes, like the Keck Telescope in Hawaii, and with instruments in orbit around the Earth, like the Hubble Space Telescope. In February 2007, the space probe New Horizons also makes detailed observations of the ring system as it passes by en route to Pluto.

Sulfurous Surprise

Linda Morabito Discovers Volcanic Activity on the Jovian Moon Io

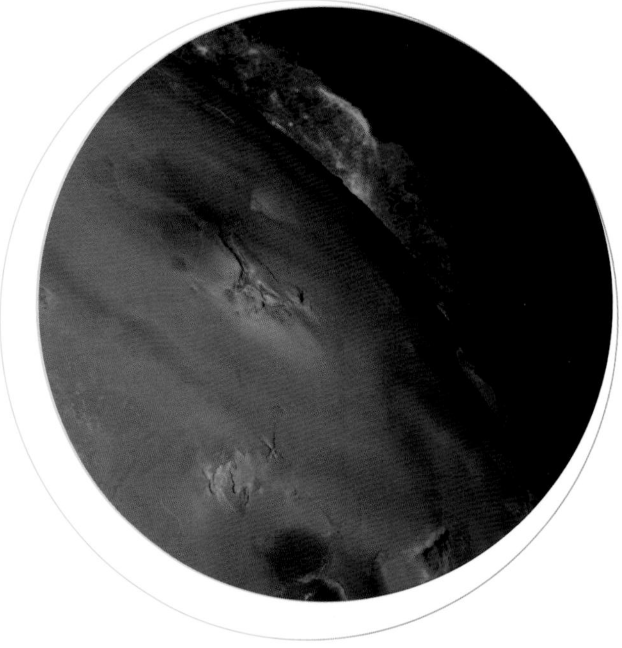

While NASA's planetary explorer Voyager 1 is racing towards the giant planet Jupiter, Stanton Peale of the University of California in Santa Barbara and Pat Cassen and Ray Reynolds of NASA come to the conclusion that Jupiter's moon, Io, is possibly volcanically active. Io is the innermost of the four large Jovian moons discovered by Galileo in 1610. According to the calculations of Peale, Cassen and Reynolds, the interior of Io is constantly kept in motion and therefore heated up by Jupiter's tidal forces which can lead to volcanic activity on the surface. They publish their astounding prediction in *Science* on March 2, 1979, three days before Voyager 1's flyby of Jupiter.

On March 5, Voyager 1 passes Io at a distance of 22,000 kilometers. The surface of this moon, which is a little larger than our own Moon, does indeed look very bizarre; Io is orange-red, has no impact craters (suggesting recent geological activity), and is covered with light and dark spots, strange halos and irregular lava-flow patterns. Planetary researchers compare Io to a pepperoni pizza that has been in the oven a little too long.

Three days after its closest flyby on March 8, Voyager 1 takes one more photograph of Io with its black and white navigation camera and from a distance of four and a half million kilometers. The following day, the photo is processed and analyzed routinely by Linda Morabito of the navigation team. When Morabito increases the contrast of the image, she sees a strange umbrella-shaped cloud on the edge of Io.

It soon becomes clear that the cloud is hanging above a heart-shaped formation on Io, which astronomers have earlier suggested may be of volcanic origin. There seems to be only one possible conclusion: Morabito has discovered the eruption of a volcano on Io. Study of other Voyager photos reveal more volcanic plumes, while infrared measurements by Voyager 1 show that there are extremely hot areas on the surface.

The discovery of Io's volcanoes is announced on March 12, 1979. It is the first evidence of extraterrestrial volcanic activity. Furthermore, Io proves to display by far the highest degree of volcanic activity of any object in the solar system. During the flyby of Voyager 2 in July 1979, a number of volcanoes are also active. Peale, Cassen and Reynolds' prediction proves to have hit the bull's eye.

Io's volcanic plumes can climb to several hundred kilometers above the surface and consist mainly of sulfur compounds. Much of the material from the plumes falls back to the surface, creating the halo-like formations, but every second more than a ton of material ends up in Jupiter's magnetosphere.

In a period of a couple of years in the 1990s, the Galileo space probe discovers more than a hundred active volcanoes on this 'pizza moon,' and at the end of 1999 even flies through one of the plumes. Io's volcanic activity is now also studied with large infrared terrestrial telescopes, like the Keck Telescope on Hawaii.

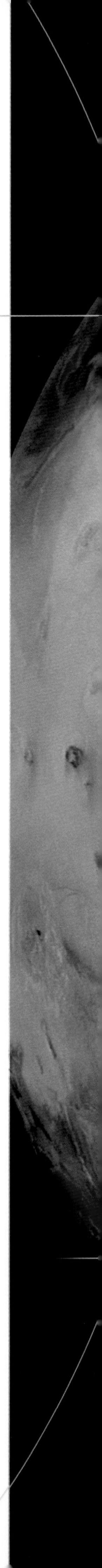

This image shows a 300-kilometer high eruption of the volcano Pele, named after the Hawaiian goddess of volcanoes, on the rim of the Jovian moon Io. Io's internal heat is caused by the tidal forces of its giant parent planet, Jupiter. (NASA/JPL/USGS)

Mosaic photograph of Io, made in the summer of 1999 by the American planetary explorer Galileo. Nearly all the dark spots on the 'pizza moon' are volcanic craters. The circular formations around some of the craters are the result of eruptions of sulfur-rich material. (NASA/JPL/University of Arizona)

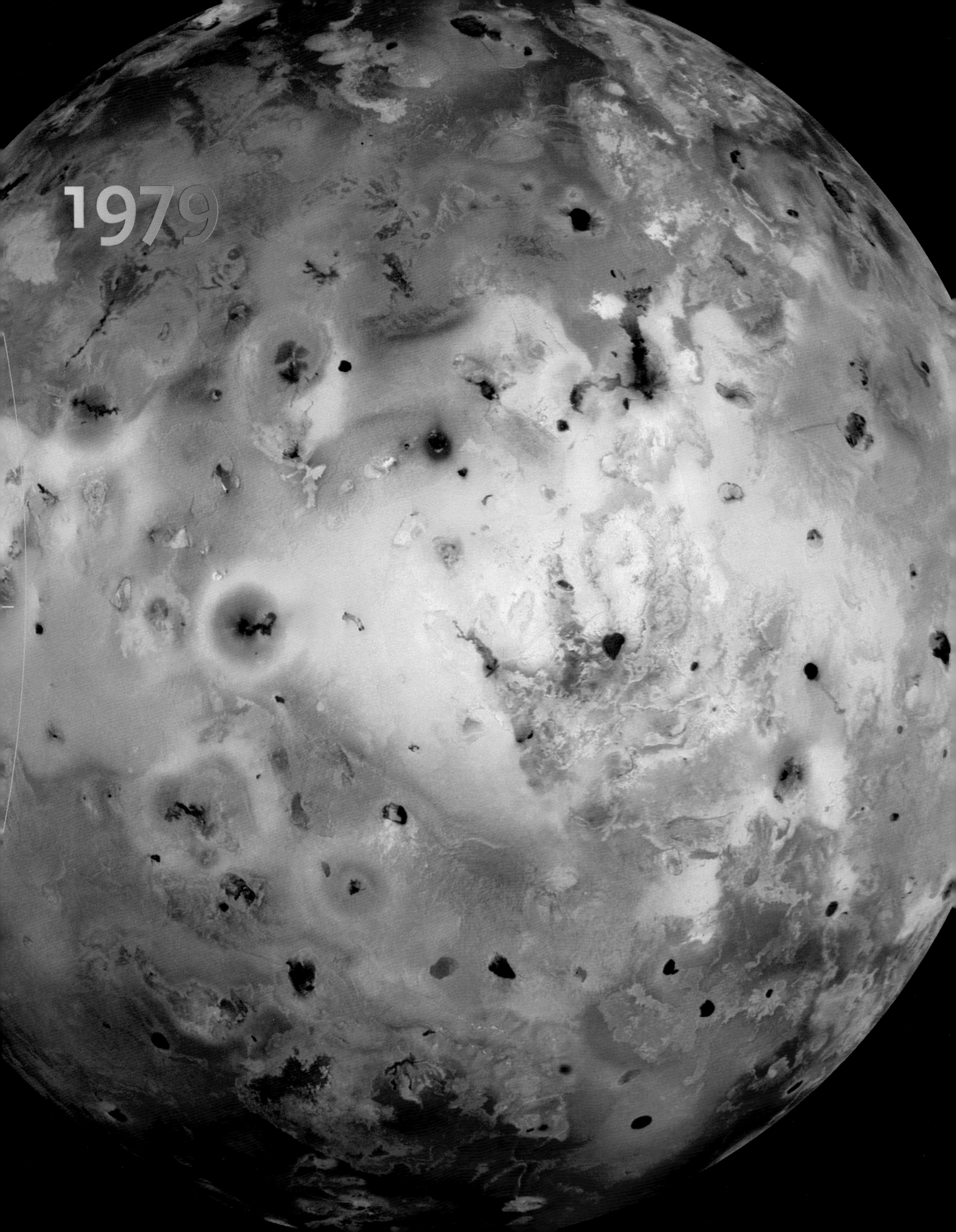
1979

1979

Curved Light

Dennis Walsh and his Colleagues Discover the First Gravitational Lens

In 1912, Albert Einstein makes his first rough notes on gravitational lenses. He has not yet published his general theory of relativity, but Einstein realizes that the bending of light in a powerful gravitational field can result in distorted or multiple images of distant celestial objects. Measurements conducted by Arthur Eddington during the solar eclipse of May 29, 1919, show that light can indeed be bent by gravity.

Einstein publishes his calculations on gravitational lenses in an article in *Science* in 1936. They could in theory be observable when, seen from the Earth, two stars are exactly aligned, one behind the other. A year later, Fritz Zwicky calculates that the chances of such a configuration occurring are much greater for galaxies than for individual stars. The first gravitational lens is, however, not discovered until more than forty years later – and only by coincidence.

In 1979, together with his compatriot Bob Carswell and the American Ray Weymann, British astronomer Dennis Walsh takes photographs with the 2.1-meter telescope at Kitt Peak Observatory in Arizona in search of optical counterparts for radio quasars. At the position of a quasar in the constellation of Ursa Major, they find two practically identical objects very close together and with exactly the same spectrum. In an article in *Nature*, they suggest that the remarkable 'binary quasar' is in fact a gravitational lens. The light from the distant quasar, more than eight billion lightyears distant, is 'bent' by a closer galaxy, resulting in a double image.

In 1980, the lens galaxy is found at a distance of 3.7 billion lightyears by Alan Stockton, using the University of Hawaii's 2.2-meter telescope on Mauna Kea. The same year, Weymann discovers a second gravitational lens, in the constellation of Leo. Discoveries then start to be made more frequently; in 1988, a four-image gravitational lens is even found in the constellation of Boötes.

Fifteen years before the discovery of the binary quasar, the Norwegian astrophysicist Sjur Refsdal publishes a method of deducing the expansion velocity of the universe from data on gravitational lenses. In 1985, Rudy Schild of Harvard University applies this method to the binary quasar. The light from the quasar arrives at the Earth via two different paths (which is why two images are visible), but these paths are not equally long. From the time difference and the geometry of the lens the speed at which the universe is expanding can be deduced.

Schild discovers that small variations in the brightness of the quasar appear in one image 417 days earlier than in the other. That produces a value for the Hubble constant (a measure of the expansion velocity of the universe) that closely concurs with the results of other methods.

Thanks to search programs using large telescopes, including the Hubble Space Telescope, many dozens of gravitational lenses have now been identified.

➋ Photograph of the renowned binary quasar in the constellation of Ursa Major, taken by the Hubble Space Telescope. Around the lower quasar, a less distant galaxy is visible that acts as a lens galaxy. (William Keel)

➌ The faint images of background galaxies are slightly deformed by the gravity of a large cluster of closer galaxies. From the distortions, it can be deduced that there is a diffuse ring of dark matter, shown here with a blue tint. (NASA/ESA/JHU)

Empty Space

Robert Kirshner Discovers the Boötes Supervoid

Galaxies are not evenly dispersed throughout the universe, which is clear from their irregular distribution across the sky. At the end of the 1950s, George Abell publishes a catalog of more than 4,000 clusters, and around the same time, Elizabeth Scott and Jerzy Neyman demonstrate the existence of even larger superclusters. But to determine the real spatial distribution of galaxies, you need to know not only their positions in the sky, but also how far away they are.

In the 1970s, astronomers measure the apparent recessional velocity of more and more galaxies. (This is the speed at which the distance from the Milky Way is increasing as a consequence of the expansion of the universe.) The higher the recessional velocity, the greater the distance. In this way, cosmologists take the first cautious steps towards producing a 3D-map of the universe. In 1978, Stephen Gregory and Laird Thompson discover that the Virgo cluster is part of an elongated supercluster, and astronomers realize that, as well as densely populated clusters, the universe must also have relatively empty voids.

Robert Kirshner of the University of Michigan decides to search for the large-scale structure of the universe, and immediately strikes lucky. Or perhaps not, because what Kirshner and his three colleagues find is basically nothing. Newspapers report on the discovery of the Great Void, a gigantic supervoid in which no more than one single galaxy has been discovered.

Kirshner's team measure the recession velocities of 133 galaxies in the direction of the constellation Boötes, to the east of the tail of the Big Dipper. There appears to be many galaxies with recessional velocities of around 8,000 kilometers per second, and a lot with speeds of around 20,000 kilometers per second, but in between the four astronomers discover only one object. In September 1981, they publish their discovery in *The Astrophysical Journal*: at a distance of about 700 million lightyears, in the constellation of Boötes, there is a supervoid about 250 million lightyears across.

If you imagine galaxies like the Milky Way and the Andromeda Galaxy as buttons one centimeter across, these large galaxies are tens of centimeters apart. On the same scale, the Boötes supervoid is tens of meters across. The discovery by Kirshner and his colleagues is initially welcomed with much disbelief and skepticism, but the existence of the supervoid is later confirmed conclusively.

Extensive search programs later show that the supervoid is not completely empty, and in the 1980s, they lead to the discovery of several dozen galaxies. Ten years later, their numbers have risen to around sixty, yet the Boötes Void remains one of the largest empty areas in the universe, though it has long been joined by others. Why the handful of galaxies that are found in the supervoid are brighter than normal, remains an unsolved mystery.

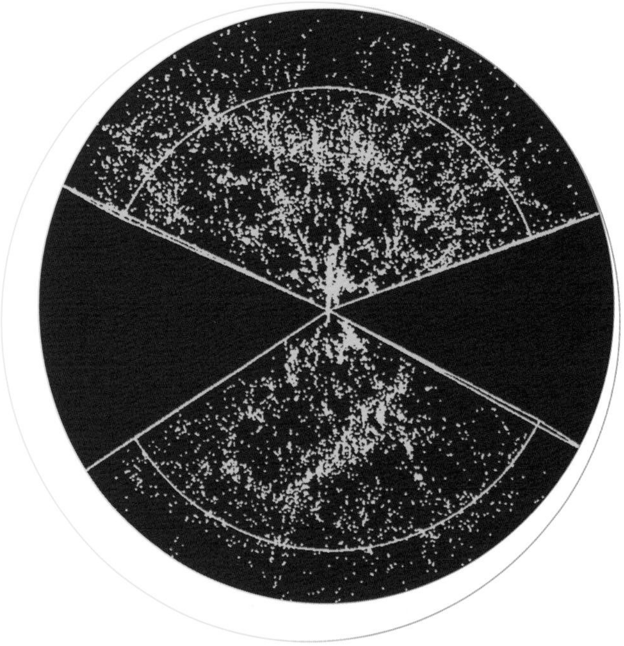

Map of the spatial distribution of galaxies in two sectors of the universe, based on redshift measurements; our own Milky Way is in the center. Similar measurements also led to the discovery of the Boötes supervoid. (CfA)

Large telescopes allow us to see countless galaxies at colossal distances in the universe. The threedimensional distribution of the galaxies cannot be determined until their distances have been calculated. It then becomes clear that they are grouped in superclusters and supervoids. (ESO)

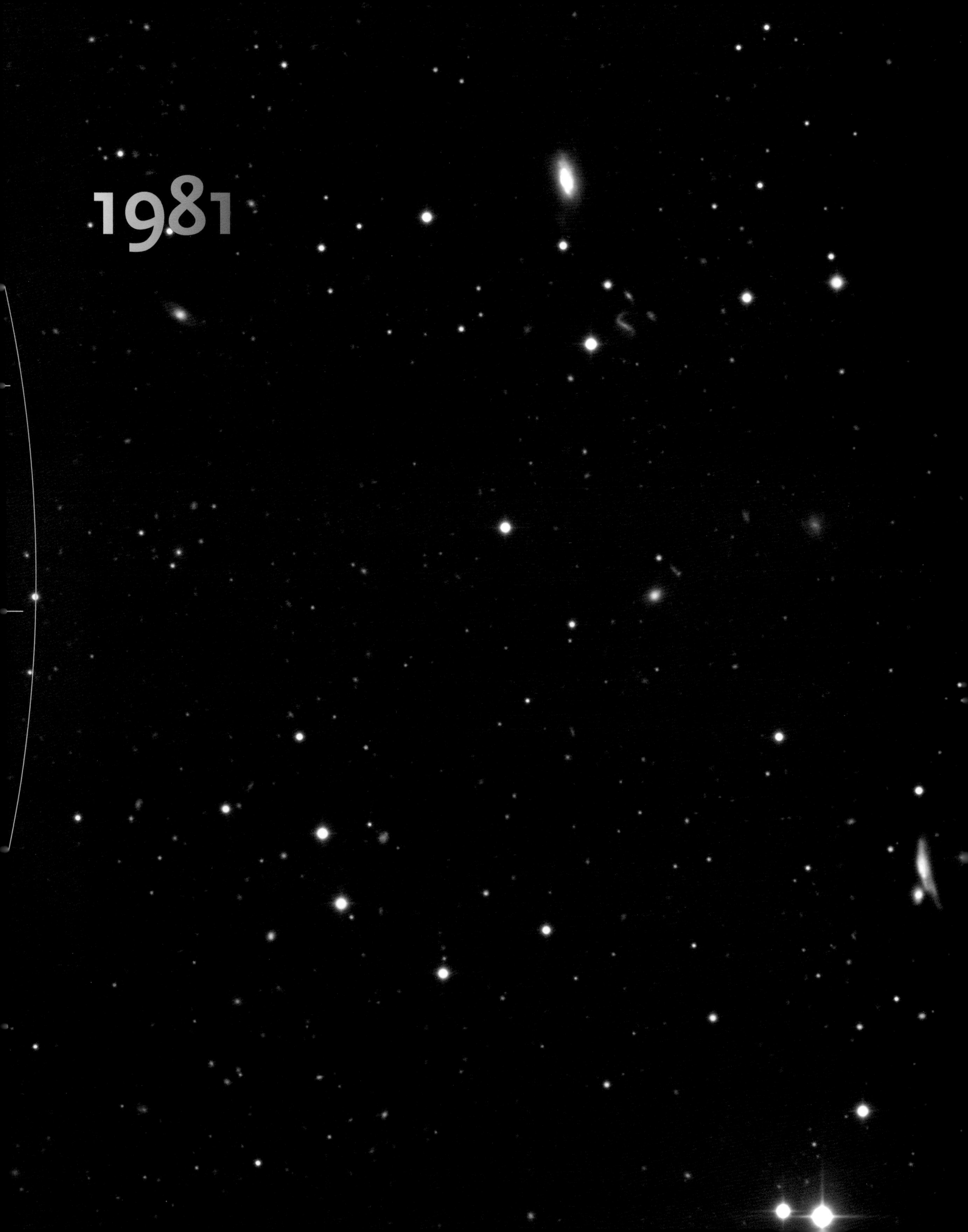
1981

1982

Nervous Beacon

Don Backer Discovers the First Millisecond Pulsar

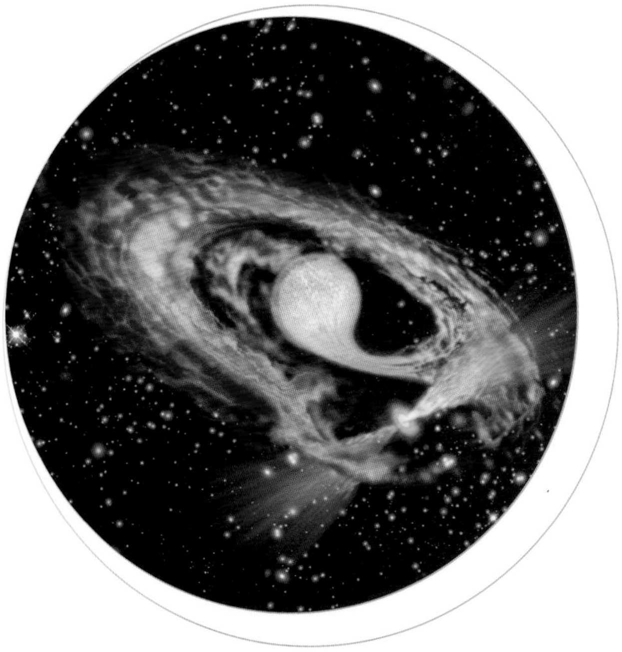

The pulsar that Jocelyn Bell discovers in 1967 spins on its axis once every 1.33 seconds, and is therefore the most rapidly rotating celestial object ever found by astronomers. That record is, however, already broken the following year with the discovery of a pulsar in the Crab Nebula, which rotates at an unbelievable 30 times a second. Because the Crab pulsar was created by a supernova explosion in 1054 and is gradually being slowed down by magnetic energy loss, astronomers assume that only younger pulsars can have high rotation speeds.

The discovery of the first millisecond pulsar in 1982, therefore, comes as a complete surprise. Don Backer and Shrinivas Kulkarni of the University of California in Berkeley find the old cosmic beacon in the small constellation of Vulpecula, only a few degrees from the location where Bell discovered the first pulsar fifteen years earlier. The data from the 300-meter Arecibo radio telescope in Puerto Rico show that PSR 1937+21, as the millisecond pulsar is called, spins on its axis 642 times a second; the rotation period of the pulsating neutron star is only 1.6 milliseconds.

A year later, astrophysicists come up with a possible explanation. According to their theory, the millisecond pulsar was originally part of a binary star system. Gas from the companion flows to the pulsar, sweeping it up like a spinning top to an incredibly high rotation velocity. During this phase of mass transfer, the binary star emits X-rays, because the gas that falls onto the neutron star is heated to a very high temperature. The pulsar is later flung into space, perhaps when the companion star explodes.

Some of the millisecond pulsars discovered after 1982 are indeed part of a binary star system. By far most of them also prove to be located, like X-ray binary stars, in globular star clusters. The theory is, however, not confirmed convincingly until 1998, with the discovery of the first millisecond X-ray pulsar. Satellite data from an X-ray binary system, conducted by Dutch astronomers Rudy Wijnand and Michiel van der Klis, demonstrate that the neutron star in the system, which emits X-rays because it sucks in gas from its companion, spins on its axis 401 times a second. The transfer of matter is indeed the cause of the high rotation speeds of millisecond pulsars.

By 2010, a little under 200 millisecond pulsars have been identified. For almost twenty-five years, the record for the most rapidly rotating celestial body is held by the first, PSR 1937+21. But at the end of 2004, Canadian astronomer Jason Hessels discovers an even faster object in the globular star cluster Terzan 5, with a rotation speed of 716 revolutions per second. In 2007, the European INTEGRAL satellite may have identified an X-ray pulsar rotating at 1,122 times a second, but the measurements have not yet been confirmed.

A millisecond pulsar has been discovered orbiting a red giant star in the globular star cluster NGC 6397. The rapid rotation of the pulsar is probably due to the transfer of matter from its companion, as shown in this artist's impression. (ESA/F. Ferraro)

Photograph of the large globular star cluster 47 Tucanae, taken by the European Very Large Telescope in Chile. The globular cluster, more than 13,000 lightyears away in the southern constellation of Tucana, contains a large number of millisecond pulsars. (ESO)

Heat Surplus

The Infrared Astronomical Satellite Discovers Dust Disks Around Stars

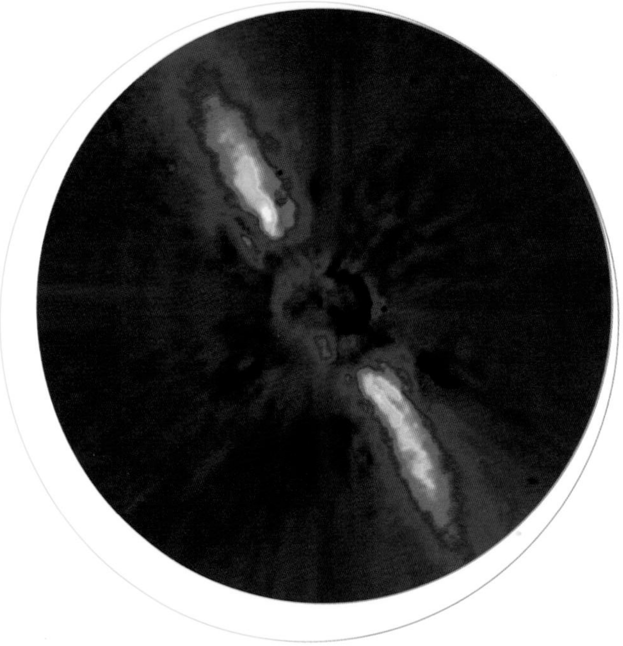

In the early 1980s, infrared astronomy is still in its infancy, but the Dutch-American Infrared Astronomical Satellite (IRAS) changes that in 1983. IRAS is a sensitive infrared telescope with a mirror diameter of 57 centimeters. The telescope is submerged in a tank containing 475 liters of liquid helium, which keeps it at a temperature of only two degrees above absolute zero enabling it to measure the extremely faint heat radiation of celestial objects.

IRAS is launched from Vandenberg Air Force Base in California on January 25, 1983. Over a period of ten months it maps almost the entire sky in detail at four different wavelengths. Hundreds of thousands of sources are detected, from asteroids and comets in our own solar system, through protostars in dusty star formation areas, to distant galaxies experiencing birth waves of new stars.

IRAS also measures the infrared radiation of regular stars, and in some cases, the radiation is much stronger than expected. One of these is Vega, the brightest star in the constellation Lyra. The hot, blue-white star, more than twenty-five lightyears from the Earth, proves to be brighter in the infrared, especially at long wavelengths, than its properties would lead to expect. That extra radiation, therefore, must come from somewhere else.

Astronomers conclude that Vega – and other stars, like Beta Pictoris and Fomalhaut – is surrounded by a flattened, rotating disk of small dust particles less than a tenth of a millimeter in size. Some of these dusty disks are later observed directly with large terrestrial telescopes. In the case of Vega, the circumstellar disk extends to a distance of several billion kilometers – much further than the distance from the Sun to the outermost planet in our solar system, Neptune.

At first, astronomers think that Vega is surrounded by a protoplanetary disk of gas and dust from which planets can form. Observations with the American Spitzer Space Telescope in 2005, however, suggest that it is a 'debris disk' and the dust particles originate from collisions between larger objects. Our own solar system has a similar extended disk called the Kuiper Belt, beyond the orbit of Neptune, in which icy objects occasionally collide.

Vega is probably a few hundred million years old – much younger than the Sun. The fact that the inner section of the dust disk is relatively empty suggests that large objects have indeed already been formed and that the star may then be surrounded by a planetary system. The discovery of circumstellar dust disks by IRAS in 1983 was, in any case, the first direct indication that there may also be planetary systems around other stars.

⮉ Beta Pictoris was one of the stars where the IRAS satellite discovered an excess of infrared radiation. The circumstellar dust disk responsible for the infrared excess was later photographed using large telescopes. From the Earth, we see it almost exactly edge-on. (NASA)

➲ The infrared Spitzer Space Telescope has found a large number of young protostars and embryonic stars in the Carina Nebula. Many of these young stars are surrounded by flat, rotating disks of dust and gas from which planets can be formed. (NASA/JPL/University of Colorado)

1983

1984

Fatal Attraction

John Tonry Discovers the First Supermassive Black Hole

Few celestial objects have as much attractive power as black holes. These bizarre and exciting knots in space-time challenge our imagination, tax the common sense and inspire scientific fantasies. Black holes are the ultimate cosmic vacuum cleaners; everything in their vicinity gets sucked in and disappears for good. In the most literal sense, they are the greatest attraction in the universe.

Black holes are by definition invisible, and hence it took a while for astronomers to be convinced of their existence. Fortunately the matter they attract is heated to extremely high temperatures before it disappears, so black holes emit enormous quantities of energetic radiation. And if a black hole is especially massive, it betrays its presence by disturbing the motion of the stars in its vicinity.

In 1984, John Tonry of the California Institute of Technology conducts measurements of the velocities of stars in the core of the nearby galaxy M32 and discovers the first supermassive black hole. M32 is only 8,000 lightyears in diameter and contains a few billion stars. In its core, the stars are a hundred million times more densely packed together than those in the vicinity of the Sun. Using large terrestrial telescopes, Tonry also shows that the stars are moving at incredible speeds.

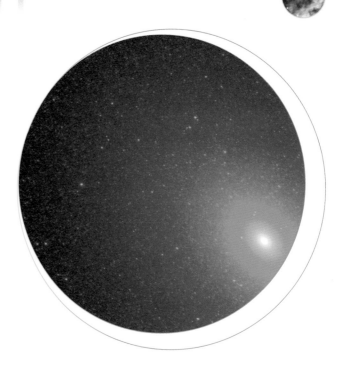

The Hubble Space Telescope has been used to discover thousands of hot, blue stars in the core of the elliptical galaxy M32, a companion of the Andromeda Galaxy. The high velocities of the stars suggest that there must be a supermassive black hole at the core of M32. (NASA/ESA/STScI)

Centaurus A, more than ten million lightyears away, is suspected to have originated in a collision between an elliptical and a spiral galaxy. At the core of Centaurus A is a supermassive black hole estimated to be a hundred million times more massive than the Sun. (ESO)

The stellar velocities that Tonry measures indicate the presence of an enormous invisible object at the core of M32, a few million times more massive than the Sun – a supermassive black hole. The existence of supermassive black holes was already suggested earlier as an explanation of the observed brightness of quasars, but this is the first observational evidence of their existence.

Black holes have since been discovered at the cores of many other galaxies, including our own Milky Way. Astronomers even think that *all* galaxies have a supermassive black hole at their centers. Some of them are only a few million times more massive than the Sun, and others have grown to become gluttonous monsters of billions of solar masses, possibly as the result of collisions and mergers with smaller black holes.

Black holes were once mathematical curiosities that mainly inspired science fiction writers, but today they are respected residents of the cosmic zoo which play an important role in the evolution of galaxies. They govern the motions of stars, eject jets of high-energy radiation into space, and create shockwaves in interstellar gas clouds, setting in motion the formation of new stars and planets.

And when supermassive black holes in the cores of galaxies swallow up gas clouds or whole stars, they generate spectacular celestial fireworks. The radio galaxies, quasars and blinding blazars driven by these cosmic monsters are visible across distances of many billions of lightyears.

Gushing Iceberg

The Space Probe Giotto Discovers What the Core of Halley's Comet Looks Like

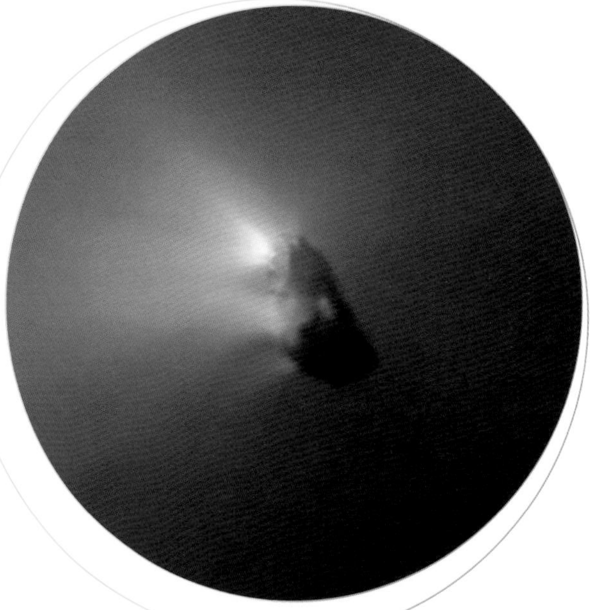

Halley's Comet is by far the most famous comet in history. It has been observed for a long period of time, and served as the Star of Bethlehem in the fresco 'Adoration of the Magi' by Giotto di Bondone, who saw the comet in the sky above Italy in 1301. In the eighteenth century the comet is named after Edmund Halley, who in 1705 is the first to predict the return of the impressive 'star with a tail' after deducing that the comets of 1531, 1607 and 1682 are probably one and the same object.

In the spring of 1986, Halley's Comet once again appears in the inner solar system. Curious space researchers launch no less than five unmanned probes to study the comet from close-up: two Russian VEGA probes (which also examine the planet Venus), the small Japanese comet explorers Sakigake and Suisei, and the European Giotto, named after the Italian artist. Giotto is the first interplanetary spacecraft to be launched by the European Space Agency (ESA).

Giotto is launched with an Ariane 1 rocket on July 2, 1985. During the night of March 13, 1986, it flies past the nucleus of Halley's Comet at the relatively short distance of 596 kilometers, at a speed of 68 kilometers per second. The drum-shaped probe is equipped with a shield to provide protection against the impact of cometary dust. A total of around 12,000 impacts with dust particles are recorded. One of them, a large grain of sand weighing about one gram, knocks the probe off balance, leading to a temporary loss of radio contact with the Earth; another impact destroys Giotto's color camera.

Fortunately, enough photographs are taken and measurements made to provide an accurate picture of the comet's nucleus. It is shaped like an irregular peanut, about 15 x 7 x 10 kilometers in size, and the nucleus is blacker than coal – probably due to the effects of ultraviolet sunlight and cosmic rays – and largely inactive. At various places on the surface, however, Giotto discovers seven powerful geysers of gas and dust, which together emit about three tons of cometary material into space per second. The material consists mainly of frozen water and carbon dioxide, with smaller quantities of methane, ammonia and more complex hydrocarbons.

In early April 1986, the Giotto flight controllers put the probe into deep hibernation, from which it is not awakened until February 1990. On July 2, 1990, it passes close to the Earth, causing its orbit to be deflected such that, on July 10, 1992, Giotto can pay a visit to a second comet. The fly-by of Comet Grigg-Skjellerup, at a distance of only 200 kilometers, is the closest a space probe has ever passed to a comet.

Photograph of the irregularly shaped core of Halley's Comet, taken by the European space probe Giotto from a distance of about 600 km. A number of active geysers of gas and dust are visible on the surface. (MPI Lindau)

On July 4, 2005, the American space probe Deep Impact fired a 300-kilogram copper projectile at the core of comet Tempel 1, after which it flew past the comet at a short distance and conducted measurements of the gas and dust particles released by the impact. (NASA/JPL/UMD)

1986

1987

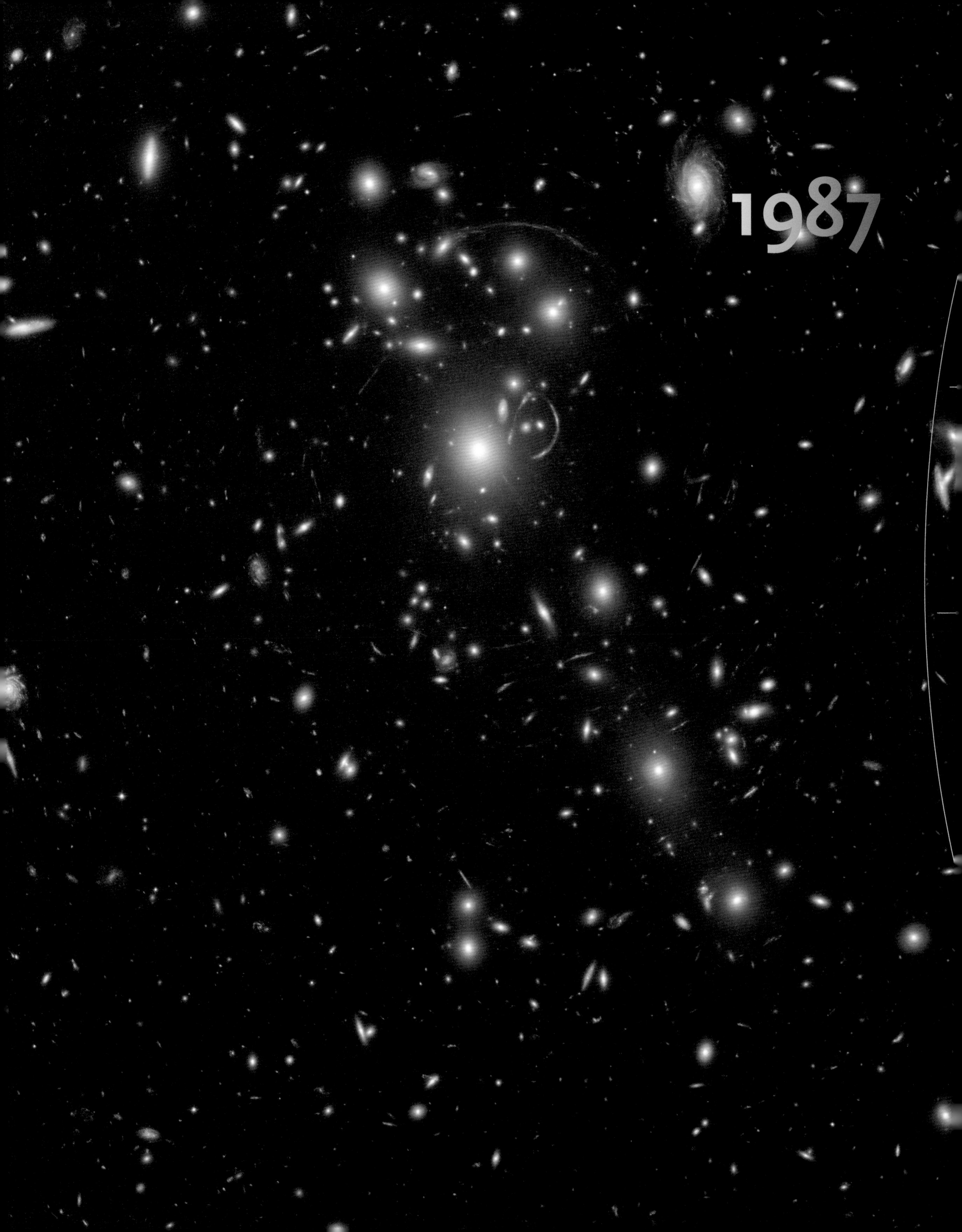

Elongated Galaxies

Geneviève Soucail and her Colleagues Discover Light Arcs

It has been clear for a long time that there is something strange about the galaxy cluster Abell 370. The American astronomer, James Westphal, photographs the cluster in 1973 with the 5-meter Hale Telescope at Palomar Observatory. An arc-like structure can be seen on the photographs, but no one notices it. Arthur Hoag first mentions the light arc in 1981, but also devotes little attention to it. Only in 1985 does Geneviève Soucail of the Observatory in Toulouse become convinced that she is on the trail of something extraordinary.

Soucail and her colleagues photograph the cluster – which is four billion lightyears from the Earth – with the Canada France Hawaii Telescope on Mauna Kea. The luminous arc traces part of a circle with its midpoint at the center of the cluster. It looks as though dozens of individual galaxies are linked together like beads on a necklace.

The French astronomers write an article on their discovery for the European journal *Astronomy & Astrophysics*. Around the time the article is published, in early January 1987, American astronomers Roger Lynds and Vahé Petrosian also announce that they have discovered light arcs, not only in Abell 370, but also in the clusters Abell 2218 and Cl 2244. Lynds and Petrosian's photographs, however, taken in 1985 during tests on a new digital camera at Kitt Peak Observatory in Arizona, show much less detail.

Lynds and Petrosian present their discovery at a meeting of the American Astronomical Society in January 1987. They have, however, no explanation for the remarkable arcs of light. They could be stars that have been ejected from the galaxies by mutual tidal forces, or elongated filaments of luminous gas for example.

The correct explanation for the light arcs is provided by Soucail and her colleagues later that year. Spectroscopic examination of the light arc in Abell 370 shows that it radiates exactly the same light as a galaxy about five and a half billion lightyears from the Earth. Furthermore each piece of the arc has exactly the same spectrum suggesting that it is an elongated (and enhanced) image of a very distant background galaxy. Through the gravitational lens effect of the cluster, the image is stretched out to form a striking arc of light.

In 1995, Warrick Couch and Richard Ellis take a spectacular photograph of the Abell 2218 cluster with the Hubble Space Telescope, at more than two billion lightyears distance from the Earth. The cluster proves to contain more than a hundred bright and fainter arcs of light – each of them the elongated images of galaxies that lie behind the cluster. Precision research into the location of the arcs makes it possible to accurately determine the distribution of dark matter in the cluster.

Photograph of the striking light arc in the cluster Abell 370, taken by a large terrestrial telescope. It shows the deformed and elongated image of a very distant galaxy. (ESO)

The cluster Abell 1703, three billion lightyears away, contains dozens of galaxies and a lot of invisible dark matter. The gravity of the cluster deforms the images of more distant galaxies, creating a plethora of light arcs. Study of the light arcs provides information on the exact distribution of matter in the cluster. (NASA/ESA/Caltech/Davide De Martin)

Mysterious Explosion

Ian Shelton Discovers a Supernova in the Large Magellanic Cloud

On February 23, 1987, at 07:36 Universal Time (UT) to be exact, a tidal wave of neutrinos washes over the Earth. Neutrinos are elusive elementary particles that are electrically neutral and have a minuscule mass. Three gigantic neutrino detectors – in the United States, Russia and Japan – register a total of twenty-five of these ghostly particles, which come from an exploding star in the southern sky. The measurements are not analyzed until a few days later, when the tsunami of neutrinos has already left the Earth long behind.

In New Zealand, on almost every clear night, amateur astronomer Albert Jones trains his telescope at the Large Magellanic Cloud, a small neighboring galaxy to our own Milky Way. From the tropics and the southern hemisphere, the galaxy can be seen as a hazy smudge of light in the sky. On the evening of February 23, 1987, around 09:30 UT, Jones sees nothing special about the cloud. An hour later, another amateur astronomer in Australia, Robert McNaught, takes a photograph of the Large Magellanic Cloud which shows a new star, just bright enough to be seen with the naked eye. McNaught does not develop the photograph until the following day.

Later that night, around 05:00 UT on February 24, staff astronomer Oscar Duhalde of the Las Campanas Observatory in Chile steps outside for a moment to enjoy the night sky. He sees a star in the Large Magellanic Cloud that should not be there but, strangely enough, devotes little attention to it. Three-quarters of an hour later, at the same observatory, 29-year-old Canadian astronomer Ian Shelton develops a photograph he made earlier that night with a small telescope. The new star in the Large Magellanic Cloud immediately attracts his attention; on a photograph from the previous night it is not visible. Shelton, too, goes outside. He sees the star in the sky with his own eyes, tells his colleagues, and shortly afterwards reports the discovery of a supernova.

Supernova 1987A is the first to be visible with the naked eye since Kepler's supernova of 1604. At a distance of 168,000 lightyears from the Earth a massive, bright giant star implodes because the nuclear reactions in its interior have stopped. The implosion of the star's core emits a tidal wave of neutrinos. A few hours later, a gigantic shock wave blows the outer layers of the star into space at more than ten percent of the speed of light. The neutrinos and the visible light from the supernova then take 168,000 years to reach the Earth.

All large telescopes in the southern hemisphere and many satellites in orbit around the Earth are aimed at the supernova. 1987A is the most closely studied supernova in history. After a few months it is no longer visible with the naked eye, but the immediate neighborhood of the explosion is still kept under close surveillance, including by the Hubble Space Telescope. The neutron star that must have been created during the supernova explosion is, however, never detected perhaps because the core of the star has imploded and become a black hole.

↑ Around 20,000 years before supernova 1987A exploded, it expelled a ring of matter into space, which glowed again when the energy from the explosion reached it. The ring is about one lightyear across. (NASA/ESA/CfA)

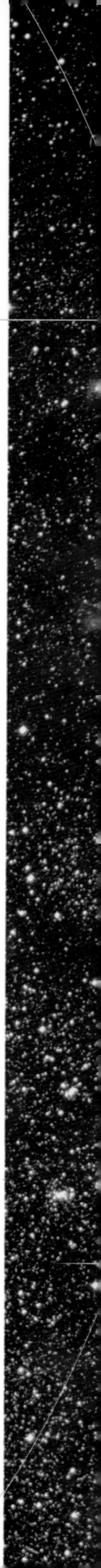

→ Supernova 1987A is visible as a bright star on this photo of the Large Magellanic Cloud. The bright area above left is the Tarantula Nebula, a gigantic star – forming region. Despite being 168,000 lightyears away, the supernova was easy to see with the naked eye. (ESO)

1987

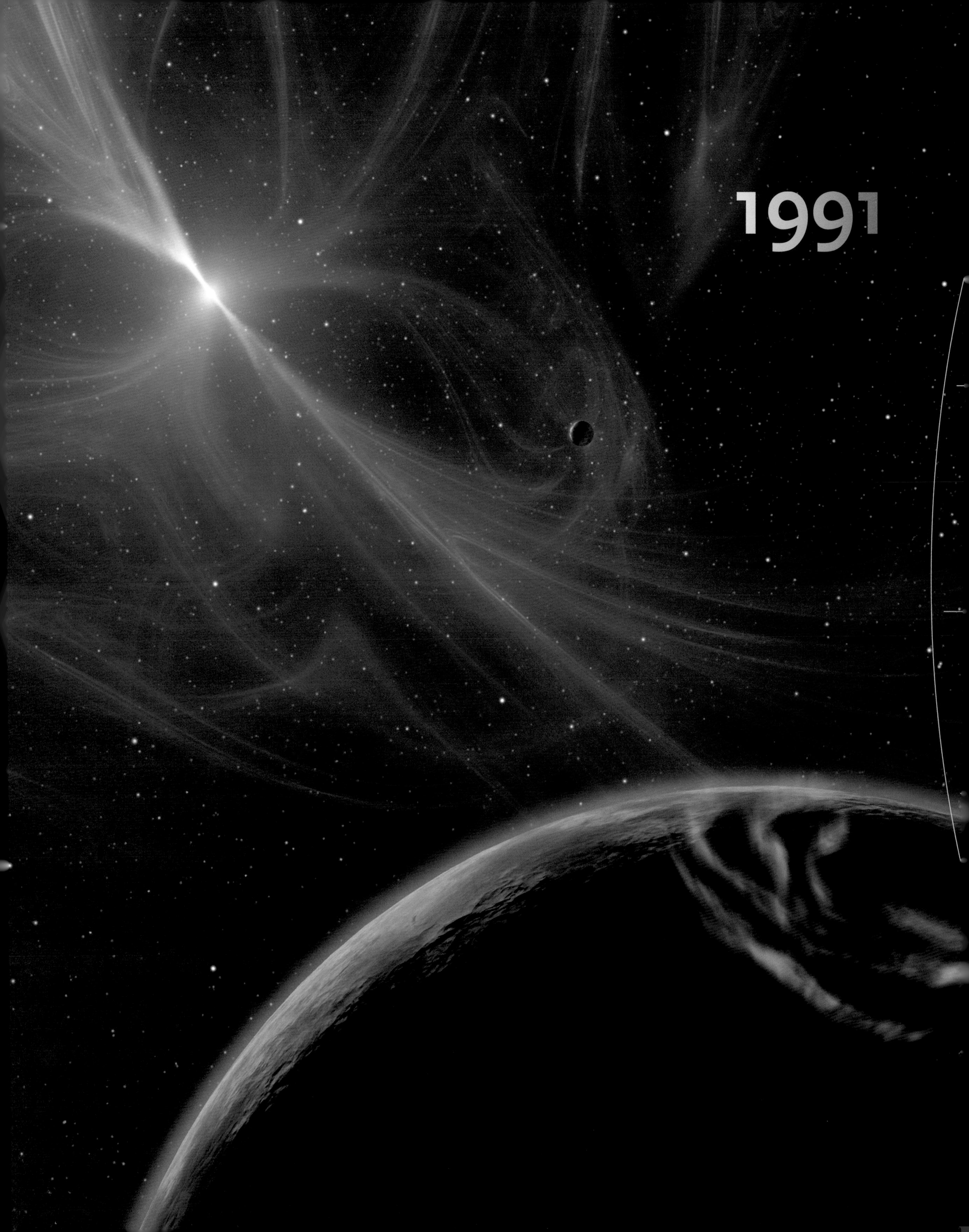
1991

Hostile Environment

Aleksander Wolszczan Discovers Pulsar Planets

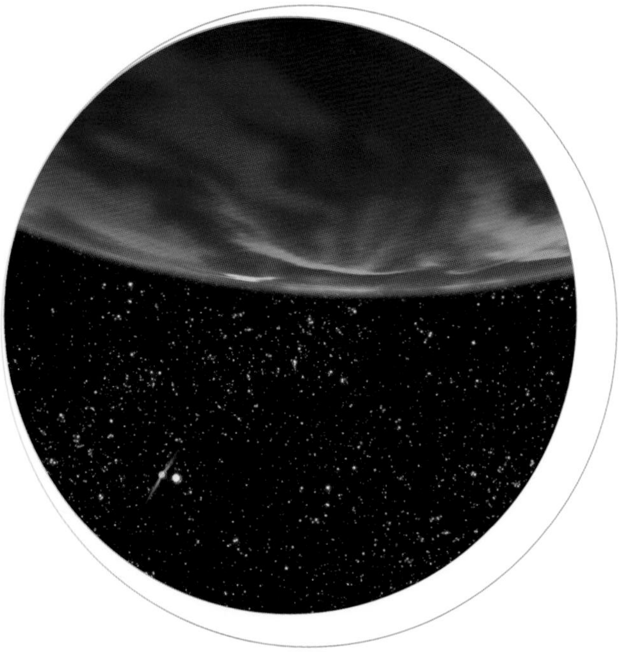

Astronomers have been searching for planets around other stars since the middle of the twentieth century, and no one expects that the first 'exoplanets' will be found not around a regular, Sun-like star but around a pulsar. Pulsars are spinning neutron stars that emit very regular, short pulses of radiation, remnants of supernova explosions. As it seems unlikely that a planetary system would survive the terminal explosion of the parent star, pulsar planets probably form after the supernova explosion.

Aleksander Wolszczan, a Polish astronomer who emigrated to the United States in 1982, searches for new pulsars with the 300-meter Arecibo radio telescope in Puerto Rico. In February 1990, at a distance of about 1,500 lightyears, he discovers a pulsar in the constellation of Virgo that spins on its axis 161 times a second. Precision measurements of the arrival times of the pulses quickly reveal that there is something extraordinary about PSR 1257+12.

Together with American astronomer Dale Frail, Wolszczan discovers that there must be at least two planets orbiting the pulsar, with orbital periods of 66.5 and 98.2 days, and at distances of 54 and 69 million kilometers respectively. The planets are a couple of times more massive than the Earth. They cause periodic changes in the position of the pulsar, which reveal themselves in minuscule variations in the arrival times of the pulses.

➋ Illustration of the extremely old planet orbiting the millisecond pulsar B1620-26. The pulsar is located in a globular star cluster, which is why so many stars are visible in the background. (NASA/STScI/G. Bacon)

➌ Pulsar PSR 1257+12 is accompanied by a small number of relatively small planets. They were probably formed from matter ejected into space during the supernova explosion that also created the pulsar. The pulsar has a strong magnetic field and blasts high-energy particles into space. (NASA/JPL)

In the course of 1991, Wolszczan and Frail become convinced that they have discovered pulsar planets. They are therefore shocked when they read an announcement by British radio astronomer Andrew Lyne in *Nature* on July 25, 1991, that he has discovered a planet in orbit around pulsar PSR 1829–10. But Lyne withdraws his claim half a year later, and Wolszczan and Frail publish their own results on January 9, 1992, again in *Nature*.

Two years later, Wolszczan announces the discovery of a third planet in the system which is much smaller and less massive than the Earth and moves in a smaller orbit with an orbital period of only 25.3 days. Furthermore, the measurements show that the pulsar planets cause small disturbances in each others orbits, exactly as predicted. That places the interpretation of the observations beyond all doubt; the variations in the arrival times of the radio pulses from PSR 1257+12 cannot be explained in any other way.

In 2002, Wolszczan announces the existence of a *fourth* planet around the pulsar. It moves in a wide circular orbit with an orbital period of 3.4 years and is smaller and less massive than the Moon. Two years earlier, in 2000, American astronomer Steinn Sigurdsson reports that he has found a planet orbiting pulsar B1620–26. This pulsar is accompanied at a short distance by a white dwarf, and the planet circles this bizarre binary system in a very wide orbit. Other pulsars, too, are suspected of having planets orbiting them.

Baby Photo

The COBE Satellite Discovers Fluctuations in the Cosmic Background Radiation

The first ideas for a 'Big Bang satellite' date from the mid-1970s. Ten years after Arno Penzias and Robert Wilson discover the cosmic background radiation, several different scientific teams propose studying the 'echo of the Big Bang' in detail, from space. Finally, work begins on building the Cosmic Background Explorer (COBE) in 1981. The space shuttle is to place COBE in orbit around the Earth in 1988.

The accident with the space shuttle *Challenger* in January 1986 frustrates these plans. Engineers, therefore, modify the design of COBE and the satellite is finally launched, using a Delta rocket, on November 18, 1989. Expectations are high; in 1987 a measuring instrument on board a rocket found indications that the spectrum of the background radiation is not in line with the Big Bang theory.

Shortly after COBE is launched, the FIRAS (Far-InfraRed Absolute Spectrophotometer) measures the intensity of the cosmic background radiation at 67 wavelengths in only 9 minutes. The resulting spectrum complies in detail with the expectations, and project scientist John Mather presents the spectrum at the national meeting of the American Astronomical Society in Arlington on January 13, 1990. Astronomers welcome the results with a standing ovation.

Analysis by COBE's DMR (Differential Microwave Radiometer) takes much longer. The DMR is designed to measure extremely small fluctuations in the temperature of the background radiation, but first it is necessary to filter out all conceivable foreground sources of cosmic microwave radiation. A team led by physicist George Smoot finally achieves this in the course of 1991. The background radiation turns out to have an average temperature of 2.73 degrees above absolute zero, but there are variations to a hundred-thousandth of a degree. These indicate minuscule variations in density in the newly formed universe, out of which galaxies and clusters were later formed.

Smoot presents COBE's map of the sky on April 23, 1992, at a physics conference in Washington. The elliptical map, with red and blue spots, is sometimes referred to as 'the baby photo of the universe'. Together with the spectrum of the cosmic background radiation presented earlier, the map provides the best evidence for the Big Bang theory. Asked about the importance of the discovery, Smoot says: 'If you're religious, it's like looking at God'. In 2006, Mather and Smoot are awarded the Nobel Prize for Physics.

From 2001 on, the background radiation is studied in much more detail by the Wilkinson Microwave Anisotropy Probe (WMAP). WMAP's results make an even greater contribution to the standard cosmological model, which considers the universe to be 13.7 billion years old and dominated by mysterious dark matter and dark energy. Yet it is the COBE measurements that mark the birth of precision cosmology.

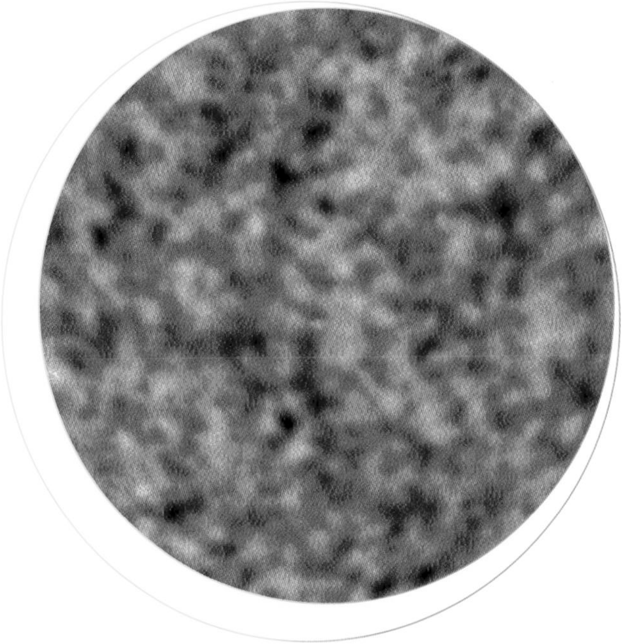

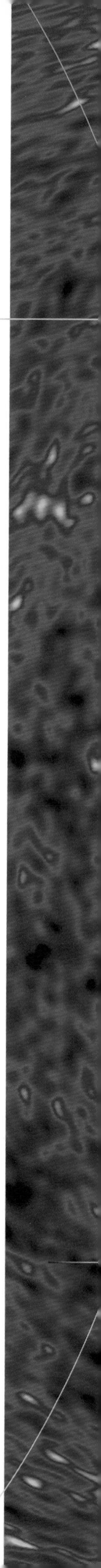

The cosmic background radiation is not only studied by satellites in orbit around the Earth. This map of temperature variations in the background radiation was made by the Boomerang Telescope, which conducted observations for ten days from a balloon. (CWRU)

The Wilkinson Microwave Anisotropy Probe (WMAP) conducted precision measurements of small temperature variations in the cosmic background radiation for many years. The variations are shown here by color differences, and the measurements provide information on the early evolution of the universe. (NASA)

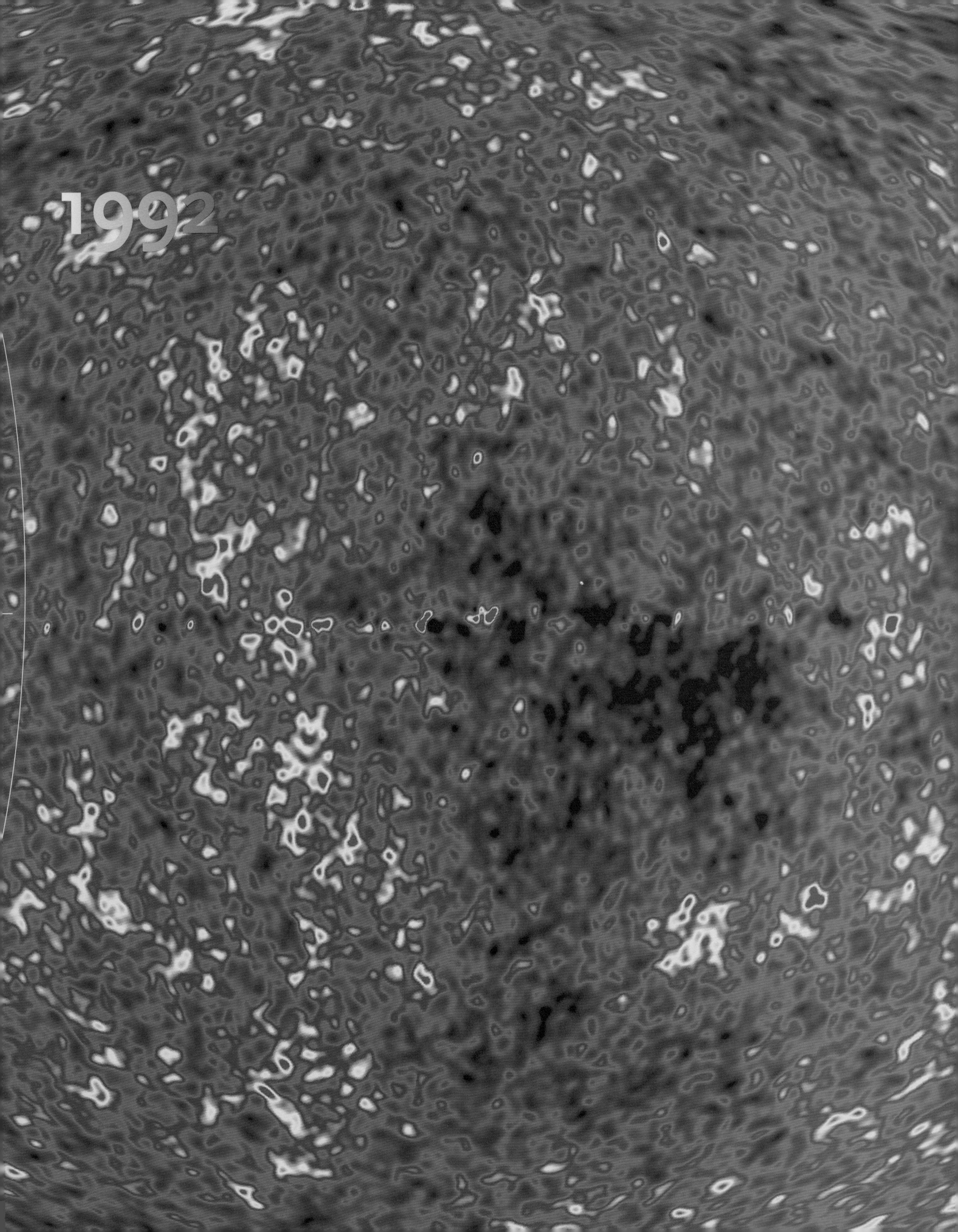
1992

1992

Beyond Pluto

David Jewitt and Jane Luu Discover the First Kuiper Belt Object

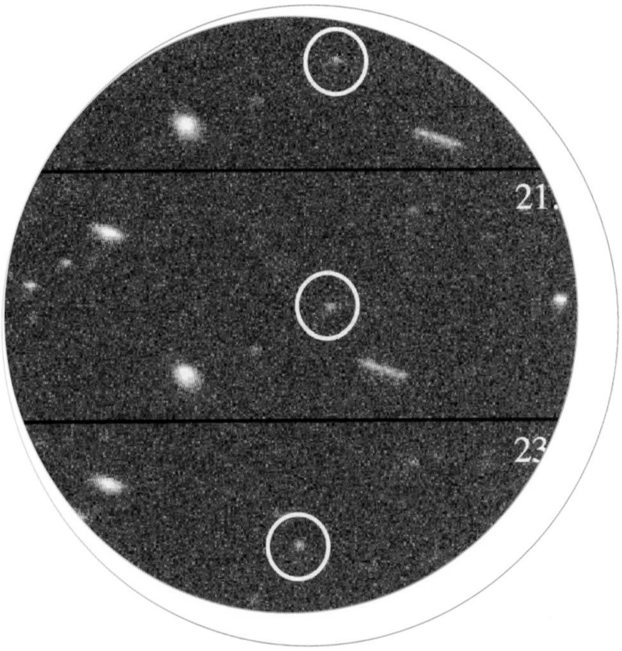

In 1951, the Dutch-American planetary scientist Gerard Kuiper publishes his theory that there must be a broad belt of small, icy objects beyond the orbit of Neptune. Computer simulations by Canadian theoreticians Martin Duncan, Scott Tremaine and Thomas Quinn at the end of the 1980s indicate that this 'Kuiper Belt' is the main source of short-period comets, with orbital periods no longer than a few hundred years. The belt is also expected to contain a large number of bigger objects, some hundreds of kilometers in size.

David Jewitt has been hunting for Kuiper Belt objects since 1985. Jewitt is born in England in 1958 and moves to the United States in 1979, where he quickly makes his name with the discovery of the small Jupiter moon Adrastea, using photographs made by the Voyager 2 planetary probe. In 1982, Jewitt is the first to identify Halley's Comet on its return to the inner solar system and achieves that using a primitive electronic camera fixed to the 5-meter Hale Telescope at Palomar Observatory.

Jewitt's photographic search for 'trans-Neptunian objects' with the Schmidt Telescope at Kitt Peak Observatory in Arizona initially produces nothing at all. In 1987, he changes to electronic CCD cameras and is assisted by Vietnamese astronomer Jane Luu. Luu continues to work with Jewitt when he moves to Hawaii in 1988 and carries on his search program using the University of Hawaii's 2.2-meter telescope on Mauna Kea. In the spring of 1992, that is updated with a new four-megapixel camera.

During the night of August 30, 1992, Jewitt compares two digital photographs of a small section of the night sky in the constellation of Pisces. On the photos, taken earlier in the evening with the new camera, there is a small speck of light that slowly but surely changes position. That same night it proves to be an object beyond the orbit of Neptune. Its brightness suggests that it must be a few hundred kilometers in diameter. Such a large object has not been found in the solar system since 1930, when Pluto was discovered by Clyde Tombaugh.

The discovery of 1992 QB_1 (the object does not yet have an official name) is announced on September 14. The ice dwarf goes around the Sun once every 289 years in a somewhat elongated orbit, in the course of which its distance from the Sun varies from 6.1 to 7 billion kilometers. In March 1993, Jewitt and Luu find a second Kuiper Belt object, followed in September 1993 by two more. After that the number rises rapidly, mainly because many other astronomers also start searching for ice dwarfs, and by 2008, more than 1,000 have been identified.

From 2002 on, large Kuiper Belt objects are also found and there is a growing awareness that Pluto is simply a large ice dwarf. In 2006, the International Astronomical Union decides that Pluto will have to continue on its way as a dwarf planet.

➋ Photographs showing the discovery of 1992 QB_1, the first ice dwarf (after Pluto) to be found beyond the orbit of Neptune. In the course of a few hours, the faint speck of light moves very slowly between the stars. The stripe is a much faster-moving asteroid. (Dave Jewitt/Jane Luu)

➌ After making a flyby of Pluto and its large moon Charon in 2015, the space probe New Horizons will visit one or two ice dwarfs in the Kuiper Belt. Astronomers hope this will provide new information on the origin of the solar system. (JHU-APL/SwRI)

Cosmic Nursery

Bob O'Dell Discovers Protoplanetary Disks in the Orion Nebula

As a small boy, American astronomer Bob O'Dell trains his self-built telescope at the Orion Nebula, an impressive star forming region about 1,500 lightyears from the Earth. In 1972, after he has been director of Yerkes Observatory for five years, O'Dell becomes project scientist of the Hubble Space Telescope. At that time the project still only exists on paper and has not yet been given its definitive name.

After many delays the Hubble Space Telescope is finally launched in April 1990. More than a year later, on August 13 and 14, 1991, O'Dell aims the space telescope at the Orion Nebula. Although as the result of a minuscule construction error in the main mirror Hubble does not provide completely sharp images, the photographs of the Orion Nebula are the sharpest ever made.

Together with his colleagues at Rice University in Houston, Texas, O'Dell discovers that many new-born stars in the nebula are not completely point-like, but seem to resemble small disks. The Space Telescope Science Institute announces the discovery at the end of 1992, and according to the researchers, the objects are 'protoplanetary disks'. As the name suggests, they are disks of gas and dust particles around young stars, which may later lead to the formation of planets.

At the beginning of December 1993, Hubble is fitted with 'contact lenses' to correct the imaging errors in the main mirror. Shortly afterwards, on December 29, O'Dell and his colleagues take new photographs of the Orion Nebula, on which the protoplanetary disks are much more clearly visible. About half of all the new-born stars studied in the nebula seem to have them.

Most of the protoplanetary disks in the Orion Nebula are bright: the gas they contain is illuminated by energetic ultraviolet radiation from extremely hot stars in the center of the nebula. Others appear dark against the bright background of the nebula. From the Hubble images, O'Dell concludes that the disks contain sufficient material for the formation of a number of planets like the Earth. Sometimes a new-born star is visible at the center of a protoplanetary disk, but in many cases the light of the star is obscured by the material in the disk.

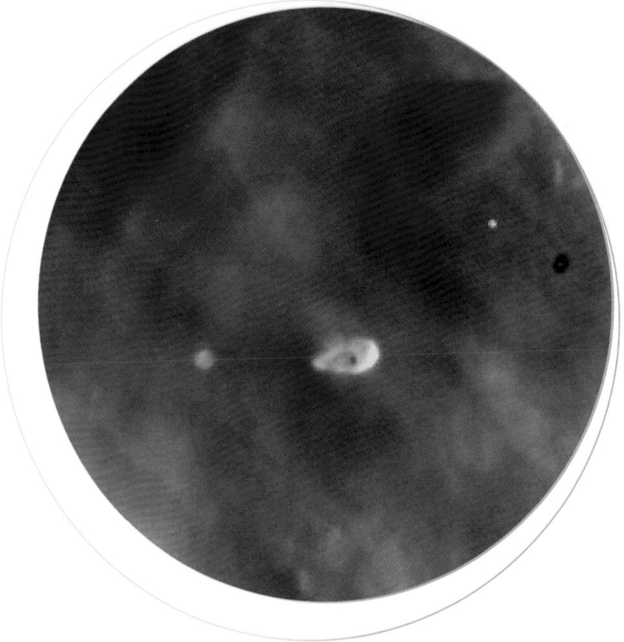

According to a Hubble press release in June 1994, the discovery of many dozens of protoplanetary disks in the Orion Nebula supports the theory that planet-formation around other stars is more likely the rule than the exception. Later, protoplanetary disks are also found in other star forming regions.

Large star forming regions like the Orion Nebula are probably not the best places for planets to form. Before the process of coagulation gets properly underway, most of the matter in the disks will be blown away by the stellar winds and the high-energy radiation of the hot giant stars in the nebula. That makes it very doubtful whether the Orion nebula will ever produce many planetary systems.

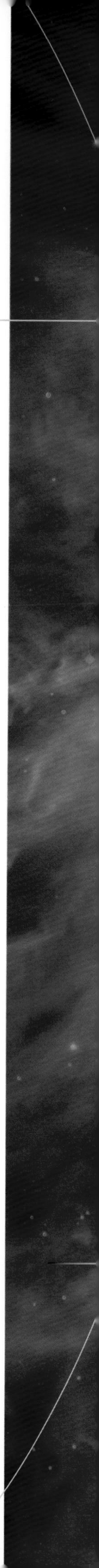

The Hubble Space Telescope photographed these protoplanetary disks around newborn stars deep in the Orion Nebula. In the future, planets may condense out of the gas and dust disks. This is how our own solar system must have formed long ago. (NASA/ESA/C.R. O'Dell)

The Orion Nebula, 1,500 lightyears from the Earth, is one of the largest star formation areas in the Milky Way. This impressive image of the cosmic nursery is composed of photographs taken by the Hubble Space Telescope and the Spitzer Space Telescope. (NASA/JPL/STScl)

1992

1993

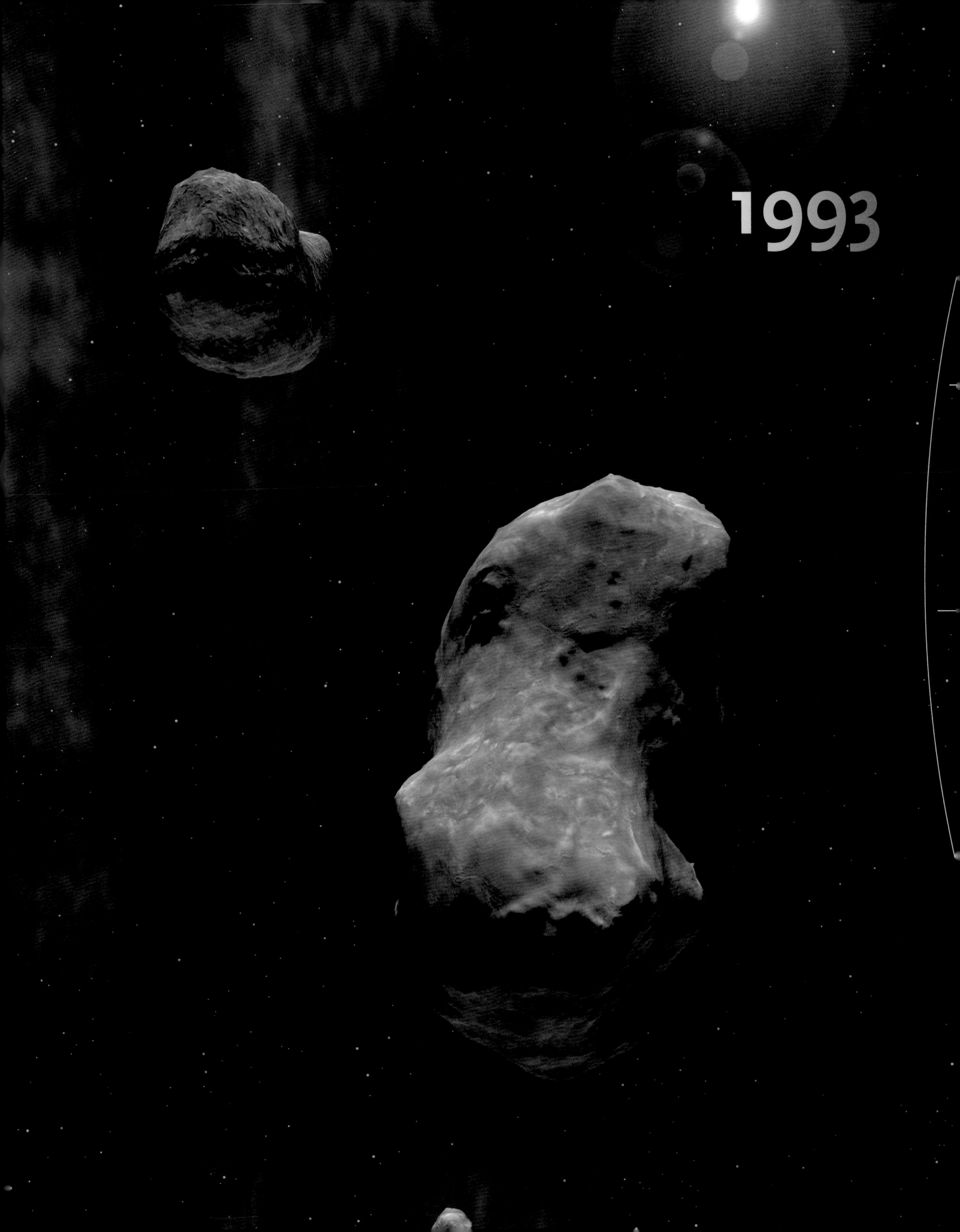

Unsightly Little Moons

The Space Probe Galileo Discovers the First Satellite of an Asteroid

Galileo Galilei's discovery of Jupiter's four large moons in January 1610 is the first indication that the Earth's Moon is not unique. Other planets, too, have their own satellites, and only Mercury and Venus, as far as we know, go on their way unaccompanied.

In 1978, for the first time in history, American observers find indications of a moon around an asteroid. The asteroid is Herculina, discovered by Max Wolf in Heidelberg in 1904. On June 7, 1978, Herculina passes in front of a star and measurements taken during the occultation provide information on the asteroid's dimensions. Many professional and amateur astronomers direct their telescopes at Herculina and some see how the star is 'occulted' a few seconds before the occultation proper. This suggests that there is a second object close to Herculina – a small satellite.

Many astronomers, however, have their doubts about this conclusion. The chance that the small satellite, if it exists, would also move directly in front of the star is, of course, very slight. Furthermore, there is no way to verify the discovery, and the existence of asteroid satellites is not confirmed with full certainty for another fifteen years until 1993. It is not confirmed by a terrestrial telescope, but by a space probe not around Herculina, but the smaller asteroid Ida.

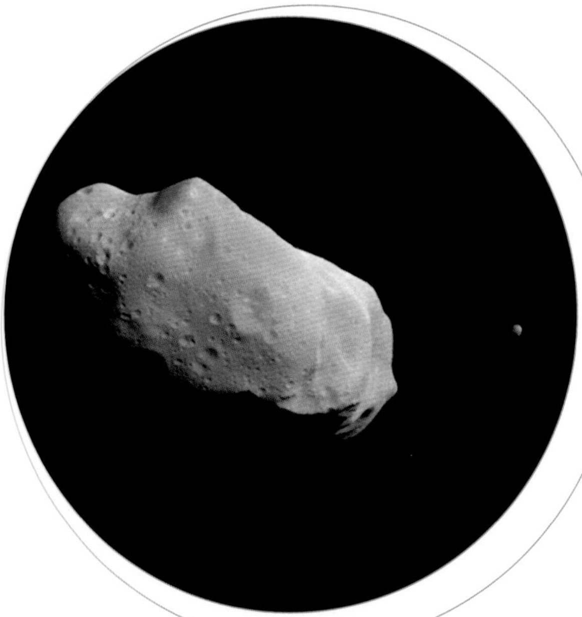

◐ Close to the asteroid Ida – an irregular lump of rock almost 60 kilometers long – the planetary explorer Galileo discovered a small moon, which was given the name Dactyl. Dactyl measures about a kilometer and a half. Small moons have since been found around many other asteroids. (NASA/JPL)

◑ Two small moons have been discovered orbiting the asteroid Sylvia, the middle object in this artist's impression. Sylvia is a large asteroid, measuring 385 × 265 × 230 kilometers. Its two moons, Romulus and Remus, measure about 18 and 7 kilometers, respectively. (ESO)

The American Jupiter probe Galileo (named after the discoverer of Jupiter's four large moons) is launched on October 18, 1989. During its long journey to the giant planet it flies past Ida on August 28, 1993, at a distance of 2,400 kilometers and takes photographs of the elongated, irregularly shaped object, but the photographs are not sent back to Earth until February 1994. In addition to Ida, Galileo has also captured a small rock on camera, about one and a half kilometers across and the small satellite is given the name Dactyl.

In the 1990s, thanks to new observation methods such as adaptive optics to compensate for the disturbing effect of atmospheric turbulence on telescopic images, astronomers discover small or even relatively large satellites around other asteroids. Radar observations show that some Earth-grazers – asteroids that fly past the Earth at short distances – are also binary objects. The Hubble Space Telescope photographs a number of asteroid satellites, but not around Herculina, and on closer inspection the asteroid proves to be making its way through space all alone.

Asteroid Sylvia even appears to have two satellites, called Romulus and Remus, which are discovered in 2001 and 2004 respectively. And the Kuiper Belt object 2003 EL$_{61}$, a large, icy object beyond the orbit of Neptune now known as Haumea, is accompanied by two satellites. Pluto, which is also part of the Kuiper Belt, actually has three moons. Asteroid satellites probably have their origins in collisions, or occur when a porous, rapidly rotating asteroid disintegrates into several pieces.

Collision Course

Astronomers Discover the Effects of a Comet Impact on a Planet

The discovery of Comet Shoemaker-Levy 9 on March 24, 1993, is pure coincidence. Husband-and-wife astronomers Carolyn and Eugene Shoemaker, together with their colleague David Levy, use a 40-centimeter Schmidt telescope at Palomar Observatory in California to hunt for Earth-grazers – asteroids that can pass close to the Earth. On one of their photographs, just a few degrees from the planet Jupiter, they see a strange little comet, which appears to have multiple nuclei.

Orbital calculations show that Comet Shoemaker-Levy 9 does not orbit the Sun, but moves in a very elongated orbit around Jupiter, with an orbital period of around two years. The comet was probably captured by the gravitational field of the giant planet in the early 1970s. Calculations show that on July 7, 1992, it flew past Jupiter's cloud tops at a distance of only 40,000 kilometer, and its porous nucleus was ripped apart into several pieces by the planet's tidal forces.

It soon becomes clear that Shoemaker-Levy 9 will not survive its next encounter with Jupiter. Between July 16 and 22, 1994, the twenty-one fragments of the comet, ranging from a few hundred meters to around two kilometers in size, bore into the planet's atmosphere at a speed of about 60 kilometers per second. For the first time in history, astronomers are witness to a collision between two celestial bodies.

Nearly all professional telescopes on Earth are directed towards Jupiter at the time of the impact, as are the Hubble Space Telescope, the German X-ray satellite Rosat, the space probe Ulysses and the planetary explorer Galileo, on its way to a rendezvous with Jupiter in December 1995. Astronomers have never before organized such a large-scale international observation campaign.

The impacts of the cometary fragments are much more spectacular than most theorists had expected. They generate gigantic fireballs with temperatures of tens of thousands of degrees, which extend to a few thousand kilometers above the atmosphere. Cometary material falling back towards the planet produces striking dark spots in Jupiter's atmosphere, which are easy to see even with amateur telescopes and remain visible for many months. The largest spot, caused by fragment G, is nearly as large as the Earth. The impact of fragment G generates an explosive energy 750 times as powerful as that of the entire nuclear arsenal on our planet.

It is estimated that an impact like that of Comet Shoemaker-Levy 9 occurs on Jupiter once every thousand years. Elongated crater chains on Jupiter's moons Callisto and Ganymede were also probably caused by the impacts of fragmented comets. The observations of the comet's impact underscore even more the risk of cosmic impacts on Earth, though they are much less common than on Jupiter.

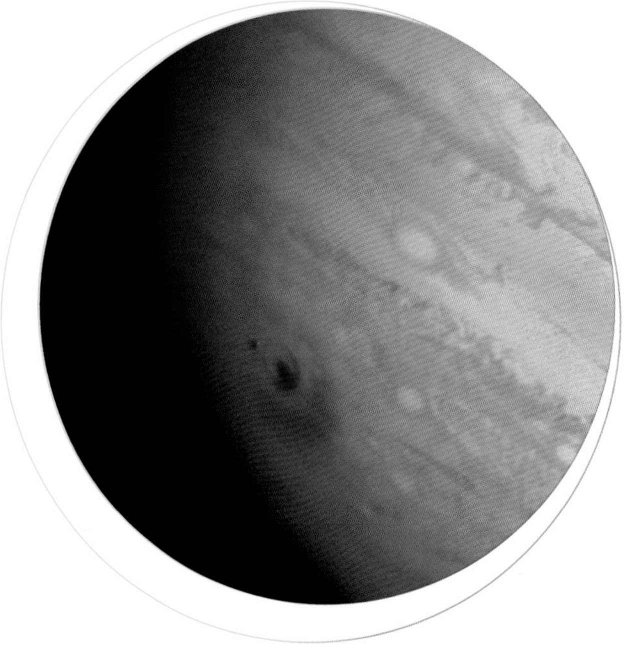

The impact of fragments of comet Shoemaker-Levy 9 on Jupiter left gigantic scars in the giant planet's atmosphere, which were even visible with small amateur telescopes on Earth. The largest spot on this photograph was caused by fragment G. (NASA/ESA/MIT/H. Hammel)

In 2006, the core of comet Schwassmann-Wachmann 3 shattered into several fragments. This photo, taken by the European Very Large Telescope, shows fragment B. The colored dots in the background are stars, photographed through different color filters. (ESO)

1994

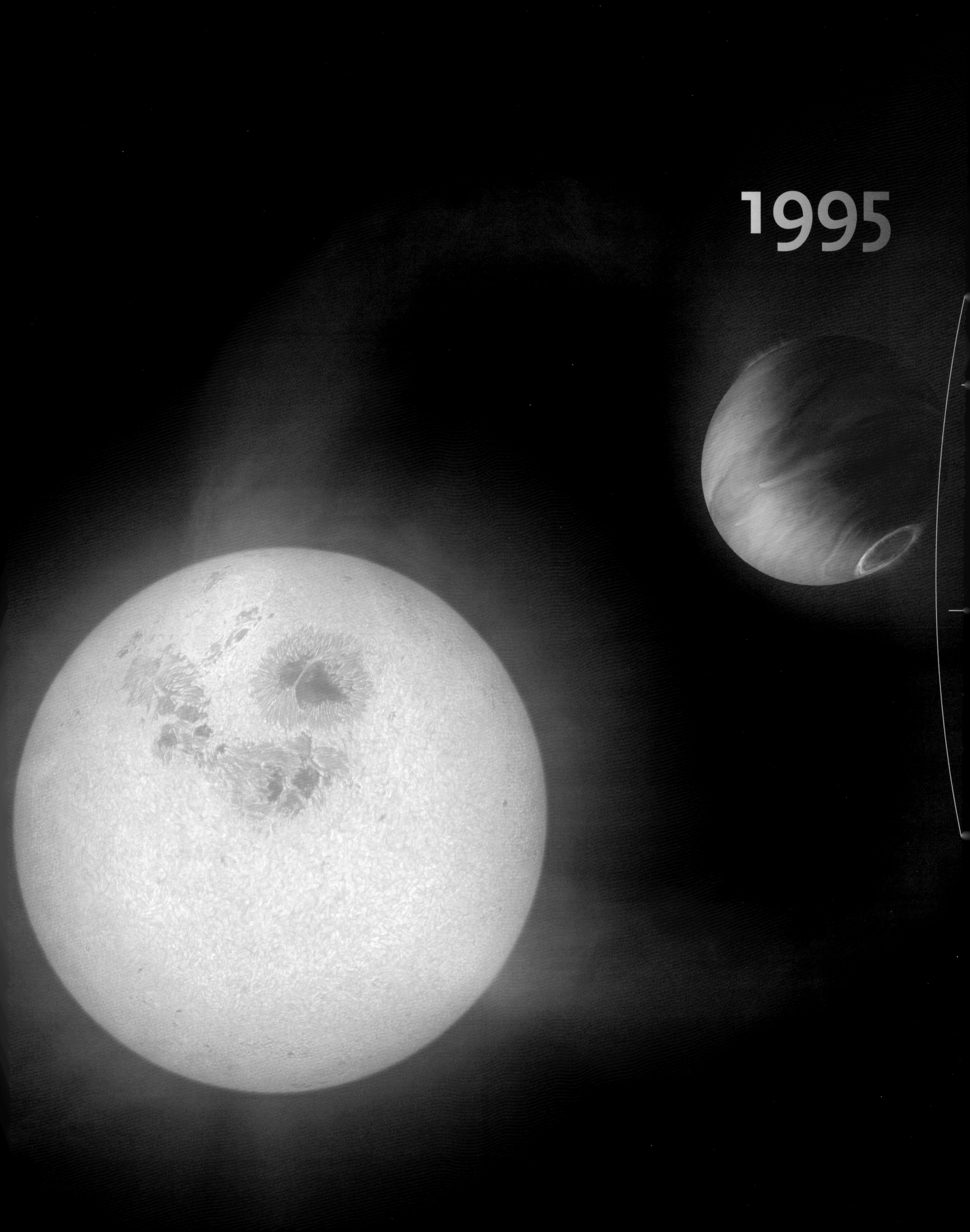
1995

Other Worlds

Michel Mayor and Didier Queloz Discover the First Exoplanet

At the end of the sixteenth century, Italian priest and philosopher Giordano Bruno is one of the first to suggest that the stars in the night sky are all suns, and are accompanied by planets like the Earth. A hundred years later, Christiaan Huygens actively starts searching for these extrasolar planets, or 'exoplanets', but soon realizes that it is a hopeless task: a planet orbiting another star will be so faint that it will be completely outshone by the light of its parent star.

In the middle of the twentieth century, Dutch-American astronomer Peter van de Kamp conducts positional measurements on Barnard's Star over a period of several decades. The star appears to wobble slightly, leading Van de Kamp to conclude that the dwarf star is accompanied by two Jovian (Jupiter-like) planets. The existence of the planets is, however, not confirmed by later precision measurements. In 1991, Aleksander Wolszczan and Dale Frail do find two planets in orbit around a pulsar, but it will be another few years before an exoplanet is discovered around a Sun-like star.

The first to strike lucky are Michel Mayor of the Geneva Observatory and his Ph.D. student Didier Queloz. In 1995, using a relatively small telescope at the Observatoire de Haute-Provence in the south of France, they discover that the star 51 Pegasi, which is about 50 lightyears from the Earth, moves a little closer to us and then away again every

4.23 days. That periodic variation in the radial velocity of the star can only be explained by the presence of a large, massive planet in a very small orbit. Mayor and Queloz present their discovery on October 6, during a workshop in Florence. On November 23, they publish the results in *Nature*.

The discovery of 51 Pegasi b, as the planet is officially called, is confirmed on October 12 by American planet hunters Geoff Marcy and Paul Butler. Shortly afterwards they find massive giant planets in small orbits around other Sun-like stars. It is not so surprising that these 'hot Jupiters' are the first to be found; small planets in wide orbits do not cause any perturbations in the motion of the parent star. The existence of the 'hot Jupiters' is initially not properly understood; the planets are probably formed at much greater distances and then spiral inwards at a later date.

At present, more than 500 exoplanets have been discovered, including a number of 'super-Earths', or planets that are only a few times more massive than the Earth. As many as five, and maybe even seven, planets have been discovered around the star HD 10180. By far the majority of exoplanets have been found using the radial-velocity method, but the transit method is yielding more and more results. If we observe its orbit exactly edge-on, the planet will move in front of its parent star once every orbit and the star will then be slightly fainter for several hours. It is estimated that at least half of all Sun-like stars are accompanied by one or more planets.

Illustration of 51 Pegasi b, the first exoplanet to be discovered around a Sun-like star. The planet is a few times more massive than Jupiter, and orbits its parent star at a distance of only 7.8 million kilometers. (Debivort)

With the Hubble Space Telescope, a giant planet has been discovered orbiting a red dwarf star at a distance of only 1.2 million kilometers. This speculative artist's impression shows possible magnetic interaction between the star and the planet, causing an aurora on the planet. (NASA/ESA/STScI/Adolf Schaller)

Failed Star

Shrinivas Kulkarni Discovers the First Brown Dwarf, Gliese 229B

Stars are nuclear power plants. In their interiors, where pressure and temperature are enormously high, spontaneous nuclear fusion of hydrogen occurs. Massive stars consume their nuclear fuel at an unimaginably high rate, and lower-mass stars like the Sun take things a little easier because conditions in their interiors are less extreme. In the smallest dwarf stars, the fusion reactions are so slow that they can keep going for billions of years on their meager stocks of hydrogen.

Theoretical calculations show that, if a star is lighter than 7.5 percent of the mass of the Sun, there will be no hydrogen fusion in its core. In 1975, American astronomer Jill Tarter calls these 'failed stars' brown dwarfs. Fusion reactions of deuterium (heavy hydrogen) can occur in a brown dwarf, but they generate almost no energy. Brown dwarfs are therefore extremely faint, and predominantly emit infrared radiation.

Almost no one doubts the existence of brown dwarfs, but it takes until 1995 for one to actually be found. In October 1994, Shrinivas Kulkarni and his colleagues use the 60-centimeter telescope at Palomar Observatory to make infrared photographs of the young dwarf star Gliese 229, which is 19 lightyears away in the constellation of Lepus. Using adaptive optics – a method of improving images by compensating for atmospheric turbulence – they find an extremely faint speck of light close to the star.

To make sure that it is not a distant background object, they observe the star again a year later. The speck of light has exactly the same proper motion as the star, proving that they belong together. The spectrum of the companion is recorded with the 5-meter Hale Telescope and proves to be an object about the same size as Jupiter but twenty to fifty times more massive, or two to five percent of the Sun's mass. It is the faintest star every observed, radiating a hundred thousand times less energy than the Sun. It is also around six billion kilometers from the parent star, about the same as the distance from the Sun to Pluto.

On November 17, Gliese 229B, as the brown dwarf is officially called, is observed by the Hubble Space Telescope. That same month Kulkarni and his colleagues publish their discovery in *Nature*. The spectrum of the brown dwarf shows lines of methane, a gas that also occurs in large quantities in the atmosphere of the giant planet Jupiter. The temperature of Gliese 229B is determined at around 1,100 degrees Celsius – mostly residual heat from the time when the brown dwarf was first formed.

Using sensitive infrared detectors and adaptive optics, astronomers discover hundreds more brown dwarfs after 1995. The closest brown dwarfs orbit the star Epsilon Indi B, less than twelve lightyears from the Earth.

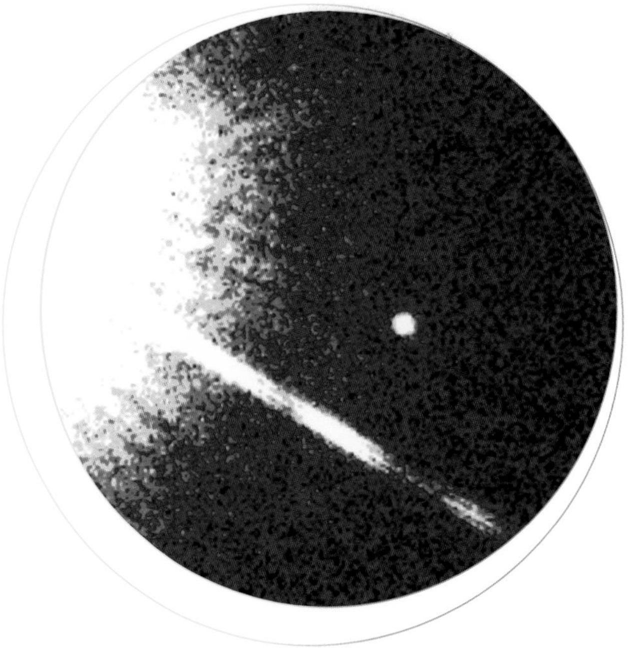

◐ Gliese 229B – a very faint brown dwarf in orbit around a young dwarf star – can be seen on this Hubble photograph as an unsightly speck of light close to the overexposed image of the dwarf star. (NASA/ESA/Caltech/JHU)

➲ With the European Very Large Telescope in Chile, a remarkable binary system has been discovered consisting of a small, hot white dwarf and a very cool brown dwarf, which orbit each other in about two hours. The white dwarf is almost half as massive as the Sun, and the brown dwarf is over fifty times more massive than the giant planet Jupiter. (ESO)

1995

1996

Fossil Remains

David McKay Discovers Signs of Life in a Martian Meteorite

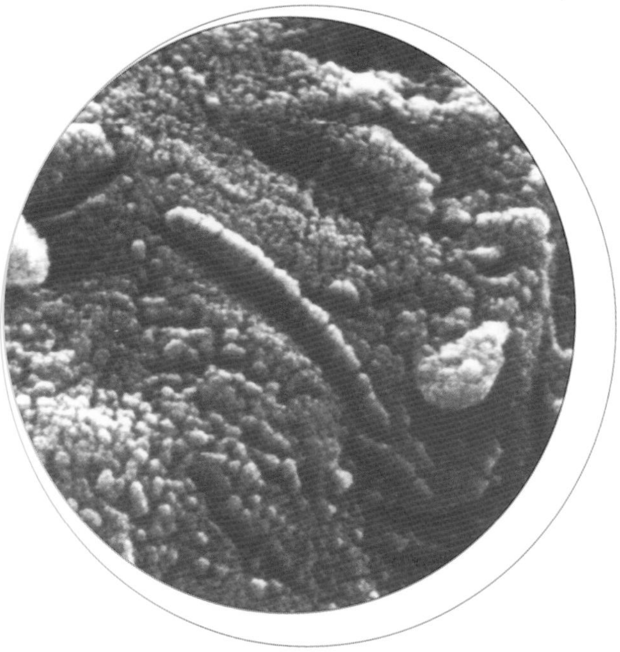

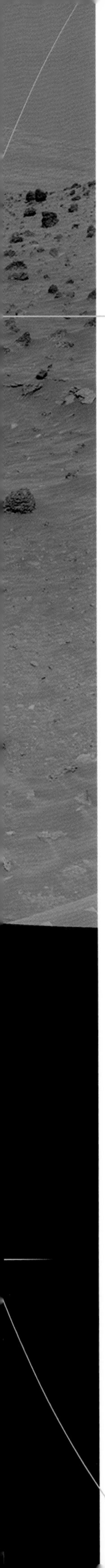

In December 1984, geologists from the American Antarctic Search for Meteorites (ANSMET) program find a large meteorite in the Allan Hills on Antarctica. The space rock is about 15 x 10 x 7.5 centimeter and weighs a little less than two kilograms and is cataloged as ALH84001.

It is not until ten years later that David Middlefehldt discovers that the rock displays many similarities with the so-called 'SNC' meteorites from the planet Mars. The gas inclusions in these meteorites have the same composition as the Martian atmosphere. One of the most famous, the Nakhla meteorite, for example killed a dog when it fell to Earth in Egypt in 1991.

Detailed geological study of ALH84001 shows that it is an igneous rock about 4.5 billion years old and that it must have been blasted into space by a severe meteorite impact – not so unlikely, given the low surface gravity of Mars. Radiological tests show that the rock then spent sixteen million years roaming around the solar system. About 11000 BC, it fell to Earth, where it remained perfectly preserved in the polar ice for thousands of years.

A team of geologists led by David McKay of NASA's Johnson Space Center in Houston discover small carbonate globules inside the meteorite, which were almost certainly formed by running water. The globules contain organic molecules, such as amino acids and polycyclic hydrocarbons. The researchers also find microscopically small magnetite crystals, which closely resemble magnetic minerals produced by some terrestrial micro-organisms. Furthermore, using a scanning electron microscope, McKay and his colleagues discover long, chain-like structures in the carbonate globules, with a diameter of twenty to a hundred nanometers, which are very similar to micro-fossils in old terrestrial rocks.

McKay draws the cautious conclusion that, long ago, there were micro-organisms living on Mars and therefore ALH84001 may contain the first evidence of extraterrestrial life. The results are published in *Science* on August 16, 1996, but because the news leaks out, the geologists present their findings at a press conference ten days earlier, at which even President Bill Clinton makes a statement. The press conference takes place several months before the launch of the planetary probe Mars Pathfinder, which is to herald a new phase in American research on Mars.

From the very beginning, there is much skepticism about the interpretation of the facts by McKay's team. The 'nanofossils' are considered too small to contain RNA molecules, organic molecules probably originate from terrestrial contamination, and magnetite crystals can also evolve in an inorganic environment. As time passes, the evidence for fossilized Martian bacteria becomes less and less credible.

➋ Electron-microscopic image of possible micro-fossils in the Martian meteorite ALH84001. Other evidence suggesting the presence of the fossil remains of organic activity has been found in the meteorite, such as carbonate globules and magnetite crystals. (NASA)

➌ The surface of Mars, photographed by the unmanned American Mars rover Spirit. Spirit and its twin, Opportunity, found convincing evidence for the theory that there was flowing water on Mars billions of years ago. Perhaps life also evolved then, too. (NASA/JPL/Cornell University)

Cosmic Fireworks

Paul Groot and Titus Galama Discover the First Afterglow of a Gamma-Ray Burst

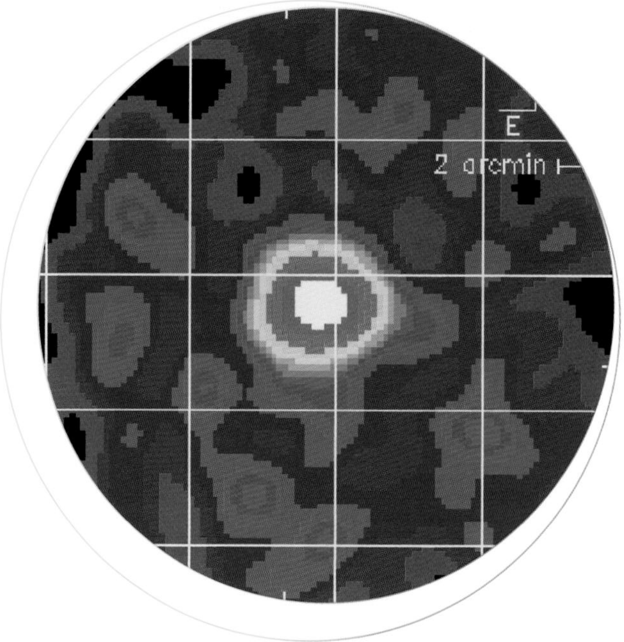

After Ray Klebesadel discovers gamma-ray bursts in 1969, it takes astronomers another thirty years to find out at what distance these explosive phenomena occur. NASA's Compton Gamma Ray Observatory may detect several thousand gamma-ray bursts in the 1990s, but it is unable to discover their true nature. They might be explosive events on neutron stars in the halo of our own Milky Way, or extremely powerful eruptions in very distant galaxies.

It is the Italian-Dutch satellite BeppoSAX that first makes it possible to determine the position of gamma-ray bursts accurately. BeppoSAX is not only equipped with a gamma-ray detector, but it also has wide-angle X-ray cameras and X-ray telescopes. Researchers suspect that gamma-ray bursts, which often last no longer than a few dozen seconds, have a longer afterglow at X-ray wavelengths. If so, BeppoSax's X-ray telescopes can determine the position of the burst.

BeppoSAX is launched on April 30, 1996, and on January 11, 1997, it succeeds in determining the position of a gamma-ray burst with reasonable precision within twenty-four hours. Yet optical and radio telescopes find nothing noteworthy at the position of the burst, in the constellation of Serpens.

On February 28, 1997, BeppoSAX strikes lucky again, registering a gamma-ray burst in the constellation of Orion, with X-ray telescopes determing the position. That same day, Titus Galama, a Ph.D. of astronomy in Amsterdam, directs the radio telescope at Westerbork at the area around the position but without any success. That evening, Galama's colleague Paul Groot passes the position on to John Telting, who uses the 4.2-meter William Herschel Telescope on La Palma to make a photograph of the area where the burst occured.

A second photograph cannot be made until March 8, when the weather is clear again on La Palma. By comparing the two images, Groot and Galama discover the optical afterglow of the gamma-ray burst: a small, faint star that is visible on the photograph made on February 28 but not on the one from March 8. Another four days later, Jorge Melnick of the European Southern Observatory in Chile makes a long-exposure photograph with the New Technology Telescope. The image shows a faint, elongated smudge of light exactly at the position of the optical afterglow: it is the distant galaxy in which the gamma-ray burst occurred.

Groot and Galama's discovery is published in *Nature* on April 17, 1997, only seven weeks after the gamma-ray burst. It is now finally clear that the high-energy explosions take place at distances of billions of lightyears away. That makes them the most powerful explosions in the universe since the Big Bang. They almost certainly occur when extremely massive, rapidly rotating stars implode to become black holes at the end of their lives, or when two neutron stars in a binary system collide and merge to become a black hole.

Shortly after the gamma-ray burst of February 28, 1997, the X-ray afterglow of the burst was recorded by the Italian-Dutch satellite BeppoSAX. That enabled the position of the burst to be determined precisely, so that an optical afterglow could also be identified. (BeppoSAX)

Gamma-ray bursts are the terminal explosions of massive, rapidly rotating stars. If the core of the star implodes into a black hole, two jets of high-energy radiation and particles are blown into space along the axis of the star. If one of the jets is more or less aimed at the Earth, we see a gamma-ray burst. (ESO)

1997

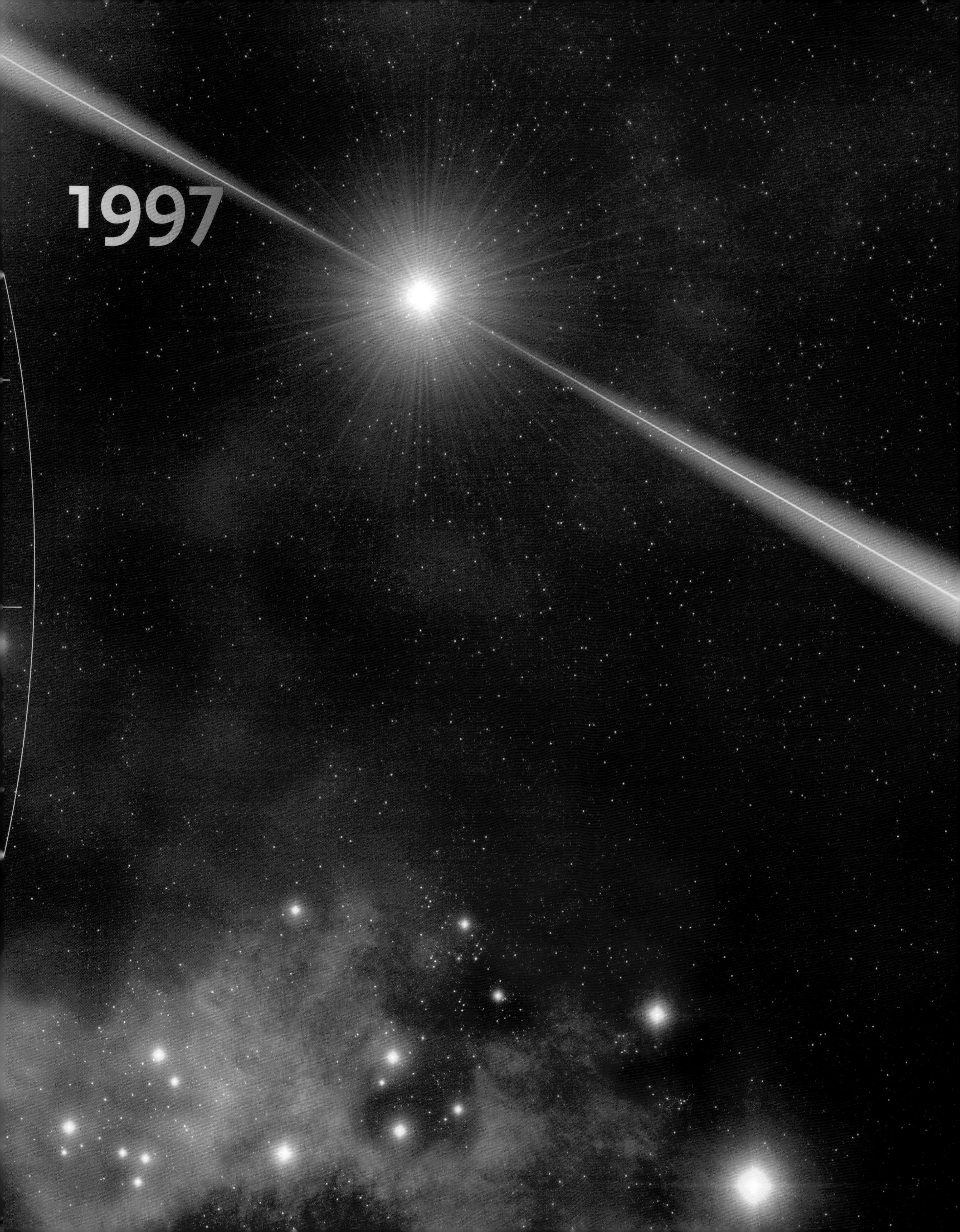

1998

Accelerating Universe

Adam Riess and Saul Perlmutter Discover Dark Energy

The discovery that the expansion of the universe is not slowing down, but actually speeding up is chosen by the American weekly *Science* as the scientific breakthrough of the year in 1998. It is a little risky so soon after the announcement of the revolutionary results, but in the years that follow more and more evidence is found to support the notion that we live in an 'accelerating universe', dominated by mysterious dark energy.

In 1917, Albert Einstein already suspects that there may be a repellent force in empty space, which he calls the cosmological constant. Einstein introduces the constant to bring his theory of relativity in line with a static, unchanging universe. After Edwin Hubble discovers in 1929 that the universe is not static, the cosmological constant becomes redundant. In a conversation with George Gamow – the 'father of the Big Bang theory'– Einstein calls the introduction of his constant the biggest blunder of his life.

For decades, astronomers believe that the gravity of all matter in the universe acts as a brake on its expansion. The idea is that in the course of time, the speed of expansion slowly but surely decreases. When, sometime in the distant future, the average density of the universe lies above a certain critical value, the expansion will turn into a contraction and the universe will end in a 'Big Crunch'.

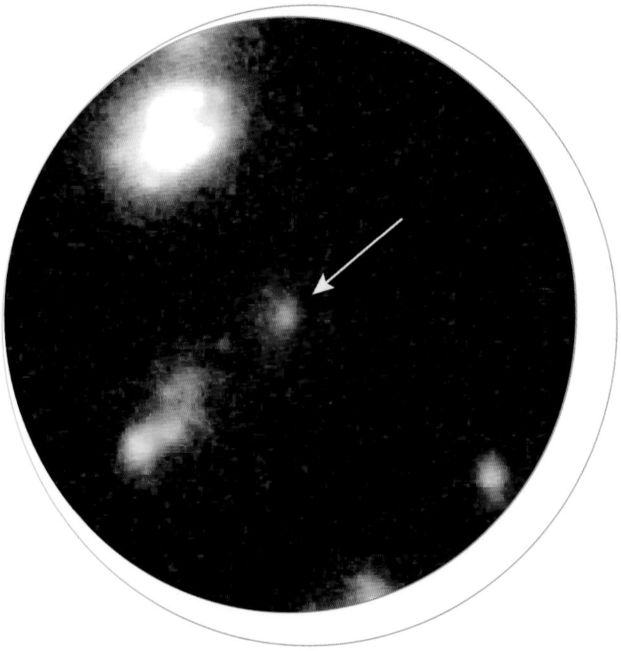

The arrow points to a supernova billions of lightyears from the Earth, discovered in the Hubble Deep Field. Measurements of these distant supernovae show that the expansion velocity of the universe used to be lower than it is now: we live in an accelerating universe. (NASA/STScI/Adam Riess)

Hubble photograph of the Spindle Galaxy (NGC 5866), 50 million lightyears away in the constellation of Draco. In the distant future all galaxies will disappear beyond our observation horizon, as a consequence of the accelerating expansion of the universe. (NASA/ESA/Hubble Heritage Team)

To discover the history of the universe's expansion, and therefore to know more about the future, you need to know the speed at which it was expanding billions of years ago. To find that out, two large international teams of astronomers use observations of distant type Ia supernovae. Because their actual luminosity is known, their distances can be calculated from their observed brightness. By comparing these distances with the redshift in the light from the supernovae, you can then determine the extent to which the expansion of the universe has slowed down in the past billions of years.

The Hi-z Supernova Search Team, led by Adam Riess of Harvard University, and the Supernova Cosmology Project, under the leadership of Saul Perlmutter of the Lawrence Berkeley Laboratory, present their astounding results at a large astronomy conference in Washington in early 1998. The supernova measurements of the two teams show that the expansion velocity of the universe has increased, rather than decreased, in the past few billion years. There appears to be a mysterious dark energy at work, as though empty space itself had some sort of repellant effect. That energy has supposedly had a greater impact on the expansion speed of the universe in the past few billion years than the combined gravity of all matter contained within it.

Precision measurements of the cosmic background radiation by the Wilkinson Microwave Anisotropy Probe (WMAP) from 2001 seem to confirm the existence of dark energy. At present, the jury is still out as to whether dark energy is identical to Einstein's cosmological constant.

Wet History

Spirit and Opportunity Discover Traces of Water on Mars

In 1972, while orbiting Mars, the American space probe Mariner 9 discovers geological structures, such as giant outflow channels and dried-up river beds, which suggest that there may once have been flowing water on the surface. The two Viking spacecraft that orbit the red planet in the late 1970s make detailed maps of the surface and discover that, in addition to frozen carbon dioxide, the polar caps also contain frozen water. Little is known, however, about the quantity of water on Mars.

In 1998, the Mars Global Surveyor space probe discovers traces of flowing water on the inner walls of craters, and in 2002, measurements by the Mars Odyssey orbiter show that there is a great deal of underground ice on the planet. This implies that, billions of years ago, Mars probably had a warmer, wetter climate, and perhaps there were seas and oceans for long periods of time. It becomes the job of NASA's robotic rovers Spirit and Opportunity to find answers to these questions.

The two Mars rovers are launched in the summer of 2003. Spirit lands in the large impact crater Gusev on January 4, 2004, and Opportunity follows 3 weeks later, landing on the other side of the planet in Meridiani Planum. The six-wheeled vehicles are equipped with stereoscopic color cameras, a robotic arm fitted with several instruments including an abrasion tool and a microscope to study Mars rocks, and various spectrometers to analyze the composition of the soil.

Only a few weeks after the landing, Opportunity finds evidence that Meridiani Planum must have been under water for a long period because rocks contain sulfur-rich salt deposits. Opportunity also finds 'blueberries'; blue-grey spherules consisting of minerals that were once dissolved in water and deposited in porous rock. Spirit, too, finds traces of water, in the form of crystallized minerals and stratified rock deposits.

The two Mars rovers keep going much longer than planetary researchers had dared to hope. Over a number of years, they cover a lot of ground on the surface of Mars where Spirit's journeys include a visit to the Columbia hills, while Opportunity descends into the impact crater Victoria. Time and time again, geological and mineralogical tests show that billions of years ago there must have been flowing water on the surface of the planet for a long period.

The Mars lander Phoenix, which makes a soft landing on the edge of the northern polar region of the planet on May 26, 2008, studies frozen soil samples and is the first space probe in history to conduct *in situ* analysis of water on another celestial body. Measurements by the European Mars Express and the American Mars Reconnaissance Orbiter also show that there are clay deposits in many locations on Mars, further indications that the planet was wet at some time in its history. The question of whether there has ever been life on Mars remains, however, unanswered.

↑ The American robotic rover Opportunity, which landed on Mars in early 2004, made this microscopic photograph of 'blueberries' – small, stratified mineral globules, which must have been created under the influence of liquid water. (NASA/JPL/USGS/Cornell University)

→ The Mars rover Opportunity spent more than a year in the immediate vicinity of the 800-meter-wide Martian crater Victoria, photographed here by the Mars Reconnaissance Orbiter. Opportunity found the strongest indications that the red planet has a wet past. (NASA/JPL/University of Arizona)

2004

2004

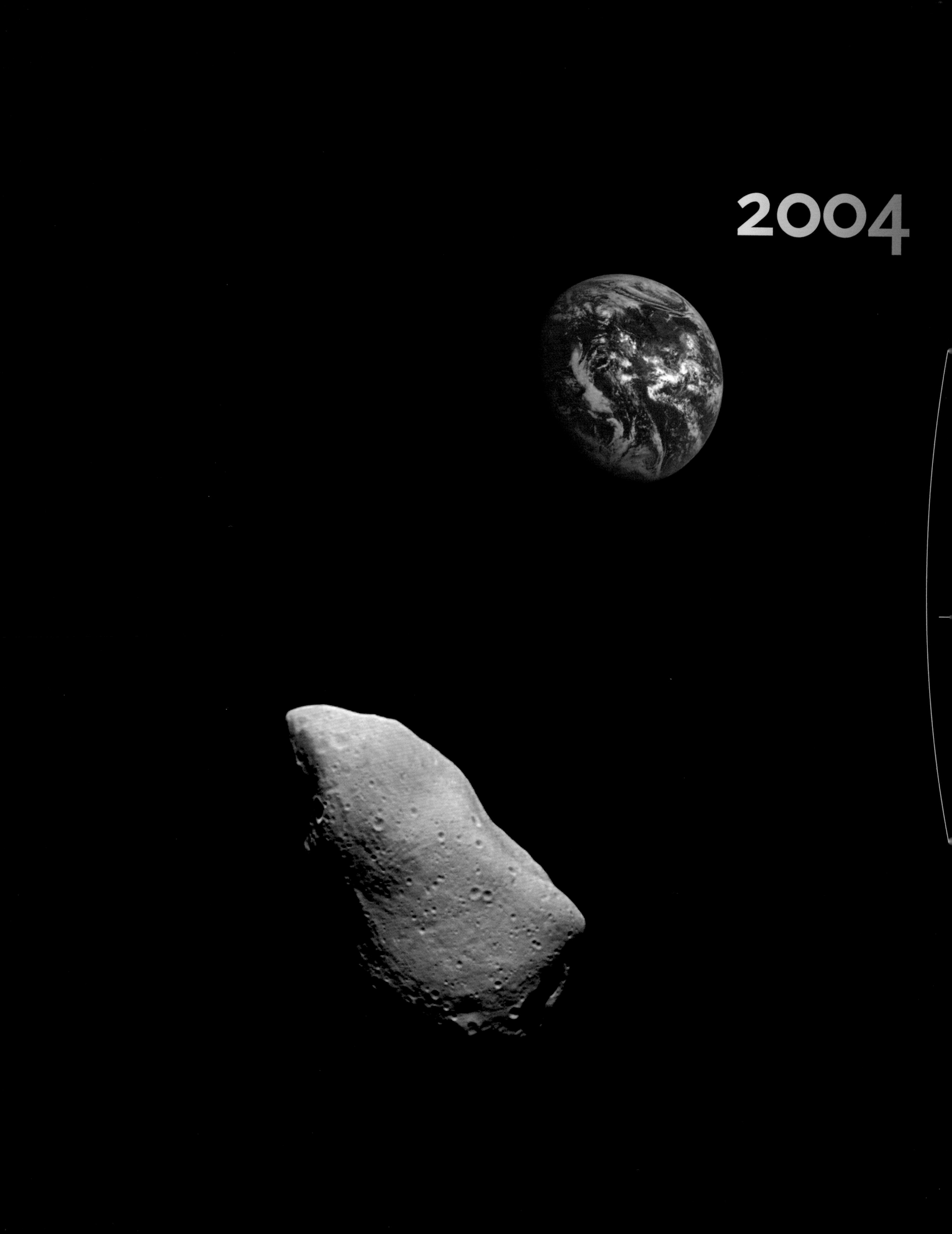

Target Earth

David Tholen and his Colleagues Discover the Earth-Grazer Apophis

More than a thousand asteroids, more than a kilometer across, are estimated to move through the inner solar system. In theory, these Earth-grazers can at some point collide with our planet. In the mid-1990s, NASA starts the Spaceguard Survey, which aims to track down at least ninety percent of these potentially dangerous projectiles.

As part of the Spaceguard project, the University of Hawaii Asteroid Survey receives a grant from NASA. David Tholen, Roy Tucker and Fabrizio Bernardi use a telescope at Kitt Peak Observatory to hunt for Earth-grazers with an orbital period of less than a year, and whose orbit lies mainly within that of the Earth. On June 19, 2004, they discover an Aten asteroid (a near-Earth asteroid named after the prototype, discovered in 1976), at a short distance from the Earth. Tholen and his colleagues photograph the chunk of rock on two successive nights, but are unable to determine its orbit accurately.

The asteroid, given the temporary designation 2004 MN_4, is re-discovered on December 18 by Gordon Garradd of Siding Spring Observatory in Australia. Now it is possible for astronomers to calculate its orbit with reasonable precision. They discover that the asteroid may collide with the Earth on April 13, 2029. At the end of December, NASA estimates the chances of a collision at 2.7 percent.

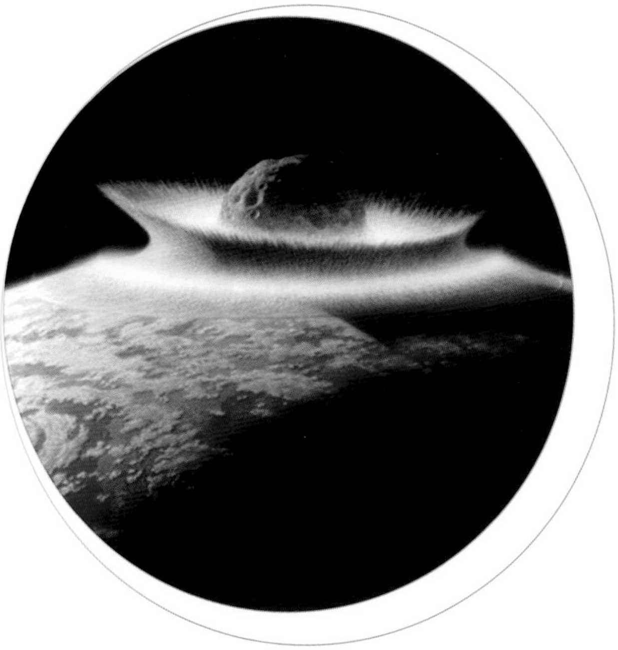

217 — 218

The media devote little attention to the announcement, largely because of the undersea earthquake and the subsequent tsunami that ravages Indonesia, Southeast Asia and Sri Lanka on December 26. Before the end of the year, 2004 MN_4 is rediscovered on photographs taken in March 2004 enabling its orbit to be calculated with even greater precision. It appears that the rock will pass very close to the Earth in 2029, but will certainly not impact.

In May 2005, 2004 MN_4 is given the definitive asteroid number 99942. Tholen and Tucker call the object, which is estimated to be 350 meter in size, Apophis – the Greek name for the Egyptian god Apep. Apophis is also the hostile alien from the television series Stargate SG-1, who wants to destroy the Earth.

During its passage on April 13, 2029, at a distance of 13,000 kilometers, Apophis will be visible with the naked eye from Europe, Africa and West Asia. It will look like a fast-moving star. There is a chance of 1 in 45,000 that the asteroid's orbit will be disturbed to such an extent by the passage that it will collide with the Earth after all, seven years later, on April 13, 2036. If that happens, the impact will occur in southern Russia, the Pacific Ocean, Central America or the Atlantic Ocean. Apophis will not be clearly visible for observation again until 2013, when it may be possible to determine its orbit even more accurately. If it then appears that the Earth will indeed be in danger in 2036, space agencies still have twenty-three years to nudge the Earth-grazer off course.

➋ Sixty-five million years ago, the Earth was hit by an asteroid about ten kilometers across. The impact led to the demise of the dinosaurs. Such massive impacts occur on Earth on average once every hundred million years. (NASA/Don Davis)

➌ Composite photograph of the asteroid Gaspra (photographed by the planetary explorer Galileo) and the Earth (photographed by the European space probe Rosetta). The Earth-grazer Apophis, estimated to be about 270 meters across in length, is fortunately much smaller than Gaspra. (NASA/JPL/USGS/ESA/Osiris Team)

Warrior Princess

Michael Brown Discovers the Dwarf Planet Eris

In the early morning of Wednesday, January 5, 2005, a shout of joy resounds through the corridors of the California Institute of Technology in Pasadena. In the constellation of Cetus, amidst thousands of immobile specks of light, Michael Brown discovers a single moving object: a celestial body in the far outer regions of the solar system that is larger than Pluto. Is this the long-sought-for Planet X?

Since the discovery of 1992 QB_1 by Dave Jewitt and Jane Luu several hundred objects have been found in the Kuiper Belt. Some of them – like Ixion, Varuna, Quaoar and Sedna – are surprisingly large. It stands to reason that there must also be a few Kuiper Belt objects that are bigger than Pluto.

Brown is determined to find these 'Pluto killers'. Just like Charles Kowal at the end of the 1970s, he photographs practically the entire sky with the 1.2-meter Schmidt Telescope at Palomar Observatory. But in 2000, after a fruitless search with relatively long-exposure photographic plates, he switches to using state-of-the-art digital cameras specially designed for this purpose. 'Planet X' is photographed on October 21, 2003, with the QUEST camera, designed and built by David Rabinowitz. Chadwick Trujillo writes the software with which the minuscule moving speck of light is found among enormous amounts of data at the beginning of 2005.

Brown, Rabinowitz and Trujillo also find Quaoar and Sedna. Once the orbit (and thereby the distance) of an ice dwarf is known, the observed brightness gives an indication of its size. You do have to make an assumption about the reflecting capacity of the object: the higher it is, the smaller the object's diameter. In the case of 'Planet X', its diameter must be larger than Pluto's, even if it reflects one hundred percent of the sunlight that reaches it.

Brown, Rabinowitz and Trujillo give their hunting trophy the nickname Xena, after the warlike leading character in the popular television series 'Xena: Warrior Princess'. When they announce the discovery in the summer of 2005, Xena is given the official designation 2003 UB_{313}. The same summer, using the 10-meter Keck Telescope on Hawaii, Brown discovers a moon around Xena, which he gives the provisional name of Gabrielle.

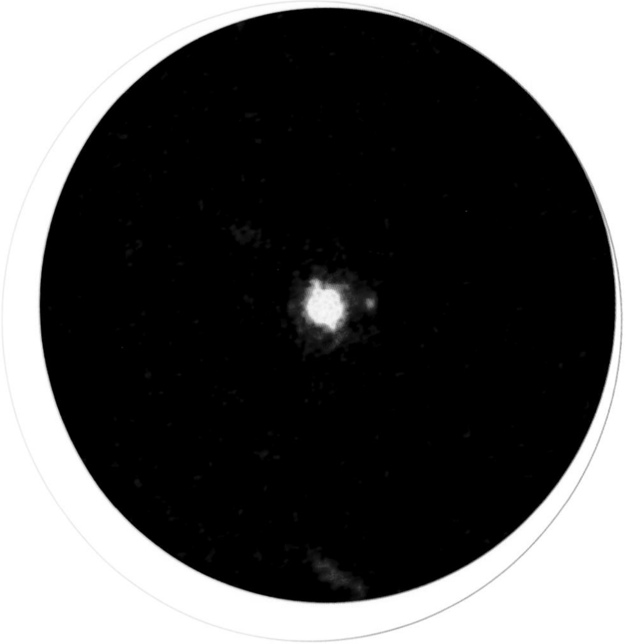

With the discovery of 2003 UB_{313}, the debate on the planetary status of Pluto flares up again. The International Astronomical Union (IAU) sets up a committee of experts to come up with a conclusive definition of what exactly constitutes a 'planet'. On August 24, 2006, during a tumultuous meeting of the IAU in Prague, it is decided that Pluto and 2003 UB313 are not 'real' planets, but dwarf planets. Later that year, Xena and Gabrielle are given the official names Eris and Dysnomia, after the Greek goddesses of strife and lawlessness.

↑ The best photograph of Eris and its moon Dysnomia was made by the 10-meter Keck Telescope on Mauna Kea, Hawaii. With a diameter of around 2,400 kilometer, Eris is the largest dwarf planet in the solar system. It is, however, not impossible that even larger ice dwarfs will be found in the future. (W.M. Keck Observatory)

→ Seen from the dwarf planet Eris, the Sun is little more than an extremely bright star. This artist's impression also shows the moon Dysnomia. Like Pluto's moon Charon, Dysnomia was probably created by a collision in the early days of the solar system. (NASA/ESA/STScI/Adolf Schaller)

2005

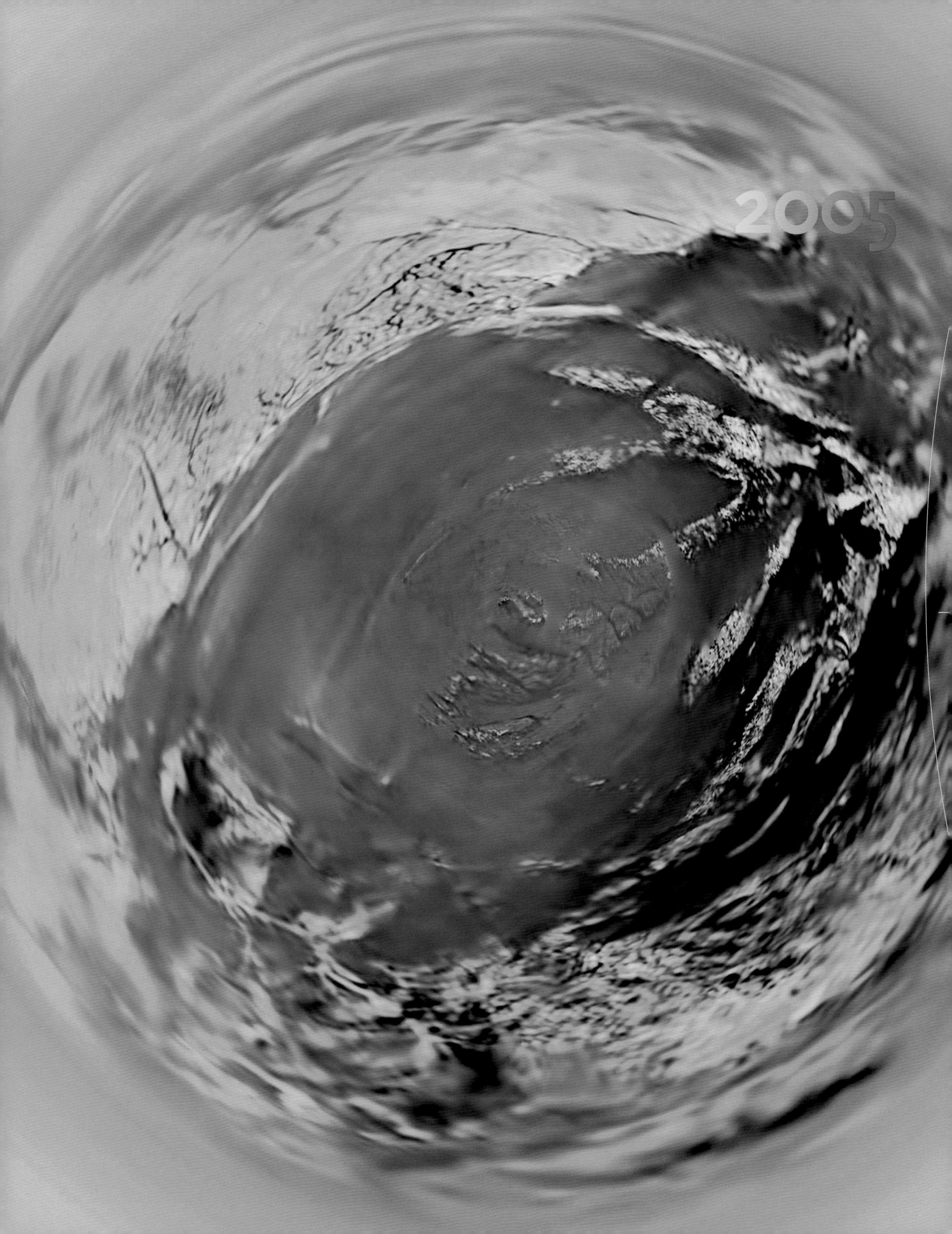
2005

Icy Landing

The Space Probe Huygens Discovers the Surface of Titan

The large Saturnian moon Titan is discovered on March 25, 1655, by the Dutch physicist and astronomer Christiaan Huygens. With a diameter of 5,150 kilometers, Titan is the second largest planetary moon in the solar system and the only one with a substantial atmosphere. Gerard Kuiper proves the existence of the atmosphere in 1944, and in 1980, the American space probe Voyager 1 discovers that Titan's atmosphere consists mainly of nitrogen.

The European space probe Huygens, named after Titan's discoverer, makes the first ever soft landing on the surface of a planetary moon on January 14, 2005. Huygens is launched on October 15, 1997, as part of the American Cassini project. After flybys of Venus, the Earth and Jupiter, the probe goes into an elongated orbit around Saturn in the summer of 2004. The Huygens capsule, weighing 318 kilograms and 1.3 meter in diameter, is launched in the direction of Titan on December 25, 2004.

Huygens is equipped with a heat shield and a large parachute. During the descent through the atmosphere it experiences strong tailwinds and conducts measurements on the pressure, temperature and wind speeds. At a height of around 20 kilometer, Huygens emerges from a thick layer of haze in the atmosphere and makes the first photographs of the surface of Titan.

Because it not known whether the probe will land on dry ground or in a sea of liquid hydrocarbons, Huygens is designed to float and finally lands on the edge of a reasonably bright area of Titan known as Xanadu. The icy landscape – the surface temperature on Titan is 180 degrees Celsius below zero – is shrouded in a somber orange glow. In the immediate vicinity of the landing site there are a number of large 'boulders' of frozen ice, and the boggy ground seems to be soaked in liquid methane gas. Huygens sends photographs and measurement data to Earth via the Cassini mothership, but an hour and a half after the landing the probe's batteries are exhausted and radio contact is lost.

On a series of flybys, Cassini charts large parts of Titan's surface using radar images that show impact craters, mountains and volcanoes. In Titan's northern hemisphere, there are extremely flat, dark areas – almost certainly lakes of liquid methane or ethane gas. Infrared observations by Cassini also show cloud structures in the southern hemisphere.

Planetary researchers suspect that Titan may have a large subterranean ocean of liquid water, like the Jovian moon Europa. In the distant future, when the Sun swells up to become a red giant, Titan will warm up and the conditions may be favorable for sustaining life.

221

222

In January 2005, the European space probe Huygens made a soft landing on the surface of the large Saturnian moon Titan. For an hour and a half, Huygens sent photographs and measurement data on the extremely cold surface of Titan back to Earth. (ESA)

As Huygens descended to Titan with the aid of a parachute, it took photographs of the landing site from an altitude of 5 kilometers. The photos were used to produce this fisheye image, in which the circular rim is the horizon. (ESA/NASA/JPL/University of Arizona)

New Earth

Stéphane Udry Discovers the Habitable Exoplanet Gliese 581c

In 1995, Michel Mayor and Didier Queloz of the Geneva Observatory discover the first planet orbiting another star. It does not take long for more of these exoplanets to be found. In most cases, they are 'hot Jupiters', gaseous giant planets which, in some way or another, have ended up in very small orbits with orbital periods no longer than a few days.

Improvements in the sensitivity of measuring instruments make it possible also to detect much smaller planets, especially if they are in orbit around low-mass dwarf stars. In 2005, the Swiss team discovers a planet approximately as massive as Neptune (sixteen times more massive than the Earth) around the red dwarf star Gliese 581. The star is 20.4 lightyears away in the constellation of Libra and, despite its proximity, is only visible with a telescope. The 'hot Neptune' is given the designation Gliese 581b. It has an orbital period of 5.4 days and is a little over six million kilometers from its parent star.

A year later, with the aid of a very sensitive spectrograph on the 3.6-meter telescope at the European Southern Observatory in Chile, teamleader Stéphane Udry and his colleagues discover another two planets around the red dwarf. One of them, Gliese 581c, is only five times more massive than the Earth. It is eleven million kilometers from the star and has an orbital period of just under thirteen days. The other planet, Gliese 581d, is eight times more massive than the Earth and three and a half times further away, and completes an orbit around the star once every twelve weeks.

The discovery of Gliese 581c is announced on April 24, 2007. The planet may be very close to its parent star, but because the star radiates so little heat and light, Udry and his colleagues believe that the surface temperature must be somewhere between 0 and 40 degrees. That means that in theory there could be liquid water on the exoplanet, making Gliese 581c the first habitable exoplanet ever discovered. If, like the Earth, it consists of rock and metals, it is one-and-a-half times larger than our home planet.

Gliese 581 is the third red dwarf star where astronomers find relatively small planets, and the discovery suggests that many more red dwarf stars have a planetary system. It is not clear whether life may be possible on such red dwarf planets because many red dwarfs display regular powerful eruptions of deadly X-rays. In addition, it is probable that the same hemisphere of the planet is facing the planet at all times, as a consequence of tidal forces.

In the fall of 2007, various teams of theorists come to the conclusion that Gliese 581c is possibly still too hot for liquid water, as the consequence of an overactive greenhouse effect, like that on Venus. Perhaps the third planet, Gliese 581d, is a more suitable candidate for the title 'Twin Earth'; thanks to clouds of frozen carbon dioxide crystals, the temperature is probably high enough for there to be flowing water on the surface.

↑ The observable universe has more than a hundred billion galaxies, each of which contains tens of billions of stars. It is highly probable that many of these stars are orbited by planets that are very similar to the Earth, with mountains, oceans, and – who knows – life. (CfA/David Aguilar)

→ Gliese 581 is a red dwarf star, around which a 'super-Earth' was discovered in 2007 – a planet with a solid surface only a few times more massive than Earth. The planet, named Gliese 581c, might be at the right distance from the star to make liquid water possible on its surface. (ESO)

2007

Four hundred years of telescopic astronomy has radically changed the way we look at the universe – and at ourselves. When Hans Lipperhey ground his first lens, science was still getting used to the idea that the Earth is not the center of the universe and that the cosmos is older than six thousand years. Four hundred years later, we now realize that mankind is a biological newcomer on a cosmic grain of sand, in an unfathomable, expanding ocean of space and

Giant Telescopes and

time, which we share with dangerous Earth-grazers, rapidly spinning pulsars, intriguing exoplanets, gluttonous black holes, and colliding galaxies. Even the atoms from which the world around us is made take a subordinate place in a cosmos dominated by mysterious dark matter and incomprehensible dark energy.

By studying objects billions of lightyears away, astronomers are looking back in time, but their telescopes are not yet capable of looking into the future. Are we on the threshold of great breakthroughs and paradigm shifts? Or will the flow of new discoveries slowly dry up, without us ever fully unraveling the secrets of the universe?

No one knows, but it seems unlikely that the role of the telescope will be played out in the near future. Astronomers are working on

the Future of Astronomy

the construction of giant telescopes with mirrors tens of meters in diameter, and on a new generation of space telescopes. There is a very good chance that we can unleash a scientific revolution, compared to which the discovery of the telescope – the most important development in the history of astronomy – will pale into insignificance.

Index

Bold page numbers refer to illustrations

3C273 (quasar) · 150, **150**
21 centimeter line · 131
21 centimeter line radiation · 142
40 Eridani (star) · 71
47 Tucanae (globular star cluster) · **181**
51 Pegasi (star) · 206
51 Pegasi b (exoplanet) · 206, **206**
55 Cancri (star) · 206
61 Cygni (star) · 59
1937 UB (Earth grazer) · 119
1987A (supernova) · 191, **191**, **192**
1992 QB_1 (ice dwarf) · **198**, 198, 219
2003 UB_{313} (ice dwarf) · 219
2004 MN_4 (Earth grazer) · 218

A

Abell 1703 (cluster) · **189**
Abell 2218 (cluster) · 190
Abell 370 (cluster) · 190, **190**
Abell 901 (cluster) · **116**
Abell 902 (cluster) · **116**
Abell catalog · 139
Abell cluster · 139
Abell, George · 139, 179
aberration of starlight · 39, **39**, 59
absolute zero · 151, 183, 195
absorption line · 86, **86**
accelerating expansion of the universe · **213**
accretion disk · **147**, **161**
achromatic lens · 58
active area on the Sun · **69**
activity maximum · **63**
activity minimum · 82
Adams, John · 67
Adams, Walter · 42, 71
adaptive optics · 202, 207
Adrastea (Jovian moon) · 198
Aerobee sounding rocket · 147, 162
afterglow · 83, 151, 211, **211**
air pump · 26
Airy, George · 67
Aitken, Robert Grant · 19
alchemy · 30
Aldebaran (star) · 38, 46
Alfonso X · 3
Algol (star) · 19
ALH84001 (Martian meteorite) · 210, **210**
Allegheny Observatory · 102
Allan Hills · 210
Almagest · 38
Alnitak (star) · **86**
Alpha Centauri (star) · 59, 98, 99, **134**
alpha particle · 74
alpha radiation · 94
Alpha Regio (region on Venus) · **154**
Alpher, Ralph · 151
Amalthea (Jovian moon) · 79, 174
American Astronomical Society · 163, 190, 195
American Science & Engineering · 147
amino acids · 210
anagram · 22
Anderson, Carl · 94
Anderson, Thomas · 83
Andromeda galaxy · **40**, **61**, **106**, 126, 142, 179, **186**
Andromeda Nebula · 102, 106, 110, 118
–, distance · 95, 106
–, radial velocity · 110
ANSMET program · 210
Antarctic Search for Meteorites · 210
Antarctica · 42, **82**, 210
antimatter · 94
Apep (Egyptian god) · 218
aperture synthesis · 146
apex · 46
Apianus, Petrus · 18, 31
Apollo (Earth grazer) · 119
Apollo 17 · **10**
Apophis (Earth grazer) · 218
Aquarius (constellation) · 47, **48**, 67
Arcturus (star) · 38, **38**, 46
Arecibo radio observatory · 119, 154, 182, 194
Ariane 1 rocket · 187
Aristoteles · 2
Armagh Planetarium · 158
Arp, Halton · 134
Asclepius (Earth grazer) · 119
asteroid belt · 51
asteroids · 47, **49**, 50, **50**, 111, 119, **119**, **120**, 167, 183, **198**, **201**, 202, **202**, 203, **217**, 218, **218**
–, moons · **201**, 202, **202**
Astronomer Royal · 43, 67, 103
Astronomia Nova · 15
astronomical unit · 27
Astronomische Nachrichten · 59, 63, 79
Astronomy & Astrophysics · 190
Astrophysical Journal, The · 143, 151, 179
astrophysics · 58, 86
Aten asteroid · 218
atmospheric refraction · 39
atomic nucleus · 118
Auriga (constellation) · 79
aurora · 70, 143, **205**

B

B1620–26 (pulsar) · 194, **194**
B^2FH · 134
Baade, Walter · 118, 126, 158
baby photo of the universe · 151, 195
Backer, Don · 182
background radiation, cosmic · 151, **151**, 163, **163**, 195, **195**, **196**, 214
balloon astronomy · 94
Barnard, Edward Emerson · 38, 79, 99, 118
Barnard's Star · 38, 99, 206
barred spiral galaxy · 142
Bayer, Johann · 19
BBC · 134
Bell Telephone Laboratories · 114, 123, 151
Bell, Jocelyn · 158, 182
Belt of Orion · **86**
BeppoSAX (satellite) · 211
Bernardi, Fabrizio · 218
Bessel, Friedrich · 59, 71
Beta Centauri (star) · **134**
Beta Pictoris (star) · 183, **183**
beta radiation · 94
Beta Regio (area on Venus) · **154**
Betelgeuse (star) · **87**
Bethe, Hans · 122, 134, 163
Big Bang theory · 110, 134, 146, 151, 163, 195, 214
Big Crunch · 214
Big Dipper (constellation) · 35, 38, 66, 179
Big Nothing · 179
binary asteroid · **119**, 202
binary quasar · 178, **178**
binary star · 19, 47, 62, 79, 86, 98, 139, 182, 194, **207**
Biot, Jean-Baptiste · 51
Birkeland, Kristian · 143
Birr Castle · 66
black hole · 147, 162, 191, 211
–, supermassive · 123, 150, 186
Blaeu, Willem Janszoon · 19
Blandford, Roger · 150
blazar · 186
blink comparator · 111
blue supergiant · 162
blueberries on Mars · 215, **215**
blueshift · 110
Bode, Johann · 43, 50
Bode's Law · 50
Bolton, Thomas · 162
Bondi, Hermann · 146
Bootes (constellation) · 38, **38**, 178, 179
Bootes Void · 179
Boyle, Robert · 26, 42
Bradley, James · 39
Brahe, Tycho · 4, 14, 15, 18, 19, 27, 79, 118
Brera Observatory · 78
brown dwarf · 207, **207**, **208**
Brown, Michael · 219
Bruno, Giordano · 206
Bubble Nebula · **80**
Buie, Marc · 171
Bulletin of the Astronomical Institutes of the Netherlands · 142
Bunsen, Robert · 58
Burbidge, Geoffrey · 122, 144
Burbidge, Margaret · 122, 144
Burney, Venetia · 111

Butler, Paul · 206
Buys Ballot, Christophorus · 62

C

calcium · 86, 134
California Institute of Technology · 115, 134, 186, 219
Callisto (Jovian moon) · 11, 203
Caloris basin on Mercurius · **153**
Cambridge catalog of radio sources · 150
camera obscura · 14
Campbell, William · 110
Canada France Hawaii Telescope · 190
canali · 78
Canes Venatici (constellation) · 66, **66**
Canis Major (constellation) · 71
Canis Minor (constellation) · 71
carbon · 122, 134
carbon dioxide · 187
–, in Venusian atmosphere · 42
carbon-14 · 82
carbonate globules · 210, **210**
Carina Nebula · **145**, **184**
Carmina Burana · 58
Carnegie Institution of Washington · 163
Carrington, Richard · 70, 143
Carswell, Bob · 178
Cassen, Pat · 175
Cassini (planetary probe) · 222
Cassini Division in Saturn's rings · 27
Cassini, Jean-Dominique · **21**, 23, 26, 27, 42
Cassiopeia (constellation) · 19, **80**
Cassiopeia A (radio soruce, supernova remnant) · **117**, 158
Celestial Police · 50
centaur · 167, **167**
Centaurus A (radio galaxy) · **185**
cepheid (type of variable star) · 95, 106, 110, 126
Cepheus (constellation) · 19
Ceres (asteroid, dwarf planet) · 50, **50**, 119
Cetus (constellation) · 14, 20, **20**, 219
Chadwick, James · 118
Challenger (space shuttle) · 195
Chandra X-ray Observatory (satellite) · 147
Chandrasekhar, Subrahmanyan · 143
Channel of Mozambique · 78
Channel, English · 78
Chapman, Sydney · 143
Chaptal, Jean-Antoine · 51
Charon (Plutonian moon) · **111**, **112**, 171, **171**, **172**
–, orbital period · 171
Chiron (centaur) · 167, **168**
–, cometary activity · 167
Chladni, Ernst · 51
Christy, Char · 171
Christy, Jim · 171
Chryse Planitia (area on Mars) · 166
Churyumov-Gerasimenko, comet · **18**
circumstellar disk · **46**, 183, **183**
Cl 2244 (cluster) · 190
Clark, Alvan Graham · 71
Clinton, Bill · 210
cluster · 115, **115**, **116**, **140**, **177**, 179, **189**, 190, **190**
CNO cycle · 122
COBE (satellite) · 151, 195
color-luminosity diagram · 87
Columbia Hills on Mars · 215
comet · **17**, 18, 31, 43, 50, 79, 127, **129**, 130, **130**, 167, 183, 187
–, geisers · 187
–, impact · 203s
–, long-period · 130
–, nucleus · 31, **187**, **188**
–, orbits · 139
–, origin · 107
–, short-period · 130, 198
–, tail · **17**, 18, **32**, **130**, 143
comets:
–, Churyumov-Gerasimenko · **18**
–, Encke · 154
–, Grigg-Skjellerup · 187
–, Halley · 31, **31**, 59, 187, **187**, 198
–, Hyakutake · **130**
–, McNaught · **27**
–, NEAT · **32**
–, of 1531 · 31
–, of 1577 · 18
–, of 1607 · 31
–, of 1682 · 31
–, Schwassmann-Wachmann 3 · **129**, **204**
–, Shoemaker-Levy 9 · 203, **203**
–, Tempel 1 · **188**
cometary reservoir · 130
Compton Gamma Ray Observatory (satellite) · 159, 211
conspiration theory · 166
constellations of the southern hemisphere · 31
continuous creation, theory of · 146
Copernicus, Nicolaus · 3, 11, 15, 59
Cordelia (Uranian moon) · 170
Cornell University · 122, 163, 170
corona · 143
Corona Australis star forming region · **52**
coronal mass ejection · **57**, 70
Cosmic Background Explorer (satellite) · 151, **151**, 195
cosmic background radiation · 151, **151**, **152**, 163, **163**, 195, **195**, **196**, 214
–, spectrum · 195
–, temperature · 195
–, temperature variations · **151**, **152**, **195**, **196**
cosmic dust · 91
cosmic rays · 94, 147
cosmological constant · 214
cosmology · 139
–, Couch, Warrick · 190
Couderc, Paul · 83
Crab Nebula · **157**, 182
Crab pulsar · 182
Crabtree, William · 42
crater chain · 203
Creti, Donato · 26
Crommelin, Andrew · 103
Curie, Marie · 94
Curie, Pierre · 94
Curtis, Heber · 102, 159
curvature of space · 103
Cydonia (area on Mars) · 166
Cygnus (constellation) · 19, 59, 123, 162
Cygnus A (radio galaxy) · 123, **123**, **124**
Cygnus X–1 (X-ray source) · **161**, 162

D

d'Arrest, Heinrich · 67
Dactyl (asteroid moon) · 202, **202**
dark energy · 195, 214, 225
dark matter · 115, **115**, **116**, **177**, **189**, 190, 195, 225
Darwin, Charles · 91
De Cometis Libelli · 18
De Revolutionibus Orbium Coelestium · 11
de Sitter, Willem · 103
debris disk · 183
Deep Impact (cometary explorer) · **188**
Deimos (Martian moon) · 75, **75**
Delta Cephei (star) · 19, 95
Delta rocket · 195
Democritus · 10
deuterium · 122, 207
Dicke, Robert · 151
differential calculus · 30
Differential Microwave Radiometer · 195
differential rotation · 70
Dione (Saturnian moon) · 27
DiPietro, Vincent · 166
diving bell · 31
DMR · 195
Domenico, Giovanni · 27
Doppler effect · 62, 110
–, for sound · 62
Doppler shift · 154
Doppler, Christian · 62
double planet · 171
Dreyer, Johann · 66
Duhalde, Oscar · 191
Dumbbell Nebula (planetary nebula) · 47
Duncan, Martin · 130, 198
Dunham, Ed · 170
Dunham, Theodore · 42
dust cloud · 83, 102
dust particles · 86
dust ring of Jupiter · 174
dust rings of Uranus · 170
dust shield · 187
Dutch Astronomers Club · 131
dwarf planet · **50**, 111, **112**, 171, **172**, 198, 219, **219**, **220**, 223
dwarf star · 115, 207
Dwingeloo radio telescope · 142
Dyce, Rolf · 154
Dysnomia (moon of Eris) · 219, **219**, **220**
Dyson, Frank Watson · 103

E

Earth · **27**
–, age · 126
–, climate · 63, 82
–, magnetic field · 31, 70

–, orbital motion • 59
–, rotational axis • 40
Earth grazer • 119, 203, **217**, 218
Echus Chasma (canyon on Mars) • **28**
Eddington, Arthur • 103, 122, 178
Eddy, Jack • 82
Eiffel Tower • 94
Einstein (satellite) • 147
Einstein, Albert • 30, 103, 110, 178, 214
electromagnetic radiation • 114
electrometer • 94
electron • 94, 131, 143
electron microscope • 210
elements, origin • 134
Elliot, Jim • 170
elliptical galaxy • 142
elliptical orbit • 15
Ellis, Richard • 190
Elysium (area on Mars) • 78
Encke's comet • 154
energy • 122
Enterprise, USS • 46
Epsilon Ring of Uranus • 170
Epsilon Indi B (brown dwarf) • 207
Eratosthenes • 2
Eris (dwarf planet) • 219, **219**, **220**
Eros (asteroid) • 27, **120**, 154
–, radar echoes • 154
Eta Carinae (star) • **79**
Euler (lunar crater) • **9**
Europa (Jovian moon) • 11, **11**, 222
European Southern Observatory • 107, 147, 211, 223
Ewen, Harold • 131
exoplanet • **144**, 194, 206, **206**, 223, 225
–, habitable • 223
expansion of the universe • 92, 110, 134, 150
–, accelerating • 214
–, asymmetric • 163
extragalactic nebula • 106
extraterrestrial life • 166

F

F Ring of Saturn • **30**
Fabricius, David • 14, 19
Fabricius, Johannes • 14
facula
failed star • 207
Far-InfraRed Absolute Spectrophotometer • 195
Ferdinand IV • 50
Feynman, Richard • 163
FIRAS • 195
fixed stars • 38
Flamsteed, John • 43, 46
Fokkes, Jan • 19
Fomalhaut (star) • 183
Ford, Kent • 163
Fowler, Willy • 122, 134
Fracastoro, Girolamo • 18
Frail, Dale • 194, 206
Franz, Otto • 171
Fraunhofer lines • 58, **58**
Fraunhofer, Joseph • 58, 59, 63, 74
Friedman, Alexander • 117
Friedman, Herbert • 162

G

Gabrielle (moon of Xena) • 219
Gaia (satellite) • 59
galactic center • 114, **114**, 123, 142, **148**
galactic halo • 102
galactic plane • 115, 142
Galama, Titus • 159
galaxy • 22, **29**, **61**, **65**, 66, **93**, 94, **101**, **105–110**, 115, 118, **118**, 123, 131, **132**, 139, 142, 146, **146**, 150, **155**, 159, 163, 179, **180**, 183, **185**, **186**, 186, 211, **213**
–, collisions • **164**
–, distances • 110, 179
–, distribution in the sky • **39**
–, recessional velocities • 163,179
–, spatial distribution • 139,179, **179**
Galex (satellite) • 19
Galilei, Galileo • 10, 11, 14, 30, 36, 75, 127, 175, 202
Galileo (planetary probe) • 11, 174, 175, 202, 203
Galle, Johann • 67
Gamma Draconis (star) • 39, **39**
gamma radiation • 159
gamma rays • 94
gamma-ray burst • 159, **159**, 211, **211**, **212**
–, afterglows • 211
–, distances • 159
Gamow, George • 122, 134, 151, 163, 214
Ganymede (Jovian moon) • 11, 124, 203
Garradd, Gordon • 218
gas inclusions • 210
gas tail of a comet • 143
Gauss, Carl Friedrich • 50
Geiger counter • 147, 162
Gemini (constellation) • 43
general relativity • 103, 110, 178
geocentric world view • 2
geomagnetic storm • 70
George III • 43
Georgium Sidus • 43
Giacconi, Riccardo • 147, 162
giant star • 87
Gill, David • 98
Giotto (cometary probe) • 31, 187
Giotto di Bondone • **31**, 187
GK Persei (star) • **83**
glass
–, production • 58
–, refractory index • 58
Gliese 229 (star) • 207
Gliese 229B (brown dwarf) • 207, **207**
Gliese 581 (star) • 223
Gliese 581b (exoplanet) • 223
Gliese 581c (exoplanet) • 223, **224**
Gliese 581d (exoplanet) • 223
globular star cluster • **60**, 102, **102**, 107, **121**, **181**, 182
–, distances • 102
–, spatial distribution • 102
Goddard Space Flight Center • 166
Gold, Thomas • 146
Goodricke, John • 95
gravitational disturbances • 111
gravitational lens • **177**, 178, 190
gravity • 8, 26, 30, 31, 67, 103
Great Comet of 1577 • 18
Great Dark Spot on Neptune • **67**
Great Red Spot on Jupiter • **12**, **25**, 26, **26**, **70**
Green Bank radio osbervatory • 123
greenhouse effect • 223
Gregory, Stephen • 179
Grigg-Skjellerup, comet • 187
Groot, Paul • 159, 211
Grubb-Parsons • 66
Guinand, Pierre Louis • 58
Guldenmann, Katharina • 18
Gulliver's Travels • 75
Gusev (Martian crater) • 215

H

Hale Telescope • 150, 190
Hale, George Ellery • 63, 106
Hall, Asaph • 75
Halley, Edmund • 30, 31, 38, 42, 46, 71, 130, 187
Halley's comet • 31, **31**, 59, 187, **187**, 198
halo of the Milky Way galaxy • 102
halo ring of Jupiter • 174
Hare (constellation) • 207
Harmonices Mundi • 15
Harrington, Bob • 171
Harriot, Thomas • 10
Hartmann, Johannes • 86
Harvard College Observatory • 95, 106, 131
Harvard University • 131, 214
Haumea (ice dwarf) • 202
Haute-Provence Observatory • 206
Hawking, Stephen • 162
HDE 226868 (star) • **161**, 161, **162**
heavy elements • 134
heliocentric world view • 4
heliometer • 59
helium • 74, 86, 122, 134
–, liquid • 183
helium nucleus • 74
Helix Nebula (planetary nebula) • **48**
Henderson, Thomas • 59, 98
Hercules (constellation) • 46
Herculina (asteroid) • 202
Herman, Robert • 151
Hermes (Earth grazer) • 119
Herschel, Caroline • 43
Herschel, Jacob • 43
Herschel, William • 36, 43, 46, 47, 50, 58, 66, 114, 170
Hertzsprung, Ejnar • 87, 95
Hertzsprung-Russell-diagram • 87
Hess, Victor • 94
Hessels, Jason • 182
Hevelius, Johannes • 19
Hewish, Antony • 146, 158
high-energy astrophysics • 147
High-Resolution Imaging Science Experiment • 166
High-Resolution Stereo Colour-camera • 166
Hipparchus • 2, 38, 46

Hipparcos (satellite) · 38, 59, 98
Hi-z Supernova Search Team · 214
Hoag, Arthur · 190
Hoagland, Richard · 166
Hodgson, Richard · 70
Holwarda, Johannes Phocylides · 19
Hooke, Robert · 26, 30
Hooker Telescope · 106, 110, 126
Horrocks, Jeremiah · 42
Horsehead Nebula · **86**
hot air balloon · 94
hot Jupiter · 206, 223
hot Neptune · 223
Howard, Edward · 51
Hoyle, Fred · 122, 134, 146, 158
Hubble constant · 178
Hubble Deep Field · **146, 215**
Hubble Space Telescope · 99, 110, 170, 171, 174, 178, 190, 191, 199, 202, 203, 207
Hubble Ultra Deep Field · **109**
Hubble, Edwin · 66, 95, 106, 110, 118, 126, 214
Humason, Milton · 110
Huygens (planetary probe) · 127, **221**, 222, **222**
Huygens, Christiaan · 22, 23, 127, 154, 206, 222
Huygens, Constantijn · 23
Hyades (open star cluster) · 103
Hyakutake, comet · **130**
Hydra (Plutonian moon) · **111**, 171
hydrocarbons · 187
hydrogen · 74, 86, 122, 131, 142, 150, 207
hydrogen nucleus · 94

I

Iapetus (Saturnian moon) · 27
ice dwarf · 47, **172**, **197**, 198, 219, **219**
Ida (asteroid) · **202**, 202
impact · 218
Infra-Red Astronomical Satellite (satellite) · 183
infrared astronomy · 183
infrared radiation · 114, 138, **183**, 207
Innes, Robert · 98
Integral (satellite) · 182
interference · 158
intergalactic matter · **124**
Internal Constitution of Stars, The · 103
International Astronomical Union · 111, 198, 219
interplanetary space · 154
interstellar dust · 102
interstellar matter · **86**
interstellar space · 130
inverse-square law · 30
Io (Jovian moon) · 11, **12**, 175, **176**
–, volcanism · **12**, 175, **175**, **176**
ionisation, degree of · 94
ionising radiation · 94
IRAS (satellite) · 183
iron · 134
iron meteorite · 51
Isaac Newton Telescope · 66
island universes · 106

J

jansky (unit of radio flux) · 114
Jansky, Karl · 114, **114**, 123, 131
Jansky's Merry-go-round · 114
Janssen, Pierre · 74
jet · 99, 124, 150, **155**, 186, **212**
Jet Propulsion Laboratory · 166
Jewitt, David · 198
Johnson Space Center · 210
Jones, Albert · 191
Juno (asteroid) · 50
Jupiter (planet) · 11, **11**, **12**, 22, 23, 25, **26**, 31, 50, 51, 67, 70, 75, 79, 98, 119, 127, 143, 154, 155, 167, 174, 175, 198, 202, 203, **206**, 207, 222, 223
–, atmosphere · 26, 203
–, cloud bands · 23
–, cometary impact · 203, **203**
–, gravitational disturbances · 31
–, Great Red Spot · 26
–, moons · 10, **11**, 155, 175, 198, 203
–, radiation belts · 174
–, rings · 22, **173**, 174
–, rotational period · 154
–, tidal forces · 175, 203

K

Kamerlingh Onnes, Heike · 74
Kapteyn galaxy
Kapteyn, Jacobus · 83, 102, 107
Kasilinkov, Andrey · 42
Keck Telescope · 174, 175, 219, **219**
Keeler, James · 110
Kepler, Heinrich · 18
Kepler, Johannes · 11, 14, 15, 15, 18, 19, 31, 42, 50, 75, 79, 143
Kepler's Laws · **16**, 23, 30
Kepler's supernova · 118, 191
Kerr, Frank · 142
Kirchhoff, Gustav · 58
Kirshner, Robert · 179
Kitt Peak National Observatory · 178, 190, 198, 218
Klebesadel, Ray · 159, 211
K-line of calcium · 86
Kosmotheoros · 23
Kowal, Charles · 167, 219
Kuiper Airborne Observatory · 170
Kuiper Belt · 167, **172**, 183, **197**, 198, 202, 219
Kuiper Belt Object · 127, 198, 202, 219
Kuiper, Gerard · 67, 127, 170, 198, 222
Kulkarni, Shrinivas · 207
Kurganov, Nikolaj · 42
Kyoto Prize · 143

L

l'Aigle, meteorite fall · 51
Lambda Herculis (star) · 46
Large Magellanic Cloud · **95**, **100**, 118, **125**, 191, **192**
large-scale structure of the universe · 151
Las Campanas Observatory · 191
Lassell, William · 67
Lawrence Berkeley National Laboratory · 163, 214
Le Verrier, Urbain · 67, 111
Leavitt, Henrietta · 95, 102, 126
Leibniz, Gottfried · 30
Leiden Observatory · 87
Lemaître, Georges · 110
Leviathan of Parsonstown · 66
Levy, David · 203
Libra (constellation) · 230
Lick Observatory · 79, 110, 139
light
–, as wave phenomenon · 62
–, bending by gravity · 103, 178
–, frequency · 62
–, velocity of · 39, 83, 155
light arc · **189**, 190, **190**
light echo · 83, **83**, **84**
light pollution · 126
light-year · 98
limb darkening · 127
Lindblad, Bertil · 107
Lipperhey, Hans · 4, 36, 225
liquid helium · 183
lithium · 122
Little Ice Age · 82
Local Group · 163
–, proper motion · 163
Lockyer, Norman · 74
Lomonosov, Mikhail · 42
long-period comet · 130
long-period variable star · 19
Lord Rosse · 66
Los Alamos National Laboratory · 159
Lowell Observatory · 78, 110, 111, 119, 171
Lowell, Percival · 67, 78, 111, 171
Lumen Frisiae · 19
Luna 1 (lunar probe) · 143
lunar eclipse · 18
Lunar Reconnaissance Orbiter (lunar probe) · 10
Luu, Jane · 198, 219
Luyten, Willem · 71, 87
Lynds, Roger · 190
Lyne, Andrew · 194
Lyra (constellation) · 46, 47, 59, 183
Lyttleton, Raymond · 130

M

M32 (galaxy) · 186, **186**
M33 (galaxy) · **132**
M51 (galaxy) · **65**, 66, **66**
M64 (galaxy) · **107**
M74 (galaxy) · **108**
M80 (globular star cluster) · **102**
M81 (galaxy) · **45**, **101**
M82 (galaxy) · **93**
M87 (galaxy) · **155**
M101 (galaxy) · **105**
Maestlin, Michael · 14
Magellanic Clouds · 95, 191
magnetite crystals · 210
main sequence · 87
Manhattan Project · 126

Marcy, Geoff • 206
Mariner 2 (planetary probe) • 42, 154
Mariner 9 (planetary probe) • 215
Marius, Simon • 11
Mars (planet) • **15**, 23, **24**, 27, **28**, 42, 50, 51, 77, 119, **165**, 166, **166, 209, 216**
–, atmosphere • 210
–, axial tilt • 23
–, blueberries • 215, **215**
–, canals • 78, **78**, 94
–, clay deposits • 77
–, craters • 79
–, diameter • 23
–, distance • 27, 78
–, dry river beds • 78
–, Face on • **165, 166**, 166
–, ice • 215
–, life • 78, 166, 210, 215
–, mass • 75
–, meteorites from • **51**, 210, **210**
–, micro-organisms • 210
–, moons • **49**, 75, **75, 76**
–, motion • 15
–, polar caps • 23, 215
–, rotation • 23
–, rotational period • 154
–, surface details • 23
–, tidal forces • 75
–, water • 215
Mars Express (planetary probe) • 166, 215
Mars Global Surveyor (planetary probe) • 166, 215
Mars Odyssey (planetary probe) • 215
Mars Orbiter Camera • 166
Mars Pathfinder (planetary probe) • 210
Mars Reconnaissance Orbiter (planetary probe) • 166, 215
Maskelyne, Neville • 43
mass transfer • 182
Massachusetts Institute of Technology • 154
Mather, John • 195
Mauna Kea Observatory • 190, 198
Maunder Minimum • 82
Maunder, Edward • 82
Maurits, Prince • 4
Maximilian IV • 58
Mayor, Michel • 206, 223
McDonald Observatory • 127
McKay, David • 210
McNaught, comet • **17**
McNaught, Robert • 191
Melnick, Jorge • 211
Mercury (planet) • 23, 63, 127, **153**, 154, 202
–, radar echos • 154
–, rotational period • 154
Meridiani Planum (area on Mars) • 215
mesa • 166
Messier, Charles • 47, 66
meteorite • 51, **51**
–, Martian • 210
–, SNC • 210
methane • 127, 207, 222
microfossils • 210, **210**
micrometer • 23
micro-organisms • 166, 210
Middlefehldt, David • 210
Milky Way • 8, **37, 97**
Milky Way galaxy • 46, 66, 86, 92, 95, 102, 106, 107, 110, 118, **141**, 142, 159, 163, 179, 186, 191, 211
–, center • 114, **114**, 123, 142, **148, 159**
–, density • 115
–, differential rotation • 107
–, dimensions • 102
–, dynamics • 107, 115, 131
–, halo • 102
–, plane • 115, 142
–, proper motion • 163
–, radio waves • 114
–, rotation • 107
–, rotational velocity • 142
–, spiral structure • 142, **142**
–, structure • 36, 107, 131
Millikan, Robert • 94
millisecond pulsar • 182, **182**
millisecond X-ray pulsar • 182
Mink, Doug • 170
Minkowski, Rudolph • 150
Mintaka (star) • 86, **86**
Mira (star) • 19, **19, 20**
–, gas tail • 19, **20**
Miranda (Uranian moon) • 43, 127
Molenaar, Gregory • 166
Molyneux, Samuel • 39
Monthly Notices of the Royal Astronomical Society • 142, 146
moon • **9**, 10, **10**, 11, **74**
–, mountains • 10
Morabito, Linda • 175
Morgan, Bill • 142
Mount Wilson Observatory • 71, 102, 106, 118, 126, 139, 167
Mullard Company • 146
Mullard Radio Observatory • 146
Muller, Lex • 131
multispectral astronomy • 114
Murdin, Paul • 162
Mutchler, Max • 171
Mysterium Cosmographicum • 15

N

Nakhla meteorite • 210
nanofossils • 210
Nature • 74, 131, 155, 158, 162, 170, 178, 194, 207, 211
Nautical Almanac • 43
NEAT, comet • **32**
nebula, planetary • 47
nebula, spiral • 66
Needle Galaxy • **213**
Neptune (planet) • 22, **67**, 94, 111, 115, 167, 171, 183, **198**, 202
–, discovery • 67
–, moons • 67, 127
–, ring arcs • 67
–, rings • 22
Nereïd (Neptunian moon) • 67, 127
neutral hydrogen • 131, 142
neutrino • 191
neutrino detector • 191
neutron • 118
neutron star • 118, **147, 158**, 162, 182, 192, 194, 211
New General Catalogue • 66
New Horizons (planetary probe) • **171**, 172, 174, **197**
New Technology Telescope • 211
New York Times, The • 87, 114
Newcomb, Simon • 79
Newton, Isaac • 7, 26, 30, 58, 66
Neyman, Jerzy • 139, 179
NGC 520 (galaxy) • **164**
NGC 1514 (planetary nebula) • 47
NGC 2440 (planetary nebula) • **47**
NGC 4526 (galaxy) • **118**
NGC 5866 (galaxy) • **213**
NGC 6397 (globular star cluster) • **60**
nitrogen • 122, 127, 222
Nix (Plutonian moon) • **112**, 171
Nobel Prize, Physics • 94, 134, 146, 147, 151, 158, 195
nova • 79, 83, 118
Nova Aurigae • 79
Nova Persei • 83
Novikov, Igor • 151
nuclear bomb
nuclear fusion • 74, 122, **134**, 207
nuclear physics • 91
nucleosynthesis • 134
nutation • 39

O

O'Dell, Bob • **199**
Oberon (Uranian moon) • 43
Object Kowal • 167
observatories
– Allegheny • 102
– Berlin • 67
– Brera • 78
– Cambridge • 103
– Cape of Good Hope • 98
– Cape Town • 98
– Dorpat • 58, 59
– Geneva • 206, 223
– Greenwich • 82, 103
– Harvard • 95, 106, 131
– Haute Provence • 206
– Heidelberg • 83
– Johannesburg • 98
– Kitt Peak • 178, 190, 198, 218
– Königsberg • 59
– La Palma • 66, 211
– Las Campanas • 191
– Leiden • 87
– Lick • 79, 110, 139
– Lilienthal • 59
– Lowell • 78, 110, 111, 119, 171
– Mauna Kea • 190, 198
– McDonald • 127
– Meudon • 74
– Milan • 78
– Mount Wilson • 71, 102, 106, 110, 118, 126, 139, 167
– Palomar • 118, **150**, 167, 190, 203, 207, 219
– Paris • 27, 67

– Potsdam • 86
– Siding Spring • 218
– Sproul • 99
– St. Petersburg • 42
– Toulouse • 190
– Transvaal • 98
– Unie
– United States Naval Observatory • 75, 171
– Yerkes • 106, 127, 199
Olbers, Heinrich • 59
Olson, Roy • 159
Omega Centauri (globular star cluster) • **121**
Omicron Ceti (star) • 19
Oort Cloud • **130**
Oort, Jan • 107, 115, 130, 131, 142, 150
open star cluster • 103
Ophelia (Uranian moon) • 170
Ophiuchus (constellation) • 99
Öpik, Ernst • 130
Öpik-Oort Cloud • 130
Opportunity (Mars rover) • **215**
opposition • 27, 78
optical afterglow • **211**
organic molecules • 210
Orion (constellation) • 2, 38, **86, 87, 199**, 211
Orion Nebula • **199, 200**
Osterbrock, Don • 142
outflow channels on Mars • 215
oxygen • 122

P

Palitzsch, Johann • 31
Pallas (asteroid) • 50
Pallas, Peter • 51
Palomar Observatory • 118, 150, 167, 190, 198, 203, 207, 219
Pandora (Saturnian moon) • **30**
Pannekoek, Anton • 142
parallax • 27, 38, 39, 42, 59, 62, 98
Paris Observatory • 27, 67
Parker, Eugene • 143
Parsons, Charles • 66
Parsons, William • 66
Peale, Stanton • 175
pen recorder • 158
pendulum clock • 22
Penthouse • 162
Penzias, Arno • 151, 195
periodic table of the elements • 122, 134
period-luminosity law • 95, 102, 126
Perlmutter, Saul • 214
Perseus (constellation) • 19, 83
Petrosian, Vahé 190
Pettengill, Gordon • 154
Philae (cometary lander) • **18**
Philosophiae Naturalis Principia Mathematica • 30
Phobos (Martian moon) • **49**, 75, **76**
–, mass • 75
–, orbital period • 75
Phoebe (Saturnian moon) • **167**
Phoenix (Mars lander) • 215
Pholus (centaur) • 167
Piazzi, Giuseppe • 50, 119
Piazzia (asteroid) • 50
Picard, Jean • 27
Pickering, Edward Charles • 95
Pickering's harem • 95
Pioneer 10 (planetary probe) • 174
Pioneer 11 (planetary probe) • 174
Pisces (constellation) • 198
Pismis 24 (star cluster) • **88**
pizza-moon • 175
Planet X • 67, 111, 167, 219
planetary migration • 203
planetary moon • 111, 222
planetary nebula • 47, **47**, 71
planetary system • 47, 183, 199, 223
Pleiades (open star cluster) • **73**
Pluto (ice dwarf, dwarf planet) • 67, 111, **111, 112**, 127, 167, 171, **171, 172**, 174, 198, 202, 207, 219
–, discovery • 111
–, mass • 171
–, moons • 171
–, planetary status • 111, 219
–, rotational period • 171
pocket watch • 26
polar lights • 70, 143
Polaris • 36, **170**
Pole Star • 38, 163
polycyclic aromatic hydrocarbons • 210
population I stars • 126
population II stars • 126
Potato Famine • 66
power • 122
precision cosmology • 195
Prince Maurits • 4
Princeton University • 151
Principe, Gulf of Guinea • 103
Principia • 30, 31
Procyon (star) • 71, 115
Procyon B (white dwarf) • 71
Prometheus (Saturnian moon) • **30**
prominence • 74, **81**
proper motion • 38, 46, 59, 62, 71, 98, 99
proton • 94, 143
proton-proton cycle • 122
protoplanetary disk • **133**, 183, 199, **199**
proto-star • 183
Proxima Centauri (star) • 98, **98**, 99
pseudo science • 166
PSR 1257+12 (pulsar) • 194, **194**
PSR 1829–10 (pulsar) • 194
PSR 1937+21 (pulsar) • 182
Ptolemaeus, Claudius • 2, 38
pulsar • **99**, 147, 158, **158**, 182, 194, **194**, 206
pulsar planet • 194, **194**
pulsating star • 19
pulsation period • 95, 102, 106, 110, 126
Purcell, Edward • 131

Q

quantum mechanics • 91
Quaoar (ice dwarf) • 219
quasar • **149**, 150, **150**, 155, **156**, 158, 178, 186
–, most distant • 150
–, triple • **149**
quasi-stellar radio source • 150
Queloz, Didier • 206, 223
QUEST camera • 219
Quinn, Thomas • 130, 198000

R

Rabinowitz, David • 167, 219
radar • 154
radar astronomy • 154, 202, 222
radar echos • 154
radial velocity • 38, 62, 110, 206
radiation pressure • 18
radio activity • 94
radio astronomy • 107, 114, 123, 131, 147
radio galaxy • 146, **185**, 186
Radio Kootwijk • 131, 142
radio quasar • 178
radio source • 146, 155
radio star • 150
radio telescope • 114, 123, 146, 155, 158
radio waves • 92, 114, 123, 150
radium • 94
Ramsay, William • 74
Reber, Grote • 123, 131
red dwarf • 38, 87, 98, 99, 223
red giant • 47, 126, 222
redshift • 134, 150, 163
Rees, Martin • 150, 155
reflecting telescope • 30
refraction • 39
refractory index • 58
Refsdal, Sjur • 178
Reinmuth, Karl • 50, 119
Reinmuthia (asteroid) • 119
relativistic expansion • 155
relativistic velocity • 155
relativity, general • 103, 110, 178
relativity, special • 155
relativity, theory of • 30, 91, 214
Remus (asteroid moon) • **201**, 202
Review of Modern Physics • 134
Reynolds, Ray • 175
Rhea (Saturnian moon) • 27
Rho Ophiuchi star forming region • **85**
Rice University • 199
Richer, Jean • 27
Riess, Adam • 214
Ring Nebula (planetary nebula) • 47
river beds on Mars • 215
RNA-molecule • 210
Roll, Peter • 151
Rømer, Ole • 155
Romulus (asteroid moon) • **201**, 202
Roque de los Muchachos Observatory • 66
Rosat (satellite) • 203
Rosetta (cometary probe) • **18**
Rosse, Lord • 66
Rossi, Bruno • 147
Royal Astronomical Society • 70
Royal Dutch Meteorological Institute • 62
Royal Greenwich Observatory • 82, 103, 162
Royal Society • 26, 30, 46, 103
Royal Statistical Society • 139
Rubin, Vera • 163
Rubin-Ford effect • 163
Rudolf II • 15
Russell Diagram • 87

Russell, Annie · 82
Russell, Henry Norris · 87, 102
Rutherford, Ernest · 74
Ryle, Martin · 146

S

Sagittarius (constellation) · 26, 102, 114
Sakigake (cometary probe) · 187
SAO 158687 (star) · 170
Sapas Mons (Venusian volcano) · **41**
Saturn (planet) · **21**, 22, **22**, 23, 27, 31, 36, 43, 50, 67, 127, 167, **168**, 174, 222
–, cloud bands · 23
–, gravitational disturbances · 31
–, handles · 8, 22
–, moons · **30**, 127, **127, 128, 167**, 222
–, rings · **21**, 22, **22**, 23, **30, 173**, 174
–, rotational period · 154
Scheiner, Christoph · 14
Schiaparelli, Elsa · 78
Schiaparelli, Giovanni · 78
Schild, Rudy · 178
Schmidt, Maarten · 150
Schmidt-telescope · 118, 167, 203, 219
Schmitt, Harrison · **10**
Schröter, Johann · 10, 59
Schwabe, Heinrich · 63
Schwassmann-Wachmann 3, comet · **129, 204**
Science · 174, 175, 178, 210, 214
Scorpius (constellation) · 147
Scorpius X–1 (X-ray source) · 147, 162
Scott, Elizabeth · 139, 179
Sedna (ice dwarf) · 219
Serpens (constellation) · 15, 19, 211
Shane, Donald · 139
Shapley, Harlow · 102, 106, 159
Sharpless, Stewart · 142
Shelton, Ian · 191
shepherd moons · 170
Shklovsky, Iosif · 75, 131
Shoemaker, Carolyn · 203
Shoemaker, Eugene · 203
Shoemaker-Levy 9, comet · 203, **203**
shooting star · 51
short-period comet · 130, 198
Sidereus Nuncius · 10, 11
Siding Spring Observatory · 218
Sidra, Gulf of · 23
Sigurdsson, Steinn · 194
Sirius (star) · 2, 38, 46, 58, 71, **72**, 115
Sirius B (white dwarf) · 71, **72**
Skiff, Brian · 119
Slipher, Vesto · 110, 111
Sloan Digital Sky Survey · 150
Sloan Great Wall · 139
Small Magellanic Cloud · 95, **96, 126**
Smoot, George · 163, 195
SNC meteorites · 210
Sobral, Brazil · 103
sodium · 74, 86
Sola, Josép Comes · 127
solar cycle · 63, 82
solar eclipse · **59**, 74, 103, **103, 104**, 143, **143**, 178
solar flare · 63, 70, **122**, 143, 159
solar particles · 143
solar system
–, dimensions · **28**, 42, 155
–, origin · 51, 130
solar telescope · 13, 14
solar wind · 18, 143
Soucail, Geneviève · 190
sounding rocket · 147, 162
space shuttle · 195
Space Telescope Science Institute · 147, 171, 199
spaceflight · 147
Spaceguard Survey · 218
spacetime · 186
–, curvature of · 103
Spacewatch-telescope · 167
spatial velocity · 38, 62, 99
special relativity · 155
spectral lines · 86, 150, 162
spectroscope · 55, 58
spectrum · 58, 86
spin state of electrons · 131
spiral arms · 126
spiral galaxy · 142
spiral nebula · 56, 66, **66**, 94, 102, 106, 142
–, distances · 106
–, extragalactic nature · 106
–, radial velocities · 110
Spirit (Mars rover) · **209**, 215
Spitzer Space Telescope · 183
Spörer Minimum · 82
Spörer, Gustav · 82
Sproul Observatory · 99
Sputnik 1 (satellite) · 147
standard model of cosmology · 151
star
–, brightness · 95
–, color · 62, **62**, 87
–, composition · 86
–, dimensions · 87
–, distance estimates · 59
–, distances · 62
–, energy source · 103, 122
–, evolution · 87
–, exploding · 79
–, gas density · 103
–, interior · 134
–, interior pressure · 103
–, interior temperature · 103
–, long-period variable · 19
–, luminosity · 87
–, nearest · 98
–, populations · 87
–, proper motion · 38, 46, 59, 99
–, pulsating · 19
–, spatial distribution · 102, 142
–, spectra · 86
–, surface temperature · 62, 87
–, temperature · 86
–, variable · 19
star catalog of Hipparchus · 38
star cluster · 47
star dust · 92, 134
star forming region · **45, 52, 85, 125, 126, 145**, 183, **184**, 199
Star of Bethlehem · 187
star streams · 107
Star Trek · 38, 46
Stargate SG–1 218
steady state theory · 146
Stella Nova · 15, 19, 79, 118
stellar black hole · 162
stellar occultation · 150, 170, 202
stellar populations · 126
Stern, Alan · 171
Stickney (crater on Phobos) · 75
Stockton, Alan · 178
Stracke, Gustav · 119
Strong, Ian · 159
Struve Telescope · 127
Struve, Wilhelm · 59
Suisei (cometary probe) · 187
sun · **13**, 14, **14, 57, 63, 64, 69, 81**, 103, **122**
–, activity cycle · 63, 82
–, apex · 46
–, corona · 143
–, death · 47
–, differential rotation · 70
–, energy source · 122
–, magnetism · 14, **64**, 143
–, motion through Milky Way galaxy · 36, 46
–, radio waves · 114
–, rotation · 14
–, spectrum · 58, **58**, 74
–, X-rays · 147
sunspot · **13**, 14, **14**, 63, **63**, 70, 82
sunspot cycle · 63, 70, 82
sunspot group · **13**, 14, **14**, 63, **63, 64**, 70
supercluster · 139
superluminal velocities · 155
supermassive black hole · 123, 150, 186
supernova · 8, 15, 19, 94, 118, **118**, 158, 159, 167, 182, 191, **191, 192**, 214, **214**
–, Kepler's · 15, 19, 79, 118, 191
–, Tycho's · 19, 79, 118
–, type Ia · 214
supernova 1987A · 118, 191, **191, 192**
Supernova Cosmology Project · 214
supernova remnant · **94, 99, 113, 117, 157**
supervoid · 179
Swift, Jonathan · 75
Sylvia (asteroid) · **201**, 202
synchronous rotation · 154
Syrtis Major (area on Mars) · 23, **24**, 78
Systema Saturnium · 22

T

Tarter, Jill · 207
Taurus (constellation) · 11, 38, 43, 47, 50, 103, **157**
telescope, invention · 7
Telting, John · 211
Tempel 1, comet · **188**
temperature variations in cosmic background radiation · **151, 152**
Terzan 5 (globular star cluster) · 182
Tethys (Saturnisn moon) · 27
Thebe (Saturnian moon) · 174

thermodynamics • 122
thermonuclear explosion • 79, 83
Tholen, David • 218
Thompson, Laird • 179
Thorne, Kip • 162
tidal forces • 190
–, of Jupiter • 175, 203
Titan (Saturnian moon) • 22, 23, 127, **127**, 222, **222**
–, atmosphere • 127, **127**, **128**, 222
–, clouds • 222
–, diameter • 127
–, impact craters • 222
–, lakes • 222
–, life • 222
–, mountains • 222
–, subterranean ocean • 222
–, surface • 222, **222**
–, volcanoes • 222
Titania (Uranian moon) • 43
Tombaugh, Clyde • 111, 167, 198
Tonry, John • 186
Toone, Mary • 31
transit technique • 206
trans-Neptunian object • 198
Transvaal Observatory • 98
tree rings • 82
Tremaine, Scott • 130, 198
Triangulum galaxy • **132**
Triangulum Nebula • 106
Triton (Neptunian moon) • 67
Trujillo, Chadwick • 219
Tucker, Roy • 218
Twin Earth • 223
Tycho Brahe • 4, 14, 15, 18, 19, 27, 79, 118
Tycho's supernova • 26, 86, 125

U

U.S. Naval Observatory • 75, 171
Udry, Stéphane • 223
Uhuru (satellite) • 147, 162
ultraviolet radiation • 138
Ulysses (space probe) • 203
unian observatory • 98
United States Naval Observatory • 75, 171
universe
–, accelerating expansion • **213**
–, age • 126
–, baby photo • 151, 195
–, density • 214
–, distance scale • 94
–, evolution • **145**
–, expansion • 91, 110, 134, 146, 214, **214**
–, expansion velocity • 178, 214, **214**
–, large-scale structure • 151, 179
–, origin • **145**
–, rotation • 163
universities:
– California • 139, 182
– Cambridge • 30
– Cornell • 122
– Erlangen • 58
– Franeker • 19
– Glasgow • 158
– Harvard • 131, 214
– Hawaii • 218
– Michigan • 179
– Oxford • 31, 66, 106
– Prague • 62
– Princeton • 151
– Rice • 199
Uranometria • 19
Uranus (planet) • 22, 43, **43**, **44**, 47, 50, 67, 167, 170, **170**, 171, 174
–, axial tilt • 43
–, discovery • 43
–, moons • **44**, 127, 170
–, orbital deviations • 67, 111
–, rings • 22, **44**, **169**, 170, **170**, 174
USS Enterprise • 46

V

V838 Monocerotis (star) • 83, **84**
vacuum • 86
Van Albada, Bruun • 134
van de Hulst, Henk • 131
van de Hulst, W.G. 131
van de Kamp, Peter • 99, 206
van der Klis, Michiel • 182
van Woerkom, Adriaan • 130
Vandenberg Air Force Base • 183
variable star • 19, **20**, 83, 95, 106, 110, 147, 162
VEGA (planetary/cometary probe) • 187
Vega (star) • 46, **46**, 59, 183
Vela (satellite) • 159
Vela supernova remnant • **94**
Venus (planet) • 23, **41**, 42, **42**, 127, 154, **154**, 202, 222, 223
–, atmosphere • 42, **42**
–, atmospheric pressure • 42
–, clouds • 23
–, distance • 154
–, live • 42
–, radar echos • 154
–, rotational period • 154
–, surface temperature • 42
–, transit • 27, 42
Verrier, Urbain Le • 37, 111
very long baseline interferometry • 155
Vesta (asteroid) • 50
Victoria (Martian crater) • 215, **216**
Viking project (Mars landers) • 166, 215
Virgo (constellation) • 194
Virgo cluster • 1, 4, 179
VLBI • 155
Void of Boötes • 179
Voltaire • 75
von Humboldt, Alexander • 63
von Weizsäcker, Carl • 122, 134
von Weizsäcker, Richard • 122
Vôute, Joan • 98
Voyager 1 (planetary probe) • 11, 127, 174, 175, 222
Voyager 2 (planetary probe) • 43, 67, 170, 174, 198
Vsekhsvyatskij, Sergej • 130
Vulcan (hypothetical planet) • 63
Vulpecula (constellation) • 47, 182

W

W5 (supernova remnant) • **152**
Walsh, Dennis • 178
War of the Worlds, The • 78
warp speed • 46
Webster, Louise • 162
Weichelsberger, Philipp • 58
Wells, Herbert • 78
Westerbork Synthesis Radio Telescope • **146**
Westerhout, Gart • 142
Westphal, James • 190
Weymann, Ray • 178
Whirlpool galaxy • **65**
Whirlpool Nebula • 66, **66**
white dwarf • 47, **47**, 56, 71, **71**, **72**, 79, 83, 87, 115, 194, **208**
White Sands Missile Range • 147, 162
Wickramasinghe, Chandra • 134
Wijnand, Rudy • 182
Wilkinson Microwave Anisotropy Probe • 195, **196**, 214
Wilkinson, David • 151
William Herschel Telescope • 211
Wilson, Robert • 151, 195
Wirtanen, Carl • 139
WMAP (satellite) • 195, **196**, 214
Wolf, Max • 83, 202
Wollaston, William • 58
Wolszczan, Aleksander • 194, 206
work • 122
world view, geocentric • 2
world view, heliocentric • 4
World War I • 103
World War II • 106, 126, 127, 131
Wulf, Theodor • 94
Würzburg radar dish • 131

X

Xanadu (area on Titan) • 222
Xena (ice dwarf) • 219
Xena: Warrior Princess • 219
X-ray afterglow • **211**
X-ray astronomy • 147, 162
X-ray binary • 147, **148**, 182
X-ray camera • 211
X-ray source • 147
X-rays • 138, 147, 150, 223

Y

Yerkes Observatory • 106, 127, 199

Z

Zel'dovich, Yakov • 151
Zephyria (area on Mars) • 78
Zwicky, Fritz • 115, 118, 158, 167, 178

Printed in the United States of America